Food
Phytates

Food
Phytates

Edited by
N. RUKMA REDDY
SHRIDHAR K. SATHE

CRC Press
Taylor & Francis Group
Boca Raton London New York

CRC Press is an imprint of the
Taylor & Francis Group, an **informa** business

CRC Press
Taylor & Francis Group
6000 Broken Sound Parkway NW, Suite 300
Boca Raton, FL 33487-2742

First issued in paperback 2019

© 2002 by Taylor & Francis Group, LLC
CRC Press is an imprint of Taylor & Francis Group, an Informa business

No claim to original U.S. Government works

ISBN-13: 978-1-56676-867-2 (hbk)
ISBN-13: 978-0-367-39647-3 (pbk)

Library of Congress Cataloging-in-Publication Data

Food phytates / edited by N. Rukma Reddy, Shridhar K. Sathe.
    p. cm.
    Includes bibliographical references and index.
    ISBN 1-56676-867-5 (alk. paper)
    1. Phytic acid—Derivatives—Physiological effect. 2. Food—Composition. I. Reddy, N. R. II. Sathe, Shridhar K.

QP801.P634 F66 2001
572'.5532—dc21   2001043780

Library of Congress Card Number 2001043780

Visit the Taylor & Francis Web site at
http://www.taylorandfrancis.com

and the CRC Press Web site at
http://www.crcpress.com

List of Contributors

Graeme D. Batten
School of Agriculture
Charles Sturt University
Locked Box #588
Wagga Wagga
New South Wales 2678
Australia
gbatten@csu.edu.au

John R. Burgess
Department of Foods and Nutrition
G-1F Stone Hall
Purdue University
West Lafayette, IN 47907-1264
burgessj@cfs.purdue.edu

Allen Cook
USDA-ARS
National Small Grain Germplasm
   Research Facility
P.O. Box 307
Aberdeen, ID 83210
phecooks@3rivers.net

Daren P. Cornforth
Department of Nutrition
   and Food Sciences
Utah State University
Logan, UT 84322
darenc@cc.usu.edu

Feng Gao
Department of Foods and Nutrition
G-1F Stone Hall
Purdue University
West Lafayette, IN 47907-1264
fengao45@hotmail.com

Elizabeth A. Grabau
Department of Plant Pathology,
   Physiology, and Weed Science
Virginia Polytechnic Institute
   and State University
Blacksburg, VA 24061-0346
egrabau@vt.edu

Mazda Jenab
Department of Nutritional Sciences
University of Toronto
150 College Street
Toronto, Ontario
Canada M5S 3E2
mazda.jenab@utoronto.ca

Srimathi Kannan
Department of Environmental
  and Industrial Health
1420 Washington Heights
University of Michigan
Ann Arbor, MI 48109-2029
kannans@umich.edu

Steven R. Larson
USDA-ARS
National Small Grain Germplasm
  Research Facility
P.O. Box 307
Aberdeen, ID 83210
stlarson@cc.usu.edu

Frank A. Loewus
Institute of Biological Chemistry
Washington State University
P.O. Box 646340
Pullman, WA 99164-6340
loewus@mail.wsu.edu

John N. A. Lott
Department of Biology
McMaster University
Hamilton, Ontario
Canada L8S 4K1
lott@mcmaster.ca

Irene Ockenden
Department of Biology
McMaster University
Hamilton, Ontario
Canada L8S 4K1
ockenden@mcmaster.ca

Brian Q. Phillippy
USDA-ARS
Southern Regional Research Center
1100 Robert E. Lee Blvd.
New Orleans, LA 70124
bqphil@commserver.srrc.usda.gov

Victor Raboy
USDA-ARS
National Small Grain Germplasm
  Research Facility
P.O. Box 307
Aberdeen, ID 83210
vraboy@uidaho.edu

N. Rukma Reddy
National Center for Food Safety
  and Technology/Food
  and Drug Administration
6502 S. Archer Road
Summit-Argo, IL 60501
rukma.reddy@cfsan.fda.gov.

Ann-Sofie Sandberg
Department of Food Science
Chalmers University of Technology
P.O. Box 5401
S-40229 Goteborg
Sweden
ann-sofie.sandberg@fsc.chalmers.se

Shridhar K. Sathe
Department of Nutrition, Food,
  and Exercise Sciences
College of Human Sciences
Florida State University
Tallahassee, FL 32306-1493
ssathe@mailer.fsu.edu

Erika Skoglund
Department of Food Science
Chalmers University of Technology
P.O. Box 5401
S-40229 Goteborg
Sweden
erika.skoglund@astrazeneca.com

Lilian U. Thompson
Department of Nutritional Sciences
University of Toronto
150 College Street
Toronto, Ontario
Canada M5S 3E2
lilian.thompson@utoronto.ca

Mahesh Venkatachalam
Department of Nutrition, Food,
    and Exercise Sciences
College of Human Sciences
Florida State University
Tallahassee, FL 32306-1493
mvenky1@hotmail.com

Connie M. Weaver
Department of Foods and Nutrition
1264 Stone Hall
Purdue University
West Lafayette, IN 47907-1264
weavercm@cfs.purdue.edu

Kevin A. Young
USDA-ARS
National Small Grain Germplasm
    Research Facility
P.O. Box 307
Aberdeen, ID 83210

Preface

FOOD crops such as cereals, legumes, and oilseeds are grown in over 90% of the world's total harvested area. Foods derived from these crops provide a major source of nutrients to mankind. Phytic acid is an important constituent of these food crops. The salt form, phytate, commonly exists in cereals, legumes, and other crops, where it serves several physiological functions, especially seed germination. Phytate is the major storage form of phosphorus and represents more than 80% of the total phosphorus in cereals, legumes, and other seed crops. Historically, phytate was considered solely as an antinutrient because it is a strong chelator of divalent minerals such as calcium, magnesium, zinc, and iron, and binds with these minerals and decreases their bioavailability. Under normal physiological conditions, phytate-mineral complexes are unavailable for absorption. However, recent investigations have focused on the possible beneficial effects of food phytates, based on their strong mineral-chelating properties. The possible beneficial effects of food phytates include lowering of serum cholesterol and triglycerides and protection against certain diseases such as cardiovascular diseases, renal stone formation, and certain types of cancer.

This book consists of fourteen chapters providing a comprehensive and succinct review of various facets of food phytates. It covers global estimates; occurrence, distribution, content, and daily intake; biosynthesis; genetics of synthesis and accumulation; phytase expression; stability of plant and microbial phytases; methods of analysis; *in vitro* and *in vivo* degradation; antioxidant effects; the use of phytate as an antioxidant in meats; phytate and mineral bioavailability; the role of phytate in cancer and disease prevention; and the influence of processing technologies on phytate removal. This publication is intended as a reference for food scientists, technologists, and nutritionists, especially those involved in the phytate research, technology, and processing of cereals, legumes, and other food crops.

The editors would like to thank Drs. Brian Q. Phillippy and John N.A. Lott, who reviewed certain chapters of this book and made several constructive comments. Special appreciation is extended to Eleanor Riemer of CRC Press.

N. RUKMA REDDY
SHRIDHAR K. SATHE

Table of Contents

CHAPTER 1

Introduction

SHRIDHAR K. SATHE
N. RUKMA REDDY

1. HISTORICAL BACKGROUND

DISCOVERY of phytate has its beginning in 1855–1856 when Hartig isolated small particles, nonstarch grains, from several plant seeds. He considered these small particles to be a source of essential reserve nutrients for seed germination and plant growth [1]. Later, in 1872, Pfeffer [2] further characterized the grains isolated by Hartig into three groups: crystals of calcium oxalate, a protein substance, and a compound that gave no reaction when tested for protein, fat, or inorganic salts. The third group was found in all of the 100 different seeds that he examined. Pfeffer [2] characterized the third group as having rounded surfaces, assuming spheroidal shapes, and frequently twining so as to present a convoluted appearance. He found that the third group was free of nitrogen but contained calcium, magnesium, and phosphorus. He named this group of grains "globoids." Organic matter was also noted in the globoids, and the suggestion was made that the substance was a phosphate combined with a carbohydrate. Later studies by Palladin [3] and Schulze and Winterstein [4] confirmed the presence and chemical composition of globoid particles in Indian mustard (*Sinapis niger*). Subsequently, Schulze and Winterstein [4] suggested "inosite-phosphoric acid" as the proper name for the "globoid," because globoid hydrolysis could yield inosite and phosphoric acid.

Extensive studies in the ensuing two decades by several researchers [5–16] indicated various chemical structures for this compound. However, the precise chemical structure of this compound remained elusive for over the next half century. The structure of phytic acid (Figure 1.1) had been a subject of controversy, which centered around the structure proposed by Anderson [17] and

a

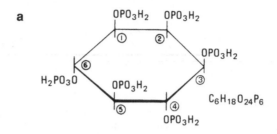

b

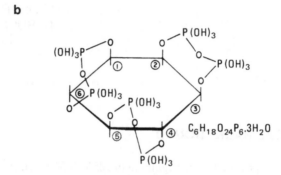

Figure 1.1 Proposed structures of phytic acid. Structure (a) was suggested by Anderson [17], and structure (b) was suggested by Neuberg [18].

the structure suggested by Neuberg [18]. Modern chemical analyses, including X-ray crystallographic and ^{31}P nuclear magnetic resonance data support the molecular structure [Figure 1.1(a)] originally proposed by Anderson [17] in 1914, which is the predominant form found in plant seeds and/or grains. The currently accepted name of phytic acid is *Myo*-Inositol 1,2,3,4,5,6 hexakis dihydrogen phosphate or *Myo*-Inositol hexakisphosphate, and phytic acid salts are generally referred to as "phytate(s)." A detailed account of the historical background of phytic acid and deduction of its chemical structure are presented earlier [19,20].

2. IMPORTANCE

Because phytic acid and its salt(s) usually occur simultaneously in many seeds, researchers often do not make a distinction between these two forms. We will collectively refer to them as "phytates" in this chapter. Phytates are the primary storage form of both inositol and phosphate in all seeds and grains [21].

Phytic acid is a strong chelator of divalent minerals such as copper, calcium, magnesium, zinc, and iron. The ability of phytic acid to chelate these minerals was recognized as a potential concern in animal and human mineral nutrition. Several *in vitro* and *in vivo* studies in animals and humans clearly indicate that phytates decrease mineral bioavailability by forming complexes with these minerals [20,22]. Many of the phytate mineral complexes are insoluble and, therefore, may become unavailable for absorption under normal physiological conditions. Because phytates are ionic in nature, they also react directly with charged groups of proteins or indirectly with the negatively charged groups of proteins mediated by a positively charged mineral ion such as calcium. The resultant phytate-protein and phytate-mineral-protein complexes may also adversely influence protein digestion and bioavailability [23,24]. Phytate can also bind with starch either directly by hydrogen bonding with a phosphate group or indirectly through the proteins [25], which may result in a decrease of starch solubility and digestibility.

Phytates are heat stable—they withstand heat, harsh field conditions, seed/grain transportation, and storage environments. Food processing (at home and in industry) includes a variety of methods: soaking, cooking, germinating, seed irradiating, extruding, milling, frying, fermenting, roasting, microwaving, and several combinations of these methods. Each of these processing methods may decrease phytate to a degree, depending upon the food to be processed and the type of processing method employed. In many instances, phytate reduction is not complete and, therefore, some phytate typically remains in the processed food. Therefore, food processing alone may not prevent the potential adverse effects of phytate. Use of exogenous enzymes for phytate reduction in monogastric nutrition has been recently reviewed, and development of thermostable phytases for the purpose of phytate reduction has been suggested [26].

Recent investigations have begun to focus on possible beneficial physiological/health effects of food phytates. The possible beneficial effects of food phytates include lowering of serum cholesterol and triglycerides and protection against certain diseases such as cardiovascular diseases, renal stone formation, and certain types of cancers [24,25,27–31]. Graf [31] has also reviewed medical and nonmedical applications of phytic acid. The primary mechanism by which the beneficial health effects of phytates may be explained is the strong mineral chelating ability of phytates. For example, because phytates can chelate copper, iron, zinc, magnesium, and several other minerals, and many of these minerals are essential cofactors for numerous oxidoreductases, several investigators have suggested that phytates may act as an antioxidant *in vivo* [32–34]. Such antioxidant activity has been suggested to be responsible for free radical scavenging action *in vivo* that may lead to anticancer activity [28]. Mineral-chelating ability of phytates has also been suggested to be responsible for antimicrobial (mold inhibition, for example) activity. Calcium and fluoride binding by phytic acid can insolubilize the minerals and, therefore, may be useful in the development of dental enamels [20,31]. Experimental data supporting many of the health benefits of phytates are inconclusive, at best. Careful, extensive, and in-depth studies must be completed before making any recommendations on beneficial health effects of phytates.

3. PURPOSE AND SCOPE

Research interest in phytic acid continues to increase, as evidenced by the number of research publications, reviews, symposia, and technical presentations at various national and international meetings. A comprehensive review was published on phytates in cereals and legumes to develop a succinct summary of what was known at that time [19]. Subsequently, we published a monograph on this subject [20]. During the intervening years, substantial advances in our understanding of this fascinating molecule have occurred. The major advances are not limited to development of analytical methodologies that permit detailed analysis of phytic acid and its various isomers and hydrolysis products, but there is a significant shift in thinking about the biological effects of this molecule. For these reasons, we organized a symposium entitled "Food Phytates: Antinutrients or Anticarcinogens" at the 1998 Annual Meeting of the Institute of Food Technologists (June 20–24, Atlanta, GA). The purpose of this symposium was to bring together leading authorities in phytate research to discuss recent advances and to identify certain future research needs. Following the symposium, we invited additional contributors to develop this book. The primary purpose of this book is to provide a comprehensive and succinct review of various facets of phytate research to date and to identify future research needs. Throughout this book, an attempt has been made to focus on recent literature in phytate research, but pertinent references to some of the older works have been retained for the purpose of continuity and perspective.

4. REFERENCES

1. Rose, A.R. 1912. A resume of the literature on inosite-phosphoric acid with special reference to the relation of that substance to plants. *Biochem. Bull.* 2:21–49.
2. Pfeffer, M. 1872. Comprehensive study of aleurone grains, identification of globoid and approximation of its chemical nature. *Johrb. Wiss. Bot.* 8:429–574.
3. Palladin, W. 1894. Discovery of inosite-phosphoric acid by chemical procedure. *Z. Biol.* 31:191–203.
4. Schulze, E. and Winterstein, E. 1896. *Physiol. Chem.* 40:120 (cited from Reference [20], Reddy et al., 1989).
5. Posternak, S. 1903. Sur la constitution de l'acide phosphorganique de reserve des plantes vertes et sur le premier produit de reduction du gaz carbonique dans l'acte de l'assimilation chlorophyllienne. *Compt. Rend. Hebdomad. des Seances de L'Acad. des Sci.* 137:439–441.
6. Posternak, S. 1903. Sur un nouveau principe phospho-organique d'origine vegetale la phytine. *Compt. Rend. Soc. Biol.* 55:1190–1192.

7. Suzuki, U., Yoshimura, K., and Takaishi, M. 1907. Uber ein enzym "phytase" das anhydro oxy methylen diphosphorsaure Spaltet. *Bull. Coll. Agric. Tokyo Imp. Univ.* 7:495–502.
8. Neuberg, C. 1908. Beziehung des cyclischen inosits zu den aliphatischen zuckern. *Biochem. Z.* 9:551–556.
9. Neuberg, C. 1909. Notiz über phytin. *Biochem. Z.* 16:406–410.
10. Levene, P.A. 1909. The conjugated phosphoric acid in plant seeds. *Biochem. Z.* 16:399–405.
11. Starkestein, E. 1908. Inosituria and the physiological significance of inosite. *Z. Exp. Pathol. Ther.* 5:378–389.
12. Starkestein, E. 1910. Die biologische bedeutung der inositphosphorsaure. *Biochem. Z.* 30: 56–98.
13. Starkestein, E. 1911. Ion action of phosphoric acids. *Biochem. Z.* 32:243–265.
14. Anderson, R.J. 1912. Phytin and phosphoric acid esters of inosite. *J. Biol. Chem.* 11:471–488.
15. Anderson, R.J. 1912. Phytin and phosphoric acid esters of inosite. *J. Biol. Chem.* 12:97–113.
16. Anderson, R.J. 1920. Synthesis of phytic acid. *J. Biol. Chem.* 43:117–128.
17. Anderson, R.J. 1914. A contribution to the chemistry of phytin. *J. Biol. Chem.* 17:171–190.
18. Neuberg, C. 1908. Zur frage der konstitution des phytins. *Biochem. Z.* 9:557–560.
19. Reddy, N.R., Sathe, S.K., and Salunkhe, D.K. 1982. Phytates in legumes and cereals. *Adv. Food Res.* 28:1–92.
20. Reddy, N.R., Pierson, M.D., Sathe, S.K., and Salunkhe, D.K. 1989. *Phytates in Cereals and Legumes*. CRC Press, Boca Raton, FL.
21. Cosgrove, D.J. 1966. The chemistry and biochemistry of inositol phosphates. *Rev. Pure Appl. Chem.* 16:209–224.
22. Krebs, N.F. 2000. Overview of zinc absorption and excretion in the human gastrointestinal tract. *J. Nutr.* 130:1374S–1377S.
23. Cheryan, M. 1980. Phytic acid interactions in food systems. *CRC Crit. Rev. Food Sci. Nutr.* 13:297–325.
24. Thompson, L.U. 1993. Potential health benefits and problems associated with antinutrients in foods. *Food Res. Intern.* 26:131–149.
25. Rickard, S.E. and Thompson, L.U. 1997. Interactions and biological effects of phytic acid. In *Antinutrients and Phytochemicals in Food*, Shahidi, F. (Ed.), ACS Symposium Series 662, American Chemical Society, Washington, DC, pp. 294–312.
26. Bedford, M.R. 2000. Exogenous enzymes in monogastric nutrition—their current value and future benefits. *Animal Feed Sci. Technol.* 86:1–13.
27. Zhou, J.R. and Erdman, J.W. 1995. Phytic acid in health and disease. *Crit. Rev. Food Sci. Nutr.* 35:495–508.
28. Shamsuddin, A.M., Vucenik, I., and Cole, K.E. 1997. IP6: A novel anti-cancer agent. *Life Sci.* 61:343–354.
29. Grases, F., March, J.G., Prieto, R.M., Simonet, B.M., Costa-Bauza, A., Garcia-Raja, A., and Conte, A., 2000. Urinary phytate in calcium oxalate stone formers and healthy people. Dietary effects on phytate excretion. *Scan. J. Urology Nephrology* 34:162–164.
30. Urbano, G., Lopez-Jurado, M., Aranda, P., Vidai-Valverde, C., Tenorio, E., and Porres, J. 2000. The role of phytic acid in legumes: antinutritional or beneficial function? *J. Physiol. Biochem.* 56:283–294.
31. Graf, E. 1983. Applications of phytic acid. *J. Amer. Oil Chem. Soc.* 60:1861–1867.
32. Graf, E. and Eaton, J.W. 1990. Antioxidant functions of phytic acid. *Free Radical Biol. Med.* 8:61–69.
33. Graf, E., Mahoney, J.R., Bryant, R.G., and Eaton, J.W. 1984. Iron-catalyzed hydroxyl radical formation. Stringent requirement for free iron coordination site. *J. Biol. Chem.* 259:3620–3624.
34. Rimbach, G. and Pallauf, J. 1998. Phytic acid inhibits free radical formation *in vitro* but does not affect liver oxidant or antioxidant status in growing rats. *J. Nutr.* 128:1950–1955.

CHAPTER 2

A Global Estimate of Phytic Acid and Phosphorus in Crop Grains, Seeds, and Fruits

JOHN N. A. LOTT
IRENE OCKENDEN
VICTOR RABOY
GRAEME D. BATTEN

1. INTRODUCTION

THIS chapter considers phytate more from the perspective of what plants deposit into the crop seeds/grains/fruits that we use than from the human and animal nutrition perspective. We provide a synopsis of the results of a global survey by Lott et al. [1] of the phosphorus (P) and phytic acid (PA) deposited each year in the world's major seed/grain/fruit crops, correct an error in those data, and extend the data to give information nutritional scientists can access more easily. It is important to point out that the beneficial and problematic aspects of phytic acid extend beyond animal nutrition aspects. Phytates, which are consumed in food by humans, swine, poultry, and other monogastric animals, often end up contributing to the eutrophication of our lakes and rivers through deposition of P-containing waste products [2]. The manure disposal problems associated with intensive animal farming operations are now widely recognized [3]. When animals are fed on grains/seeds, most of the total P consumed is PA-P, and thus, a large amount of the P in manure must be from phytate phosphorus that was not absorbed by the farm animals [4,5].

Phytate occurs in protein bodies and often, but not always, assembles into small, dense spheres called globoids [6–8]. Phytate concentration may vary in different tissues of a seed/grain/fruit [9,10]. Phytate is used by all seeds/grains/fruits as a mineral nutrient store for the growing seedlings. Generally, phytate accounts for one to several percent of the dry weight of grains and other dry seeds [1,11]. It provides substantial *myo*-inositol, P, K, and Mg to the growing seedling, and perhaps Ca, Fe, Zn, and Mn [11–15]. The synthesis and roles of phytate in plants have been reviewed by Greenwood [16], Lasztity and

7

Lasztity [17], Raboy [18], and Raboy and Gerbasi [19]. Thus, from the plant nutrient perspective, phytate is important for seedling establishment and fosters good crop yields, even though from an animal nutrition point of view, phytate may be an antinutritional substance [20]. The low phytic acid mutants recently isolated by several groups have normal levels of seed total P and have potential for improved bioavailability of P and other nutrients in humans and livestock.

The large growth of the human population makes the study of phytate significant. The importance of P and K in crop yields and the fact that a very significant amount of P, K, and Mg is removed with seeds/grains as phytate, links our understanding of phytate to the global food supply [1].

Given the results of the "green revolution" that resulted in a major increase in food production, there may be a perception in some minds that increases in food production will be sustainable almost without limit and that good nutrition is merely a matter of choice. It is important that we point out that the impending crisis of human population growth potentially will drive many agendas—political, economic, and environmental. The world population of humans reached 6 billion in 2000. It is projected to increase to 7.2 billion by 2010 and likely will reach 11 billion by the year 2100 [21,22]. The widening gap between human population in developed and developing countries is a serious problem for human nutrition. If the predictions given above are realized, then by 2100, only 1.5 billion of the total human population of 11 billion will be living in countries now regarded as prosperous. Many of the areas of the world that have high human populations are located in areas with low soil P availability to plants [23]. In the long term, this problem is compounded by the fact that high available P rock phosphate is a limited commodity [24].

2. MAKING A GLOBAL ESTIMATE

Global estimates are very useful but not perfect. Oerke et al. [22] in discussing this point, concluded that because the data can always be improved, the only alternative to an imperfect estimate is no estimate at all. The global estimates of P and PA in crop grains/seeds/fruits [1], parts of which are summarized here, made detailed notations of where data came from, what assumptions were used to fill in missing information, and what calculations were used to calculate phytic acid concentrations from phytic acid-phosphorus (PA-P) values. These data can clearly be improved in many places, but at least, this is a start.

Using the Raboy and Dickinson [25] procedure based on a formula weight of 659.02 for PA, the PA values were calculated from published values for PA-P. The PA-P value was multiplied by 3.55 to obtain the PA amount. Because grains produced on field grown plants generally have lower P concentrations than those grown in greenhouses [26], our estimate used data from field grown plants, wherever possible. Lott et al. [1] estimate contains information on how the total seed/grain/fruit tonnage was corrected for tissues that do not contain PA (for example, the hard shell of a nut). Many of these corrections can be improved by study of larger samples. Where possible, our estimate used the average of several published values for the % P and % PA. In terms of reliability, it was fortunate that those seeds/fruits that humans produce in the largest tonnages were also those for which four or more published values from different laboratories were available for averaging to calculate a value for % PA for a given species. Thus, the errors are likely to be smaller with regard to our major seed/grain/fruit crops estimates than with more minor ones.

However, the readers will recognize that a mean % PA cannot take into account all the possible variables that can lead to differences in % PA in a crop seed/grain/fruit. Some of the possible sources of such variation include the following:

(1) Climatic factors such as the amount and timing of rainfall [27]
(2) Fertilizer applications and soil type [28,29]
(3) Procedures to measure PA-P that cannot distinguish between PA (Ins P_6) and less substituted inositol phosphates such as Ins P_5, Ins P_4, and Ins P_3 (Because only Ins P_6 and Ins P_5 effectively inhibit uptake of Ca and Zn in monogastric animals, the contribution of lower substitutions of inositol phosphates may be misleading [30]. There is a general consensus that in mature seeds, the *myo*-inositol phosphates are almost exclusively in the form of Ins P_6, but because of the methods used to identify PA, some of this data may also not be as accurate as one would like.)
(4) Cultivar differences [31–34]
(5) Differences in moisture content of the crop seeds/fruits at the time of weighing for sale
(6) Fertilizer applications and the starting soil conditions [28,29]
(7) Choice of years for production tonnage [In our estimate, we averaged three years (1994–1996) of data from the Food and Agriculture Organization (FAO) of the United Nations (UN). Values would be slightly different if data from different years were used.]

In making our global estimate, we recognized that total production of crop seeds/grains/fruits was underestimated, and correction factors were put in to raise the total production. These correction factors, which total 45%, are our educated guesses as to what these correction factors should be, but they could be too high or too low. At least by mentioning these correction factors, we have highlighted the need for such adjustments. The FAO production estimates used in our survey are low because they do not consider such items as (1) losses of seeds/grains/fruits during harvesting, shipping, and storage; (2) losses due to pests or environmental conditions; (3) under-reporting of seed/fruit production where farmers grow grain to feed livestock but only report the sale of animals; (4) production of seeds/fruits that are consumed locally; (5) production of seeds/fruits not reported by the UN; and (6) production of seeds/fruits used to plant many types of crops.

3. RESULTS OF A GLOBAL ESTIMATE

A global estimate [1] of annual production of PA and sequestration of P in the world's seeds/grains/fruits is presented in Table 2.1. Details of the % total P, % PA, and corrections for tissue not containing PA can be found in the original estimate. Production estimates for dry cereal grains, dry legume seeds, other dry seeds/fruits, and fleshy fruits with seeds are listed in Table 2.1. The largest tonnages of food come from maize (corn), wheat, and rice, each with over 550 million metric tonnes (1000 kg = 1 metric tonne) per year (Table 2.1). Of these three largest crops, wheat produces the most PA (over 5.6 million tonnes) and rice the least (over 4.0 million tonnes), without corrections for production underestimates (Table 2.2). The next largest producers of PA are soybean (over 2 million tonnes without corrections) and barley (over 1.5 million tonnes). Other crops in which over 300,000 tonnes of PA are produced without corrections for production underestimates are rapeseed (canola), sorghum, cotton, peanut, and oats. When corrections for production underestimates are applied, the total PA synthesized is higher (Table 2.2).

The totals for the four categories are listed in Table 2.3. Dry cereals account for 69.5% of the total global crop seed/grains/fruit production each year but synthesized 77.3% of the total PA. Legumes account for 7.6% of the annual global production of crop seeds/grains/fruits and 13.0% of the total PA. The fleshy fruits with seeds have a high water content, up to 95% in some cases, and as a result, this category accounts for 17.2% of the tonnage produced each year but less than 1% of the total PA. In many fleshy fruits listed in Table 2.1, the seeds are not eaten as food (e.g., apricot, mango, papaya). The largest consumption of PA by animals eating fleshy fruits occurs from fruits where mature or nearly mature seeds are eaten (e.g., tomatoes). At least 22 million metric tonnes of PA are produced by crop seeds/grains/fruits each year, and with corrections for underestimates, the total is likely to be over 33 million tonnes (Table 2.3).

Table 2.4 lists the crops that produce a high concentration of PA, on a percent of weight basis of the seed portions eaten by humans. Sesame, pumpkin/squash, and flax (linseed) have 3.7 to 4.7% PA on a dry weight basis. Rapeseed (canola), sunflower, and mustard are in the 2% PA range, and many others listed in Table 2.4 exceed 1.0% dry weight. Percent PA concentrations do not take into account that PA exists in nature mainly as phytate, which is commonly a K and Mg salt. Other elements binding with PA in varying, but usually trace amounts, include Ca, Fe, Zn, and Mn [7,11]. A model K and Mg salt of PA was used to approximate phytate in crop seeds/fruits. While recognizing that this is a

TABLE 2.1. Estimate of Annual Production of Phytic (PA) Acid and Sequestration of Phosphorus (P) in the Crop Seeds/Grains/Fruits of the World.

Taxonomic Name	Common Name	Annual Global Production[a] (Metric Tonnes)	Annual Global Sequestration of P[b] (Metric Tonnes)	Annual Global Synthesis of PA[b] (Metric Tonnes)
A) Cereals (dry grain only)				
Hordeum vulgare	barley	153,099,000	581,776	1,561,610
Zea mays	maize	553,968,000	1,606,507	4,764,125
Digitaria sp.	millets	28,287,000	107,491	260,240
Echinochloa frumentaceae				
Eleusine coracana				
Eragrostis abyssinica				
Panicum sp.				
Pennisetum sp.				
Avena sativa	oats	31,155,000	130,851	317,781
Oryza sativa	rice	550,084,000	1,398,314	4,059,620
Secale cereale	rye	22,792,000	86,610	221,082
Sorghum bicolor	sorghum	61,795,000	247,180	655,027
Other cereals	may include	16,671,000	66,684	148,372
Triticale secale	triticale, wild			
Zizania sp.	rice, buckwheat			
Fagopyrum esculentum				
Triticum sp.	wheat	552,480,000	2,044,176	5,635,296
B) Legumes (dry seeds only)				
Phaseolus sp.	beans	18,247,000	89,410	257,283
Vicia sp.	broad beans	3,421,000	23,605	37,973
Cicer arietinum	chickpeas	8,287,000	27,347	45,579
Lens esculenta	lentils	2,820,000	10,716	19,740

(continued)

TABLE 2.1.   (continued).

Taxonomic Name	Common Name	Annual Global Production[a] (Metric Tonnes)	Annual Global Sequestration of P[b] (Metric Tonnes)	Annual Global Synthesis of PA[b] (Metric Tonnes)
Other beans	may include cow	11,775,000	44,745	102,443
Vigna sp.	peas, pigeon peas,			
Cajanus indicus	jack bean, lupines			
Canavalia sp.				
Lupinus sp.				
Arachis hypogaea	peanut (groundnut)	28,898,000	93,052	343,886
Pisum sativum	peas	12,248,000	56,341	122,480
Glycine max	soybean	130,761,000	889,175	2,026,796
C) Other dry seeds/fruits				
Prunus dulcis	almond	1,175,000	2209	6674
Anacardium occidentale	cashew	682,000	1228	5374
Ricinus communis	castor bean	1,463,000	6952	22,753
Castanea sp.	chestnuts	497,000	725	443
Theobroma cacao	cocoa, coco beans	2,800,000	10,114	25,043
Cocos nucifera	coconut	45,711,000	21,393	61,801
Coffea arabica	coffee	5,821,000	11,060	27,359
Gossypium sp.	cotton	34,851,000	228,274	671,126
Corylus avellana	filberts, hazelnuts	647,000	1101	2738
Corylus cornuta				
Cannabis sativa	hemp	36,000	252	626
Linum usitatissimum	linseed (flax)	2,489,000	36,837	91,844
Brassica sp.	mustard	182,000	1401	3640
Other tree nuts	includes Brazil nuts,	497,000	1118	4473
Bertholletia excelsa	macadamia nuts,			
Macadamia ternifolia	pecans, pili nuts,			
Carya illinoinensis	sapucaia nuts			

TABLE 2.1. (continued).

Taxonomic Name	Common Name	Annual Global Production[a] (Metric Tonnes)	Annual Global Sequestration of P[b] (Metric Tonnes)	Annual Global Synthesis of PA[b] (Metric Tonnes)
Canarium sp.				
Lecythis sp.				
Elaeis guineensis	palm nut (oil palm)	4,872,000	658	1900
Pistachia vera	pistachio	395,000	1106	2726
Brassica sp.	rapeseed (canola)	31,577,000	318,928	789,425
Carthamus tinctorius	safflower	818,000	2290	5726
Sesamium indicum	sesame	2,434,000	16,551	114,641
Helianthus annuus	sunflower	24,414,000	100,097	256,347
Camellia sasanqua	teaseed, tung	495,000	1114	4455
Aleurites sp.				
Juglans sp.	walnuts	1,004,000	2024	3036
D) Fleshy fruits with seeds				
Malus x domestica	apple	50,528,000	5871	2077
Prunus armeniaca	apricot	2,323,000	906	585
Persea americana	avocado	2,082,000	866	167
Vaccinium sp.	blueberries, cranberries	260,000	26	26
Prunus sp.	cherries (sweet and sour)	1,595,000	391	215
Citrus sp.	citrus including oranges, lemons, tangarines, limes, grapefruit, etc.	90,333,000	19,164	2927
Cucumis sativus	cucumber, gherkins	22,442,000	6773	240
Cucumis anguria				
Ribes sp.	currants, gooseberries	667,000	200	84
Phoenix dactylifera	date	2,435,000	1683	1013
Solanum melongena	eggplant	11,460,000	4584	6347

(continued)

TABLE 2.1. (continued).

Taxonomic Name	Common Name	Annual Global Production[a] (Metric Tonnes)	Annual Global Sequestration of P[b] (Metric Tonnes)	Annual Global Synthesis of PA[b] (Metric Tonnes)
Ficus carica	figs	1,500,000	300	396
Vitus sp.	grape	55,806,000	10,268	2079
Phaseolus sp.	green beans including as well	3,552,000	2131	7459
Dolichos lablab	lablab but not others			
Pisum sativum	green peas, immature fruits	5,072,000	2556	8947
Actinidia sp.	kiwifruit	1,020,000	306	396
Mangifera indica	mango	18,919,000	1892	10,525
Cucumis melo	melons, cantaloupes	15,788,000	4269	6783
Olea europaea	olive	11,318,000	2807	2173
Carica papaya	papaya	5,835,000	5088	10,083
Prunus persica	peaches, nectarines	10,414,000	3832	4999
Pyrus sp.	pears	12,337,000	1360	318
Capsicum sp.	peppers, chilies	13,663,000	5547	3798
Prunus sp.	plums	6,593,000	1437	475
Cucurbita sp. and others	pumpkins/squashes, gourds	9,602,000	7355	5798
Rubus sp.	raspberry	329,000	66	221
Fragaria x ananassa	strawberry	2,602,000	520	437
Zea mays	sweet corn (maize)	5,540,000	5540	13,850
Lycopersicon esculentum	tomatoes	83,545,000	25,064	10,401
Citrullus lanatus = (*Citrullus vulgaris*)	watermelon	39,824,000	4072	5958

[a]Almost all of the production estimates given here are an average calculated from the FAO Yearbook on Production 1996 [35] using data from 1994, 1995, and 1996. A metric tonne equals 1000 kilograms. The latest values for cherries are from 1971–73 given in the FAO Yearbook on Production 1973 [36]. Mustard production figures are from Ensminger et al. [37]. Teaseed and tung were estimated from values in Vaughan [38]. *Vaccinium* and kiwifruit estimates are based on Macrae et al. [39]. Date and sweet corn production is based on Considine and Considine [40]. Fig production values are based on Janick et al. [41].
[b]These values, without any correction for production underestimates, are from Lott et al. [1].

TABLE 2.2. Annual Synthesis of Phytic Acid (PA) in Seeds/Grains/Fruits.

| Plant | Annual Synthesis of PA (Metric Tonnes) | |
	Without Corrections	With Corrections[a]
Wheat	5,635,296	8,171,179
Maize	4,764,125	6,907,981
Rice	4,059,620	5,886,449
Soybean	2,026,796	2,938,854
Barley	1,561,610	2,264,335
Rapeseed (canola)	789,425	1,144,666
Cotton	671,126	973,133
Sorghum	655,027	949,789
Peanut	343,886	498,635
Oats	317,781	460,782
Millets	260,240	377,348.
Dry beans	257,283	373,060
Sunflower	256,347	371,703

[a]The United Nations crop production data, upon which the Lott et al. [1] estimate was based, are clearly underestimates of the tonnage of a given crop that is actually produced by plants. For example, losses during harvesting, shipping, and storage are not considered and losses due to spoilage, pests, or weather are also not considered. The FAO production estimates do not cover many species used as food, do not consider production in small gardens as occurs widely in many countries, and often do not consider the seeds/fruits used to feed livestock. In many cases, seeds/grains are consumed by poultry and livestock on the farm and are not recorded in production figures. The 45% correction factors applied by Lott et al. [1] and shown here are educated guesses that we assume are conservative.

simplification of the complexities found in nature, it is still a useful estimate. Without and with corrections for production underestimates, the phytate produced each year by crop seeds/grains/fruits is certainly over 33 million metric tonnes and may well be near 50 million metric tonnes (Table 2.5). Within this amount of phytate are over 6.4, 8.1, and 2.5 million tonnes of P, K, and Mg, respectively, if presented without corrections for production underestimates. Correction factors increase those values considerably.

4. DISCUSSION AND SUMMARY

The negative effects of phytate in food on human health are likely to be most pronounced in people on marginal subsistence diets that consist mainly of seeds/grains/fruits. Over two billion humans, largely located in developing countries, have micronutrient deficiencies [42–44]. Considerable scientific literature indicates that the strong chelation capacity of PA can decrease the uptake of elements such as Ca, Fe, and Zn in monogastric digestive tracts [20,45–47].

TABLE 2.3. Total Phosphorus (P) and Phytic Acid (PA) Sequestered in One Year in Seeds/Grains/Fruits of Crop Plants without and with Correction Factors for Underestimates.

Category	Annual Global Production (Metric Tonnes)	Total Production (%)	Annual Global Sequestration of P (Metric Tonnes)	Annual Global Production of PA (Metric Tonnes)	Total PA (%)
A) Cereals (dry grains only)	1,970,331,000	69.5	6,269,589	17,623,153	77.3
B) Legumes (dry seeds only)	216,457,000	7.6	1,234,391	2,956,180	13.0
C) Other dry seeds/fruits	162,860,000	5.7	765,432	2,102,150	9.2
D) Fleshy fruits with seeds in different stages of formation	487,385,000	17.2	124,874	108,777	0.5
Subtotal (no corrections)	2,837,033,000		8,394,286	22,790,260	
Estimated underestimate 45%[a]	1,276,664,850		3,777,429	10,255,617	
Total with corrections	4,113,697,850		12,171,715	33,045,877	

[a]Refer to the note under Table 2.2.

TABLE 2.4. Seeds/Grains/Fruits with Seeds that Are Commonly Eaten by Humans that Contain Phytic Acid (PA) Concentrations of One Percent or More on a Weight Basis.

Plant[a]	Structure	% PA
Sesame	dry seed	4.71
Pumpkin/squash	embryo	4.08
Flax (linseed)	dry seed	3.69
Rapeseed (canola)	dry seed	2.50
Sunflower	embryo	2.10
Mustard	dry seed	2.00
Cashew	embryo	1.97
Brazil and other tree nuts	embryo	1.80
Hemp	dry fruit	1.74
Peanut	seed in shell	1.70
Tomato	seed only	1.66
Soybean	dry seed	1.55
Almond	dry embryo	1.42
Eggplant	seed only	1.42
Beans	dry seed	1.41
Pistachio	embryo	1.38
Watermelon	seed only	1.36
Kiwi fruit	fleshy fruit	1.34
Broad beans	dry seed	1.1l
Cucumber	immature seed	1.07
Sorghum	dry grain	1.06
Coco beans	dry seed	1.04
Barley	dry grain	1.02
Oats	dry grain	1.02
Wheat	dry grain	1.02
Peas	dry seed	1.00

[a]Data obtained from Table 2.1 in Lott et al. [1]. A number of species were excluded from this table, including (a) cases where the seeds generally are not eaten by humans (e.g., apple, pears, melons, papaya), (b) cases where the seeds constitute a very small portion of the total weight (e.g., strawberry), and (c) cases where humans eat very little of the seeds (e.g., cotton, castor bean, teaseed/tung).

While reductions of bioavailability of minerals by phytate is the most prominent negative aspect of phytate ingestion, there are other impacts that may be beneficial. Possible positive effects of PA in food are likely to be most pronounced in humans with high red meat consumption, because there is a link to iron intake and certain kinds of cancer [48]. Antioxidant effects of phytate may help reduce colorectal cancer and mammary cancer [49–51].

The P, K, and Mg that are common in phytate in seeds are important mineral macronutrients for humans and animals [52]. Those elements come from many food sources, not just seeds/fruits/grains [53]. However, it is worth pointing

TABLE 2.5.   Estimates of the Weight of Phosphorus (P), Potassium (K), and Magnesium (Mg) Associated with Phytate that Are Removed Annually Worldwide by Crop Seeds/Fruits. [a]

Category	Estimated Weight of Phytate (Metric Tonnes)	Estimated Weight of P in Phytate (Metric Tonnes)	Estimated Weight of K in Phytate (Metric Tonnes)	Estimated Weight of Mg in Phytate (Metric Tonnes)
Major crops (A to D in Table 2.3)	33,501,682	6,432,323	8,140,909	2,512,626
Corrected for underestimates (Table 2.3)	15,075,757	2,894,545	3,663,409	1,130,682
Total	48,577,439	9,326,868	11,804,318	3,643,308

[a]The data has been recalculated from Lott et al. [1]. The model phytate, which is a simplified approximation of what is found in seeds, has complete substitution of the twelve negative binding sites per molecule of PA. The model phytate has six potassium and three magnesium per molecule. PA values from Table 2.3 were converted to phytate by multiplying by 1.47, a factor based on atomic weights. The proportions of P, K, and Mg in the model phytate are 19.2%, 24.3%, and 7.5%, respectively.

out that for humans, the recommended dietary allowance for P, a level that is designed for maintenance of good nutrition, varies from 300–500 mg per day for infants to 800–1200 mg per day for children and adults [53]. A reasonable average dietary requirement for P is thus 1 gram (1000 mg) per person per day or 365 g per person per year.

If we assume the following: (a) all the P found in seeds/grains/fruits was bioavailable for humans, (b) the entire global production of seeds/grains/fruits was used to feed humans, (c) this P supply is distributed equally to all people across the globe, and (d) no P was coming from other food groups, then each of the 6 billion humans alive today would receive about 1399 grams of P per year, based on the total uncorrected P value from Table 2.3. This is 3.8× the needed 365 g of P per year. If the anticipated population increase occurs, then in a century, the 11 billion humans would equally get 763 g of P per year, just two times above the yearly requirement. Clearly, the assumptions made above do not reflect current realities. However, this calculation shows that even for a nutrient like P, that is commonly considered not to be a problem with regard to deficiencies in man, there are limits that we are approaching more rapidly than some would anticipate. The proportion of nutrients coming from seeds/grains/fruits is higher for humans on subsistence or vegetarian diets than for those living in wealthier countries, where food variety is high. The Oerke et al. [22] estimate predicts that by 2100, the human population will be about 11 billion and that about 85% of that population will be in developing countries. It is in these countries where food shortages are likely to be most acute. Low soil P in those areas also will result in minimal P in the seeds/grains/fruits produced [23].

One of the assumptions made above is that all the P found in seeds/grains/fruits is bioavailable for humans. Realistically, at present, if people obtain most of their P from grains and legumes, a significant fraction of the total P consumed is PA-P and, thus, may be largely unavailable. Clearly, the bioavailability of P is complex and depends upon many factors including cooking and other food preparation procedures as well as the initial composition of the grains/seeds/fruits. If the calculation done above is corrected for 50% and 75% of the P being not bioavailable, the amounts for 6 billion humans fall to 700 g and 350 g, respectively, and for 11 billion humans fall to 382 and 191 g per person per year, respectively. This calculation highlights the need for higher available P from our cereal and legume crops, because several of these calculated values are close to or lower than the 365 g per person per year required for good nutrition, and several other of the assumptions above are not likely to occur.

We estimate that the P in the phytate in the world's crop seeds/grains/fruits is about 9.3 million metric tonnes per year (Table 2.5). Total global usage of elemental P from mineral sources for fertilizer for all purposes is in the 13–16.6 million tonnes per year range [1]. Thus, a weight of elemental P, equivalent to 56–71% of the total elemental P applied globally each year as fertilizer, is removed in the form of phytate in seeds/grains/fruits. Keep in mind that P is one of the most widely used plant fertilizers. Not only do fertilizers add significantly to the cost of producing food [54], but the removal of these elements from soils in the form of phytate in harvested seeds/grains reduces the sustainability of crop production by that soil.

The development of low phytic acid (*lpa*) mutants of several important crop plants (see Chapter 5) offers a particularly interesting nutritional opportunity. About 75% of the total P in grains is PA-P that is not readily bioavailable to nonruminant animals. In the *lpa* mutants, the higher inositol phosphates are greatly reduced, so the available P is much greater [55–57]. If more of the phosphorus in seeds/grains was absorbed by humans and animals, the environmental problems related to P pollution would be reduced. One option is the use of *lpa* varieties to improve bioavailability of P and other minerals in crop seeds/grains/fruits. The world total tonnage for PA in crop seeds/grains/fruits is about 22.8 million metric tonnes, without corrections (Table 2.3). Currently, *lpa* mutants are available for wheat, rice, maize, soybean, and barley. If we assume that 25% of all the world cereal and legume crops were replaced by *lpa* mutants with 66% PA reduction, that would reduce the world total for PA in seeds/grains/fruits by about 3.4 million metric tonnes or about 15% of the total.

Some major findings are listed below.

(1) We report here a summary of the Lott et al. [1] estimate of the P and PA in most of the world's crop seeds/grains/fruits. We made some changes in the way data were presented and corrected small errors. Without and with corrections for underestimates in production, the tonnage of crop seeds/grains/fruits was in the 2.8–4.1 billion metric tonne range. The P content was in the 8.3–12.1 million tonne range, and the PA content was in the 22.7–33.0 million metric tonne range. Calculations showed that use of low phytic acid mutants of important crop plants could bring about considerable reductions in the PA in crop seeds/grains/fruits.

(2) The cereals represented 69.5% of the production but accounted for 77.3% of the total PA. The legumes were the next largest category with 7.6% of the production and 13.0% of the PA. Fleshy fruits with immature or mature seeds formed 17.2% of the production, but because water content is high and concentration of seed tissue is generally low, this category accounted for only 0.5% of the PA. Other dry seeds/fruits accounted for 5.7% of the production but 9.2% of the PA. This reflects the fact that a number of high-phytate-containing seeds are in this category.

(3) The crop seeds/grains/fruits with the concentration of PA over 2% in the edible portion on a weight basis were sesame, pumpkin (squash), flax, rapeseed (canola), and sunflower.

(4) Using a model K + Mg salt of PA, one that is an approximation of the

phytate found in seeds/grains/fruits in nature, we estimated that at least 33.5 and perhaps close to 50 million metric tonnes of phytate are produced each year globally in crop seeds/grains/fruits. The higher values are based on correction factors for production underestimates, but these may be too low or too high. We believe that at least 8.1 and perhaps up to 11.8 million tonnes of K is likely sequestered in phytate in crop seeds/grains/fruits each year and 2.5 to 3.6 million tonnes of Mg.

5. SUGGESTIONS FOR FUTURE WORK

We have begun additional estimates to more closely relate P and PA in crop seeds/grains/fruits in relation to total production, fertilizer usage, and hectares farmed in different countries and continents. Because the *lpa* mutants have considerable potential to provide better nutrition, there is a need for many structural, compositional, nutritional, and agronomic aspects of these mutants to be better characterized. Development of additional *lpa* mutants is continuing, and these will require ongoing study. For example, whether or not these *lpa* mutants have altered levels of Mg, Ca, Fe, Zn, and other nutritionally significant elements needs to be ascertained, because trace element deficiencies are nutritionally very important issues.

6. REFERENCES

1. Lott, J. N. A., Ockenden, I., Raboy, V. and Batten, G. D. 2000. Phytic acid and phosphorus in crop seeds and fruits: a global estimate. *Seed Sci. Res.* 10:11–33.
2. Loehr, R. C. 1979. Potential pollutants from agriculture—an assessment of the problem and possible control approaches. *Prog. Wat. Tech.* 11:169–193.
3. Poggi-Varaldo, H. M. 1999. Agricultural wastes. *Water Environment Research* 71:737–784.
4. Das, P. C. and Netke, S. P. 1979. Utilization of phytin and non-phytin phosphorus by laying pullets. *Indian J. Anim. Sci.* 49:557–562.
5. Jongbloed, A. W. and Kemme, P. A. 1990. Apparent digestible phosphorous in the feeding of pigs in relation to availability, requirement and environment: I. Digestible phosphorous in feedstuffs from plant and animal origin *Neth. J. Agric. Sci.* 38:567–575.
6. Greenwood, J. S. and Bewley, J. D. 1984. Subcellular distribution of phytin in the endosperm of developing castor bean: a possibility for its synthesis in the cytoplasm prior to deposition within protein bodies. *Planta* 160:113–120.
7. Prattley, C. A. and Stanley, D. W. 1983. Protein-phytate interactions in soybeans. I. Localization of phytate in protein bodies and globoids. *J. Food Biochem.* 6:243–253.
8. Lott, J. N. A., Randall, P. J., Goodchild, D. J. and Craig, S. 1985. Occurrence of globoid crystals in cotyledonary protein bodies of *Pisum sativum* as influenced by experimentally induced changes in Mg, Ca and K contents of seeds. *Aust. J. Plant Physiol.* 12:341–353.
9. Wada, T. and Lott, J. N. A. 1997. Light and electron microscopic and energy dispersive X-ray microanalysis studies of globoids in protein bodies of embryo tissues and the aleurone layer of rice (*Oryza sativa* L.) grains. *Can. J. Bot.* 75:1137–1147.

10. O'Dell, B. L., de Boland, A. R. and Koirtyohann, S. R. 1972. Distribution of phytate and nutritionally important elements among the morphological components of cereal grains. *J. Agric. Food Chem.* 20:718–721.

11. Lott, J. N. A. 1984. Accumulation of seed reserves of phosphorus and other minerals. In *Seed Physiology,* Volume I. Murray, D. R. (Ed.), Sydney, Orlando, San Diego, Petaluma, New York, London, Toronto, Montreal, Tokyo, Academic Press. pp. 139–166.

12. Batten, G. D. and Lott, J. N. A. 1986. The influence of phosphorus nutrition on the appearance and composition of globoid crystals in wheat aleurone cells. *Cereal Chem.* 63:14–18.

13. Johnson, L. F. and Tate, M. E. 1969. Structure of "phytic acid." *Can. J. Chem.* 47:63–73.

14. Loewus, F. A. 1990. Structure and occurrence of inositols in plants. In *Inositol Metabolism in Plants.* Morré, D. J., Boss, W. F., Loewus, F. A. (Eds.), New York, Chichester, Brisbane, Toronto, Singapore, Wiley-Liss, Inc. pp. 1–11.

15. Lott, J. N. A., Greenwood, J. S. and Batten, G. D. 1995. Mechanisms and regulation of mineral nutrient storage during seed development. In *Seed Development and Germination.* Kigel, J. and Galili, G. (Eds.), New York, Basel, Hong Kong, Marcel Dekker, Inc. pp 215–235.

16. Greenwood, J. S. 1989. Phytin synthesis and deposition. In *Recent Advances in the Development and Germination of Seeds.* Taylorson, R. B. (Ed.), New York, Plenum Press. pp. 109–125.

17. Lásztity, R. and Lásztity, L. 1990. Phytic acid in cereal technology. In *Advances in Cereal Science and Technology,* Volume X. Pomeranz, Y. (Ed.), St. Paul, Minnesota, The American Association of Cereal Chemists, Inc. pp. 309–371.

18. Raboy, V. 1990. Biochemistry and genetics of phytic acid synthesis. In *Inositol Metabolism in Plants.* Morré, D. J.; Boss, W. F.; Loewus, F. A. (Eds.), New York, Chichester, Brisbane, Toronto, Singapore, Wiley-Liss, Inc. pp. 55–76.

19. Raboy, V. and Gerbasi, P. 1996. Genetics of myo-inositol phospate synthesis and accumulation. In *Subcellular Biochemistry.* Biswas, B. B.; Biswas, S. (Eds.), New York, Plenum Press. pp. 257–285.

20. Thompson, L. U. 1993. Potential health benefits and problems associated with antinutrients in foods. *Food Res. Internat.* 26:131–149.

21. Alexandratos, N. 1995. *FAO Study, World Agriculture: Towards 2010* (Food and Agriculture Organization of the United Nations). Chichester, New York, Brisbane, Toronto, Singapore, John Wiley & Sons.

22. Oerke, E.-C., Dehne, H.-W., Schönbeck, F. and Weber, A. 1994. *Crop Production and Crop Protection. Estimated Losses in Major Food and Cash Crops.* Amsterdam, Lausanne, New York, Oxford, Shannon, Tokyo, Elsevier Science B.V.

23. Lynch, J. P. 1998. Introduction. In *Phosphorus in Plant Biology: Regulatory Roles in Molecular Cellular, Organismic, and Ecosystem Processes (Current Topics in Plant Physiology,* Volume 19). Lynch, J. P. and Deikman, J. (Eds.), Rockville, Maryland, American Society of Plant Physiologists. pp. vi–viii.

24. Stevenson, F. J. 1986. *Cycles of Soil, Carbon, Nitrogen, Phosphorus, Sulfur, Micronutrients.* New York, Chichester, Brisbane, Toronto, Singapore, John Wiley & Sons.

25. Raboy, V. and Dickinson, D. B. 1984. Effect of phosphorus and zinc nutrition on soybean seed phytic acid and zinc. *Plant Physiol.* 75:1094–1098.

26. Batten, G. D. 1986. Phosphorus fractions in the grain of diploid, tetraploid, and hexaploid wheat grown with contrasting phosphorus supplies. *Cereal Chem.* 63:384–387.

27. Horvatic, M. and Balint, L. 1996. Relationship among the phytic acid and protein content during maize grain maturation. *J. Agron. Crop Sci.* 176:73–77.

28. Miller, G. A., Youngs, V. L. and Oplinger, E. S. 1980. Effect of available soil phosphorus and environment on the phytic acid concentrations in oats. *Cereal Chem.* 57:192–194.

29. Bolland, M. D. A., Jarvis, R. J., Coates, P. and Harris, D. J. 1993. Effect of phosphate fertilisers on the elemental composition of seed of wheat, lupin, and triticale. *Commun. in Soil Sci. Plant Anal.* 24:1991–2014.

30. Lönnerdal, B., Sandberg, A.-S., Sandström, B. and Kunz, C. 1989. Inhibitory effects of phytic acid and other inositol phosphates on zinc and calcium absorption in suckling rats. *J. Nutr.* 119:211–214.

31. Batten, G. D. 1994. Concentrations of elements in wheat grains grown in Australia, North America, and the United Kingdom. *Aust. J. Exp. Agric.* 34:51–56.

32. Miller, G. A., Youngs, V. L. and Oplinger, E. S. 1980. Environmental and cultivar effects on oat phytic concentration. *Cereal Chem.* 57, 189–191.

33. Ockenden, I., Falk, D. E. and Lott, J. N. A. 1997. Stability of phytate in barley and beans during storage. *J. Agric. Food Chem.* 45:1673–1677.

34. Raboy, V., Noaman, M. M., Taylor, G. A. and Pickett, S. G. 1991. Grain phytic acid and protein are highly correlated in winter wheat. *Crop Sci.* 31:631–635.

35. *FAO Yearbook on Production 1996*, Volume 50. 1997. Rome, Food and Agriculture Organization of the United Nations.

36. *FAO Production Yearbook 1973*, Volume 27. 1974. Rome, Food and Agriculture Organization of the United Nations.

37. Ensminger, A. H., Ensminger, M. E., Konlande, J. E. and Robson, J. R. K. 1994. *Food & Nutrition Encyclopedia*, 2nd Edition. Boca Raton, Ann Arbor, London, Tokyo, CRC Press, Inc.

38. Vaughan, J. G. 1970. *The Structure and Utilization of Oil Seeds*. London, Chapman and Hall Ltd.

39. Macrae, R., Robinson, R. K. and Sadler, M. J. (Eds.). 1993. *Encyclopaedia of Food Science, Food Technology and Nutrition*, London, San Diego, New York, Boston, Sydney, Tokyo, Toronto, Academic Press, Harcourt Brace Jovanovich, Publishers.

40. Considine, D. M. and Considine, G. D. 1982. *Foods and Food Production Encyclopedia*. New York, Van Nostrand Reinhold Company Inc.

41. Janick, J., Schery, R. W., Woods, F. W. and Ruttan, V. W. 1974. *Plant Science. An Introduction to World Crops*, 2nd Edition. San Francisco, W. H. Freeman and Company.

42. FAO/WHO. 1992. Preventing specific micronutrient deficiencies. Theme Paper 6, *International Conference on Nutrition*. Rome, Food and Agriculture Organization of the United Nations/World Health Organization.

43. WHO (World Health Organization). 1992. National strategies for overcoming micronutrient malnutrition. Geneva.

44. Graham, R. D. and Welch, R. M. 1996. *Breeding for Staple Food Crops with High Micronutrient Density*. Washington, D.C., International Food Policy Research Institute.

45. Nwokolo, E. N. and Bragg, D. G. 1977. Influence of phytic acid and crude fibre on the availability of minerals from four protein supplements in growing chicks. *Can. J. Animal Sci.* 57:475–477.

46. Sandberg, A.-S., Brune, M., Carlsson, N.-G., Hallberg, L., Rossander-Hulthen, L. and Sandström, B. 1993. The effect of various inositol phosphates on iron and zinc absorption in humans. In *Proceedings of the International Conference on Bioavailability 1993–Nutrition Chemical and Food Processing Implications of Nutrient Availability*. pp. 53–57.

47. Zhou, J. R., Fordyce, E. J., Raboy, V., Dickinson, D. B., Wong, M.-S., Burns, R. A. and Erdman, J. W. 1992. Reduction of phytic acid in soybean products improves zinc bioavailability in rats. *J. Nutr.* 122:2466–2473.

48. Weinberg, E. D. 1994. Association of iron with colorectal cancer. *BioMetals* 7:211–216.

49. Thompson, L. U. and Zhang, L. 1991. Phytic acid and minerals: effect on early markers of risk for mammary and colon carcinogenesis. *Carcinogenesis* 12:2041–2045.

50. Ullah, A. and Shamsuddin, A. M. 1990. Dose-dependent inhibition of large intestinal cancer by inositol hexaphosphate in F344 rats. *Carcinogenesis* 11:2219–2222.

51. Shamsuddin, A. M., Ullah, A. and Chakravarthy, A. K. 1989. Inositol and inositol hexaphosphate suppress cell proliferation and tumor formation in CD-1 mice. *Carcinogenesis* 10, 1461–1463.

52. Groff, J. L. and Gropper, S. S. 2000. *Advanced Nutrition and Human Metabolism*, 3rd Edition. Belmont, CA, Wadsworth/Thompson Learning.

53. Pennington, J. A. T. 1998. *Bowes & Church's Food Values of Portions Commonly Used*, 17th Edition. Philadelphia, PA, Lippincott-Raven Publishers.

54. Batten, G. D. 1992. A review of phosphorus efficiency in wheat. *Plant and Soil* 146:163–168.

55. Ertl, D. S., Young, K. A. and Raboy, V. 1998. Plant genetic approaches to phosphorus management in agricultural production. *J. Environ. Qual.* 27:299–304.

56. Mendoza, C., Viteri, F., Lönnerdal, B., Young, K., Raboy, V. and Brown, K. H. 1998. Effect of genetically modified low-phytate maize on absorption of iron from tortillas. *Amer. J. Clin. Nutr.* 68:1123–1127.

57. Sugiura, S. H., Raboy, V., Young, K. A., Dong, F. M. and Hardy, R. W. 1999. Availability of phosphorus and trace elements in low-phytate varieties of barley and corn for rainbow trout (*Oncorhynchus mykiss*). *Aquaculture* 170, 285–296.

CHAPTER 3

Occurrence, Distribution, Content, and Dietary Intake of Phytate

N. RUKMA REDDY

1. INTRODUCTION

PHYTATE (*myo*-inositol hexakisphosphate, InsP$_6$) widely occurs in plant seeds and/or grains [1–4], roots and tubers [1,3,5,6], fruits and vegetables [3,5,6], nuts [3,5], pollen of various plant species [7–9], and organic soils [10,11]. The phytate fraction of organic soil contains a mixture of phosphorylated derivatives of *myo*-, *chiro*-, *scyllo*-, and *neo*-inositol [12]. Inositol phosphates with fewer than six phosphate groups, such as *myo*-inositol 1,3,4,5,6-pentakisphosphate, have been isolated and identified from the nucleated erythrocytes of birds, turtles, and freshwater fish [13–17].

2. OCCURRENCE

Phytate occurs primarily as a salt of mono- and divalent cations in discrete regions of grains and seeds [18–20]. It rapidly accumulates in grains and seeds during their ripening period and maturation, accompanied by other substances such as starch, proteins, and lipids [21–25]. The accumulation site of phytate in grains and seeds is within the subcellular single membrane particles, aleurone grains or protein bodies [18,20,26,27]. The aleurone grains are located in the aleurone cells of monocotyledonous seeds such as cereals. The aleurone grains of rice are composed of at least two major parts: high phytate-containing particle and surrounding coat that consists of protein and carbohydrate [28,29]. The aleurone grains of rice are spherical, about 1–3 μm in diameter [30].

In dicotyledonous seeds such as legumes (beans) and many other seeds, globoids are located within the proteinaceous matrix of protein bodies [31–33]. Globoids are present in the cotyledons of dicotyledonous seeds but not their seed coats. They vary in size and number depending on the species. For example, Prattley and Stanley [34] reported that isolated soy globoids varied in size from 0.1–1.0 μm and were comparatively small in relation to protein bodies (2–20 μm). Some dicotyledonous seeds, namely peas, lack globoids within the protein bodies but still contain phytate [32]. Occurrence and biogenesis of globoids within the protein bodies may be controlled by the calcium, magnesium, and potassium contents [19,33]. The presence of phytate within the globoids has been shown for a wide range of cereals such as oats [35], barley [36,37], wheat [26,38], rice [28,30], sorghum [39], and corn [40], and dicotyledonous seeds such as peas [27,33], soybeans [28,34], peanuts [41,42], and broad beans [43,44].

Phytate content can range up to 60–80% of the dry weight of globoids [45,46]. The chemical composition of phytate-rich particles of rice, globoids of cottonseed, soybean, and peanuts, and phytate-rich isolated particles of Great Northern beans is presented in Table 3.1. These were isolated by different isolation methods. The chemical composition of these isolated particles or globoids is characterized by high phytate, potassium, magnesium, and calcium concentrations. The isolated globoids of cottonseed had low amounts of protein, carbohydrate, and lipid and 60% and 10%, respectively, of phytate and metals (potassium, magnesium, and calcium) [46]. Major components of globoids from peanuts [42] were protein (35.1%), phytate (28%), and metals (5%). The isolated phytate-rich particles of Great Northern beans [47] contained 34.3% protein, 30% carbohydrate, 26.6% phytate, and 3% metals. Over 90% of the compounds of the isolated particles of rice bran were reported to contain phytate, potassium, and magnesium (Table 3.1). The isolated particles of rice bran and embryo had low amounts of protein and carbohydrate.

Phytate occurs primarily as a potassium-magnesium salt in rice [18,28], wheat [26], broad beans [27], and sesame seeds [48], and as a calcium-magnesium-potassium salt in soybeans [27,34] and Great Northern beans [47]. Prattley and Stanley [34] indicated that phytate in protein bodies of soybeans is present in water-soluble as well as water-insoluble forms as salts of metals and proteins. Great Northern beans contain phytate in the water-soluble and water-insoluble forms [47,49]. Reddy et al. [49] found that phytate is present as a water-soluble salt with a molecular weight <1000 daltons and water-insoluble complex with molecular weight >1000 daltons in Great Northern beans. However, the exact water-soluble forms of phytate in Great Northern beans have not been identified. Lott et al. [32] suggested that most phytate in peas is water soluble and present as potassium phytate. Further studies are needed to identify the chemical form(s) in which phytate occurs in many grains and seeds.

TABLE 3.1. Chemical Composition of Phytic Acid-Containing Particles[a] Isolated from Cottonseed, Peanuts, Soybeans, Great Northern Beans (GNB), and Rice.

Composition (Weight %)	Globoids of Cottonseed	Globoids of Peanuts[b]	Globoids of Soybeans	Isolated Particles of GNB	Isolated Particles of	
					Rice Bran	Rice Embryo
Nitrogen	0.70	—	—	—	—	—
Protein	—	35.10	—	34.30	0.66	0.84
Carbohydrate	1.35	N.S.[c]	—	30.00	0.72	0.78
Phytic acid	—	28.00	23.80	26.58	66.68	70.30
Organic phosphorus	13.85	7.30	—	—	—	—
Inositol	13.21	7.00	—	—	—	—
Minerals						
Potassium	6.40	2.00	3.61	0.26	14.55	22.83
Magnesium	1.70	2.50	1.59	2.05	11.97	9.50
Calcium	1.30	0.50	0.86	0.64	0.83	0.73
Moisture	9.71	8.60	—	—	—	—

[a]For details of isolation procedures, see References [28,29] for rice, [46] for cottonseed, [34] for soybeans, [42] for peanuts, [47] for Great Northern beans.
[b]Globoids of peanuts were obtained after centrifugation at 20,000× g.
[c]N.S. indicates that result is not significant.

3. PHYTATE DISTRIBUTION AND CONTENT

In many seeds and grains, phytate accumulates during seed development and reaches its highest level at seed maturity [18,21–23,50,51]. Yao et al. [50] reported that during soybean maturation, phytate content increased from 0.87 to 1.26% on a dry weight basis (Table 3.2). Phytate content increased from 1.0 to 2.2 mg between stage 1 and stage 3, when calculated on a per bean basis. Welch et al. [52] found that during maturation of peas, the phytate content increased from 0.16 to 1.23%. The proportional increase of phytate at four developmental stages of seed maturity in soybeans and winged beans has also been reported [23,51].

3.1. PHYTATE DISTRIBUTION

In monocotyledonous seeds such as cereals, phytate is associated with specific components or parts within the grain and can be preferentially separated with those components. The starchy endosperm of wheat and rice grains is almost devoid of phytate, as it is concentrated in the germ and aleurone layers (pericarp) of the cells of the grain (Table 3.3). Corn differs from most other cereal grains, as 88% of phytate is concentrated in the germ portion of the grain [53]. Corn endosperm has small amounts of phytate (3.2% of total phytate). Rice and wheat germ portions contain appreciable amounts of phytate, but a major portion of phytate is found in the aleurone layers or pericarp. Of the total phytate, 84–88% has been reported to be present in the bran part of the rice [54]. In pearl millets, the majority of the phytate appears to be present in germ and bran fractions [55]. However, the distribution of phytate in different fractions of pearl millets has not been determined.

In dicotyledonous seeds, such as beans and other seeds, phytate is distributed throughout the cotyledon and located within the subcellular inclusions of protein bodies [19,20]. Ferguson and Bollard [56] found that 99% of the phytate in dry peas was in the cotyledons, and 1% was in the embryo axis. Phytate phosphorus represents about 65% of the total phosphorus in the pea cotyledons and 20% of the total phosphorus in the pea embryo axis. Beal and Mehta [57] reported

TABLE 3.2. Phytate Content of Soybean Seeds at Three Stages of Maturity.[a]

Stage of Maturity	Harvest Date	Phytate (%)
Stage 1	September 20	0.87
Stage 2	September 26	1.08
Stage 3 (mature)	October 3	1.26

[a]Soybeans were planted on June 16, 1980.
See Reference [50] for details.

TABLE 3.3. Phytate Concentration and Distribution in Morphological Components of Cereals and Legumes.

Cereal or Legume	Morphological Component	Phytate (%)	Distribution[a] (%)
Commercial hybrid corn	Whole	0.89	—
	Endosperm	0.04	3.20
	Germ	6.39	88.00
	Hull	0.07	0.04
High lysine corn	Whole	0.96	—
	Endosperm	0.04	3.00
	Germ	5.72	88.90
	Hull	0.25	1.50
Soft wheat	Whole	1.14	—
	Endosperm	0.004	2.20
	Germ	3.91	12.90
	Aleurone layer (bran)	4.12	87.10
Brown rice	Whole	0.89	—
	Endosperm	0.01	1.20
	Germ	3.48	7.60
	Pericarp	3.37	80.00
Pearl millet	Whole	0.89	—
	Endosperm	0.32	—
	Germ	2.66	—
	Bran	0.99	—
Peas	Whole	0.79	—
	Cotyledon	0.78	88.70
	Germ	1.23	2.50
	Hull	0.01	0.10

[a]Percentage of phytate in the component part; see References [53,55,57] for details.

that more than 88% of the phytate is present in the pea cotyledon (Table 3.3). The hull or seed coat fractions contain little or no phytate.

3.2. PHYTATE CONTENT

3.2.1. Cereals and Cereal Products

The phytate contents of cereals, cereal products, and cereal-based foods are presented in Table 3.4. The amount of phytate varies from 0.06 to 2.22% in cereals, 0.08 to 6% in cereal-milled fractions and protein products, 0.03 to 2.41% in various types of breads and other products, 0.05 to 3.29% in ready-to-eat cereal products, and 0.06 to 1.38% in infant cereals. Among all cereals, polished rice contains the lowest amounts (<0.25%) of phytate. Some of the ready-to-eat cereals such as wheat cereals (100% Bran, Shredded Wheat, Wheaties, Raisin Bran), and infant cereals have the highest phytate content (Table 3.4). White

TABLE 3.4. Phytate Contents of Cereals, Cereal Products, and Cereal-Based Foods.

Cereal/Cereal Products	Phytate (%)	References
Cereals		
Wheat	0.39–1.35	[62,64,65,72–78,223]
Hard wheat	0.84	[63]
Soft wheat	0.94–1.13	[53,78]
Durum wheat	0.88–1.16	[66,73,74,79]
Corn	0.75–2.22	[6,66,73,74,76]
Yellow corn	0.72	[80]
White corn	0.76	[80]
Corn (high lysine)	0.97–0.99	[66]
Popcorn (unpopped)	0.62–0.68	[3]
Triticale	0.50–1.89	[59,76,78]
Oat	0.42–1.16	[62,67,72,74–76,80,81]
Barley	0.38–1.16	[62,65,72,73,76,83–84]
Rye	0.54–1.46	[65,72,73,76,85,86]
Sorghum	0.67–1.35	[6,67,76,87,88]
Sorghum (low tannin)	0.57	[69]
Sorghum (high tannin)	0.96	[69]
Common millets	0.50–0.70	[6,89]
Proso millets	0.60–1.67	[6,90]
Pearl millets	0.18–0.99	[6,55,91–94]
Pearl millets (dehulled)	0.30–0.52	[94]
Ragi	0.55–0.67	[6,70,95]
Wild rice	2.20	[73]
Brown rice (long-grain)	0.84–0.99	[6,66,74]
Polished rice (long-grain)	0.34–0.60	[6,60,83]
Polished rice (medium-grain)	0.14–0.19	[61,74]
Polished rice (short-grain)	0.14	[61]
White rice (enriched)	0.23	[96]
Glutinous rice (Vietnam)	0.23	[97]
Rice (Basmati)	0.06	[98]
Rice (Egyptian)	0.17	[98]
Cereal milled fractions and protein products		
Wheat flour	0.25–1.37	[3,78,99,100,222]
Wheat flour (India)	0.50–0.55	[101]
Wheat bran	2.02–5.27	[64,65,73,76,81,102-104]
Wheat gluten	2.13	[105,106]
Wheat germ	0.08–1.14	[97,222]
Corn meal	0.79–1.07	[3,222]
Corn germ	1.94	[3]
Rye flour	0.33–1.08	[3,78,222]
Triticale flour	0.18–0.48	[3,59]
Oat meal	0.89–2.40	[3,73,222]
Oat white flour	0.40	[67]
Oat bran	0.60–1.42	[67,76,104]
Oat groats	1.37	[76]
Rice bran	2.59–6.00	[64,65,76]
Wheat protein concentrate	1.88–2.70	[107,108]

TABLE 3.4. **(continued).**

Cereal/Cereal Products	Phytate (%)	References
Cereal-based foods		
Wheat breads		
Whole wheat breads	0.43–1.05	[3,71,74,109,110]
Whole wheat bread (flat)	0.43–0.83	[71,74,111]
Wheat breads	0.28–1.00	[71,96]
Honey wheat berry bread	0.73	[71]
Wheat bread (high fiber)	0.37	[3]
White breads		
White bread	0.03–0.23	[3,73,109,110]
Other breads		
Corn bread	1.36	[110]
Rye bread	0.03–0.41	[5,110]
French bread	0.03	[110]
Raisin bread	0.10	[110]
Mixed flour breads	0.03–0.19	[86]
Sour rye bread	0.03	[86]
Sour buckwheat bread	0.03	[86]
Coffee bread	0.08	[86]
Crisp breads	0.08–0.68	[5,86]
Pumpernickel bread	0.16	[110]
Plain hamburger bun	0.12	[3]
Pita bread	0.12–0.16	[3,112]
Indian chapalies (flat bread)	0.25–0.56	[101,111]
Norwegian flat bread	0.68	[3]
Iranian flat breads		
Bazari (leavened)	1.17	[58]
Sangak (leavened)	1.38	[58]
Tanok (unleavened)	2.41	[58]
Kuwaiti breads		
Brown pita bread	0.23	[98]
White pita bread	0.14	[98]
Brown toast	0.27	[98]
Iranian type bread	0.30	[98]
Pakistani flat breads and other foods		
Flat breads	0.67–0.68	[113]
Roti	0.47–0.52	[113]
Nan	0.04	[113]
Puri	0.03–0.04	[113]
Roasted corn	0.55	[80]
Popped corn	0.60	[80]
Corn chapati	0.36	[80]
Wheat porridge	0.16	[113]
Italian cereal foods		
Breads	0.06–0.26	[114]
Crackers	0.37–0.58	[114]
Biscuits	0.11–1.05	[114]

(*continued*)

TABLE 3.4. (continued).

Cereal/Cereal Products	Phytate (%)	References
Other foods		
Wheat bran muffins	0.77–1.27	[102,115]
English muffins	0.12	[3]
Wheat germ pancakes	1.76	[3]
Doughnut cake, sugar-coated	0.48	[3]
Brown rice, parboiled	1.60	[116]
Barley, parboiled	0.54	[3]
Corn chips	0.24–0.66	[3]
Corn tortillas	0.11–0.95	[117,220]
Crackers	0.09–0.34	[3]
Rabadi (Indian fermented food)	0.11–0.15	[118]
Kenkey (fermented corn)	1.60	[119]
Ready-to-eat-cereals		
Wheat cereal (40% Bran)	1.12	[3]
Wheat cereal (100% Bran)	3.29	[3]
Wheat cereal (Bran Chex)	1.39	[3]
Wheat cereal (Bran Flakes)	1.10	[3]
Wheat cereal (Raisin Bran)	0.72–1.83	[3,73]
Wheat cereal (Shredded Wheat)	0.95–1.53	[3,73,76]
Wheat cereal (Wheaties)	1.52	[3]
Other wheat cereals	0.06–0.69	[3,73]
Corn cereal (Corn Flakes)	0.05–0.09	[3,73,86,109]
Corn cereal (Corn Bran)	0.24	[3]
Corn cereal (Corn Pops)	0.10	[3]
Mixed grain cereals	0.18–0.52	[3]
Rice cereal (Rice Krispies)	0.18–0.24	[3,73]
Other rice cereals	0.14–0.19	[3]
Quick oats	0.99–1.03	[102,121]
Old fashioned oats	0.64	[121]
Instant oats with bran and raisins	0.99	[121]
Puffed oats	0.66	[86]
Puffed rice	0.07–0.11	[86]
Rolled oats	0.80–1.03	[5,222]
Instant porridge	0.50–0.68	[5]
Infant cereals		
Infant cereal	1.38	[73]
Barley cereal	1.00	[3]
Rice cereal (dry form)	0.98	[3]
Mixed grain cereal (dry, instant)	0.81	[3]
Weaning foods[a]	0.06–0.70	[120]
Children's biscuits	0.13	[5]

[a]Prepared with cereal and legume mixtures using processes such as roasting, malting, and fermentation.

breads, French, rye, raisin, mixed flour, sour rye, sour buckwheat, coffee, crisp, pita, pumpernickel, Kuwaiti, and some Pakistani flat breads have low amounts of phytate when compared to certain wheat breads and Iranian flat breads. The differences in phytate contents of various breads could be due to degree of fermentation during preparation and use of different extraction rate flours. These factors influence the phytate content in breads [58]. During fermentation of bread dough, some of the phytate in the dough is hydrolyzed by wheat and bacterial and yeast phytases, which may result in reduced phytate content, especially in breads made with white flours. As most of the phytate in cereals is located in the aleurone layers (bran), milling of cereals and subsequent separation of bran results in a significant reduction of phytate in flours [59]. Singh and Reddy [59] reported that the bran fractions of triticale, wheat, and rye had higher amounts of phytate than the corresponding flours.

Phytate phosphorus accounts for the major portion (>80%) of total phosphorus in cereals and cereal products. Of the total phosphorus, phytate phosphorus represents 73.7–81% in brown rice, 51–61% in polished rice [6,60,61], 60–80% in wheat [53,62,65], 55–70% in barley [62,65], 48.7–70.9% in oats [62,67], 38–66% in rye [59,62,65], 18–73% in triticale [59,65], 71–88% in corn [6,20,64,66], 87.1% in high lysine corn [20], 63.9–90.5% in sorghum [6,68], 88.9% in high tannin sorghum [69], 58.3–78% in ragi (common millets) [6,70], 70.4% in foxtail millets [6], 64–85.7% in rice bran [6,64,65], 49.6–93% in wheat bran [6,64,65,121], 59–76% in wheat middlings [65], 54% in oat bran [67], 80% in oat white flour [67], 84% in Quaker instant oats with bran and raisins [121], 34% in Quaker® old fashioned oats [121], 15–33% in white breads [71,121], 55% in brown bread [71], and 38–66% in whole wheat breads [71].

3.2.2. Beans and Bean Products

Phytate content ranges from 0.17 to 9.15% in whole beans, 0.58 to 4.20% in bean flours and bean protein products, 0.05 to 5.20% in bean-based foods, and 0.004 to 0.03% in crude soybean oil (Table 3.5). Among all whole beans, dolique beans contain the highest amount of phytate, i.e., 5.92–9.15% Crude soybean oil and soy milk appear to contain the lowest amounts (<0.12%) of phytate. The wide variations reported for phytate content within the same type of bean may be due to differences in cultivars, growing conditions, and locations. Because most of the phytate in beans is distributed in the cotyledons, removal of the hull or seed coat typically leads to a higher phytate content of beans (Table 3.6) [57,111,122–124].

In beans and bean products, phytate phosphorus content constitutes a major portion of the total phosphorus. For example, of the total phosphorus, phytate phosphorus accounts for 50–70% in soybeans [6,62,125], 27–87% in lentils [6,125], 40–95% in chickpeas [6,125,126], 39.5–95% in broad beans [20,125],

TABLE 3.5. **Phytate Content of Whole Beans, Bean Flours and Protein Products, and Bean-Based Foods.**

Bean/Bean Products	Phytate (%)	References
Whole beans		
Adzuki beans	0.86	[77]
African yam beans	0.10	[133]
Dolichos beans	1.00–1.40	[130]
Winged beans	0.63–2.67	[6,51,131,134–136]
Soybeans	1.00–2.22	[6,62,66,87,99,137,138]
Broad beans (faba beans)	0.51–1.77	[97,123,125,139–143,222]
Dolique beans	5.92–9.15	[125]
Peas	0.22–1.22	[3,52,83,125,139,143–144]
Dwarf grey peas	0.40	[145]
Early Alaskan Pea	0.67	[145]
Pigeon pea	0.22–7.00	[6,111,125,146–149]
Pigeon pea (early maturing)	1.25–1.64	[150]
Pigeon pea (medium maturing)	1.10–1.60	[150]
Pigeon pea (late maturing)	1.10–1.63	[150]
Lentil	0.27–1.05	[6,83,97,99,111,139,151–155]
Linseed	2.15–2.78	[125]
Peanuts, Spanish	1.88	[96]
Peanuts, Bambara	0.29	[148]
Peanuts	1.05–1.76	[73,222]
Chickpeas	0.28–1.26	[3,6,83,125–126,139,152,156]
Cowpeas	0.37–1.45	[3,6,126,128,138,152,157–158]
Moth bean	0.85–0.92	[159]
Green gram	0.59–1.10	[6,111,147,160]
Black gram	0.72–1.46	[6,111,122,147,155–156,161]
Lima beans	0.23–2.52	[63,73–74,127,142,162,222]
Lima beans (immature, raw)	0.70	[3]
Blackeye peas	0.91–1.38	[3,109,111]
Lupine	0.20–1.20	[64–65,163]
Kidney beans	0.89–1.57	[109,111,119,149]
California small white beans	0.26–1.03	[63,74,142,164]
Pinto beans	0.61–2.38	[3,96,129,142,165–166,185]
Navy beans	0.74–1.78	[6,73,129]
Great Northern beans	0.50–2.70	[47,124,129,167,222]
Small white beans	0.55–1.80	[124,129,185]
Red Mexican beans	0.54–1.10	[129]
Red kidney beans	1.20–2.06	[129,165]
Small red beans	1.30–2.07	[124,185]
Sanilac beans	2.75	[166]
Cranberry beans	2.63	[166]
Velvet beans	0.60–2.48	[168]
Viva pink beans	2.16	[166]
Black beauty beans	1.04–2.93	[124,142]
Light red kidney beans	1.20–2.63	[124,164]
Dark red kidney beans	2.86	[124]
Yellow peas, split	0.54–1.68	[3,116,151]

TABLE 3.5. (continued).

Bean/Bean Products	Phytate (%)	References
Green peas, split	0.85	[96]
Sweet peas	0.38	[98]
Rice bean	0.17	[169]
Indian *Bauhinia* bean	0.50	[170]
Indian tribal bean	0.50–0.63	[171]
Bean flours and protein products		
Soy flour, full-fat	1.24–1.30	[132,172]
Soy flour, defatted	1.52–2.25	[102,116,132]
Soy flakes, defatted	1.52–1.84	[66,96–97]
Soy protein concentrate	1.24–2.17	[81,102,108,115,172–173]
Soy protein isolate	1.40–2.11	[81,102,108,132,173]
Soy beverage	1.24	[172]
Soybean oil, crude	0.004–0.03	[174]
Peanut flour	1.50–1.94	[81]
Peanut meal, defatted	1.70	[41]
Linseed meal, defatted	4.20	[175]
Pea protein concentrate	0.58–1.90	[99,143,151]
Faba bean protein concentrate	1.48–3.39	[99,176]
Lentil protein concentrate	1.02	[151]
Bean-based foods		
Soy milk	0.05–0.11	[177]
Tempe	0.67–1.08	[132,137,178]
Tofu	1.46–2.90	[76,87,117,132]
Calcium-tofu	1.60–1.91	[179]
Magnesium-tofu	1.49–1.77	[179]
Idli	0.54	[60]
Wadies[a]	1.00	[161]
Khaman	5.20	[180]
Dhokla	1.80	[181]
Oncom	0.70	[182]
Peanuts, toasted and salted	1.00	[3]
Ewa-Ibeji[b]	1.00	[183]
Moin-Moin[c]	1.00	[183]
Falafel	0.11	[221]
Foul Mudames	0.17	[221]
Balila	0.18	[221]
Foul Jerra	0.39	[221]
Chicken analog	0.27	[184]
Ham analog	0.12	[184]
TVP[d] pork	1.42	[184]
TVP bacon	0.95	[184]
TVP ham	1.26	[184]
TVP beef	1.36	[184]
TVP beef chunks	1.36	[184]

[a]Fermented black gram product.
[b]Product made from cowpeas.
[c]Product made from cowpeas.
[d]TVP = texturized vegetable protein from soybeans.

TABLE 3.6. Phytate Content (%) of Whole Beans and Bean Cotyledons.

Cultivar	Whole Beans	Bean Cotyledon
U.S. dry beans		
Great Northern	2.04	3.26
Small white	1.16	1.63
Sanilac	2.75	2.94
Cranberry	2.63	3.39
Viva pink	2.16	2.91
Pinto	2.38	2.56
Light red kidney	2.63	3.47
Dark red kidney	2.86	3.67
Small red	2.07	3.05
Black beauty	2.93	3.61
Other beans		
Chickpeas	0.56	0.92–1.05
Black gram	1.46	1.40–1.70
Broad beans (faba beans)	0.76–1.62	0.92–1.90
Lentils	0.44–0.50	0.49–0.53
Lupine	0.30–0.35	0.18–0.24
Peas	0.79	0.78

Compiled from References [57,111,123,124,139,142,143,186].

36–53% in peas [65,125], 75–76% in pigeon peas [6,125], 70–87% in linseed [125], 31–60% in lima beans [127], 63.2–69% in green gram [6,63,126], 37–54% in dolique beans [125], 29.8–71.8% in cowpeas [6,128], 74.4–79% in black gram [6,122], 57–81% in navy beans [129], 68–72% in red kidney beans [129], 55–80% in Great Northern beans [47,129], 73.2–93.3% in dolichos beans [130], 57–81.6% in peanuts [6,20], 44–73% in winged beans [16,51,131], 70% in California small white beans [63], 20–54.7% in lupine [64,65], 58.6% in velvet beans [6], 77% in tempe and 94.5% in tofu [132], 60–60.9% in soybean meals [6], 87% in defatted soy flour [132], and 62% in soy protein isolate [132].

3.2.3. Tubers, Fruits, Leafy Vegetables, Nuts, and Other Foods

Among all leafy products and vegetables, tomatoes and okra have high amounts of phytate (Table 3.7). Most fruits contain low amounts of phytate. However, some fruits, namely apples, apricots, bananas, pears, peaches, and grapefruit, contain no detectable phytate [3,5]. Nuts are reported to contain high amounts of phytate compared with tubers, fruits, and leafy products and vegetables. In tubers and some fruits, phytate phosphorus accounts for 20–34% of total phosphorus [6].

TABLE 3.7. Phytate Content of Tubers, Fruits, Leafy Vegetables, and Other Foods.

Products	Phytate (%)	References
Tubers		
Potato	0.01–0.18	[5,6,223]
Yam	0.04–0.29	[6,119,223]
Sweet potato	0.07–0.32	[6,119]
Taro	0.32	[6]
Cassava	0.10–0.19	[3,6,119,223]
Sugar beet	0.01	[76]
Gari (fermented cassava product)	0.12	[119]
Fruits		
Plantain (unriped)	0.10–0.18	[6,119,223]
Dates	0.14	[6]
Bread fruit	0.14	[6]
Jack fruit	0.14	[6]
Mango raw/ripe	0.14	[149]
Avacado	0.51	[220]
Strawberry	0.13	[5]
Others		
Coconut meal (extracted)	1.17	[6]
White sesame seeds	1.44	[77]
Sesame seeds (dehulled powder)	5.36	[76]
Coriander seeds	1.28–1.31	[187]
Leafy products and vegetables		
Spinach	0.01–0.07	[5,6]
Chinese cabbage	0.08	[149]
Red cabbage, raw	0.03	[5]
Okra	0.29	[3]
Cauliflower, raw	0.05	[5]
Rhubarb, raw	0.01	[5]
Carrots, raw	0.09	[5,223]
Cucumber	0.05	[5]
Tomato	0.04–0.31	[5,223]
Lettuce	0.04	[5]
Kale, raw	0.13	[3]
Raddish, fresh	0.11	[3]
Nuts		
Hazelnut	1.91	[5]
Brazll nut	1.97–6.34	[3,5]
Walnut	0.65–2.38	[5,222]
Almond	1.35–3.22	[3,5]
Cashew nuts	0.63–1.97	[3,222]

3.3. EFFECTS OF ENVIRONMENTAL AND OTHER FACTORS ON THE PHYTATE CONTENT

Environmental fluctuations, growing locations, irrigation conditions, types of soils, various fertilizer applications, and year during which a cultivar or variety is grown influences phytate content of seeds and grains. Bassiri and Nahapetian [188] observed that wheat varieties grown under dry land conditions had lower concentrations of phytate compared with the ones grown under irrigated conditions. Nahapetian and Bassiri [189], Singh and Reddy [59], Miller et al. [190,191], Feil and Fossati [192], and Simwemba et al. [55] reported variations in the phytate content of triticales, wheat, rye, oats, and pearl millet grown at different locations and in different years. A variation in phytate content of navy beans was observed by Proctor and Watts [193] as a result of variety and location effects. Griffiths and Thomas [194] reported that the phytate phosphorus content of broad beans, when calculated as a percent of total phosphorus, increased significantly from 39.5 to 57.7% for beans grown under greenhouse conditions as opposed to field conditions. Application of different fertilizer (nitrogen and phosphorus) regimes to field crops during their growth is reported to increase phytate content of their seeds and/or grains [82,191,195–198].

4. DIETARY INTAKE OF PHYTATE

Estimates of phytate daily intake in different parts of the world are presented in Table 3.8. These estimates are compiled from various nutritional studies based on different methods of data collection and analysis. Harland and Peterson [199] suggested that the average American (weighing 75 kg) consumes about 750 mg phytate per day. However, several studies [200–203] conducted in the United States indicate a wide variation in daily intake of phytate (Table 3.8). These variations could be due to differences in data collection methods and consumption of foods rich in phytates. Harland et al. [202] conducted a nutrition assessment of a lacto-ovo-vegetarian Trappist monk community in 1977 and 1987 and found a significant decrease in intake of phytate from 4569 mg/day in 1977 to 972 mg/day in 1987. They attributed this variation to decreased intake of phytate-containing foods such as cereals and increased consumption of milk, milk products, and others. In general, vegetarians consume a higher amount of phytate compared to nonvegetarians. Phytate intake also varies with season. For instance, Ellis et al. [200] reported that phytate intake in self-selected diets of omnivorous females varied from 585 mg/day in spring to 734 mg/day in winter and from 781 mg/day in spring to 762 mg/day in winter for omnivorous males. Cereals were the major source of phytate for the omnivorous males consuming self-selected diets.

TABLE 3.8. Dietary Intake of Phytates in Various Countries.

Country	Groups	Phytate (mg/day) Mean	Phytate (mg/day) Range/S.D.[a]	References
U.S.	Non-lactating young women (18–24 yrs)	395	±14	[201]
	Lacto-ovo-vegetarian Trappist monks (1977)	4569	615–5,770	[202]
	Lacto-ovo-vegetarian Trappist monks (1987)	972	58–3186	[202]
U.S.	College students (19–35 yrs)	1293	198–3098	[203]
U.S.	Omnivorous females	631	585–734	[200]
	Omnivorous males	746	714–781	[200]
Canada	Preschool girls (4–5 yrs)	250	132–318	[209]
	Preschool boys (4–5 yrs)	320	203–463	[209]
Mexico	Toddlers (18–30 months)	1666	±650	[210]
Guatemala	Pregnant women (15–37 yrs)	2254	877–4708	[213]
U.K.	Students and faculty staff	670	500–840	[205]
U.K.	—	—	600–800	[204]
U.K.	Infants (1–18 months)	—	0–200	[219]
Sweden	—	180	—	[206]
Italy	—	219	112–1367	[207]
Egypt	Toddlers (18–30 months)	796	±248	[210]
Taiwan	Graduate students and faculty members (20–60 yrs)	780	±260	[218]
Nigeria	—	2100	2000–2200	[119]
Malawi	Preschool girls (4–6 yrs)	1675	1621–1729	[211]
	Preschool boys (4–6 yrs)	2010	1857–2161	[211]
Ghana	Omnivorous children	578	±161	[212]
Papua New Guinea	Children	569	±561	[212]
Kenya	Toddlers (18–30 months)	1066	±324	[210]
India	Faculty families	670	596–742	[216]
India	Children (4–9 yrs)	940	720–1160	[215]
	Adolescents (10–19 yrs)	1565	1350–1780	[215]
	Adults (20–45 yrs)	2030	1560–2500	[215]
	Adults (>45 yrs)	1685	1290–2080	[215]
India	Lacto-vegetarian young women (16–20 yrs)	840	—	[214]
	Non-vegetarian young women (16–20 yrs)	848	—	[214]
Gambia	Infants (1–17 months)	—	10–560	[219]

[a]Standard deviation.

The estimates of daily phytate intakes in the U.K. range from 600–800 mg/day [204,205]. About 70% of this phytate is derived from cereal products, 20% from fruits, and the remainder from vegetables and nuts. On the other hand, Swedish people appear to consume diets containing very low (180 mg/day) levels of phytate [206]. Analysis of 13 Italian diets indicated a wide range for daily phytate intake from 112–1367 mg/day [207]. This large variation is due to the consumption of a variety of refined cereal and pasta products. Carnovale et al. [207] reported that a value of 219 mg/day may be indicative of the average national phytate intake in Italy. Average phytate intake in Finland has been estimated to be 370 mg/day, mainly from cereal products [86,208].

In countries such as Mexico and Kenya, the phytate intake of toddlers and preschool children is much higher than in Western countries such as Canada (Table 3.8). The estimated intake of phytate for toddlers (18–30 months) is reported to be high in Mexico and Kenya and low in Egypt, where the main diet of toddlers consisted of yeast-leavened bread [210]. The phytate intake estimates among Malawi children consuming a vegetarian and fish diet appear to be three times higher than Ghana and Papua New Guinea children [211,212]. Fitzgerald et al. [213] reported that pregnant Guatemalan women on average consume high amounts (2254 mg/day) of phytate with 68% of it from corn tortillas.

Indian and Nigerian diets also provide high phytate intake because they consist mainly of cereals and beans. The large variation reported for intake of phytate in India may be due to the differences in the socioeconomic groups (urban vs. rural) involved in these studies [214–216]. The diets of urban populations were reported to be varied and contained more vegetables, fruits, milk, and leafy vegetables. This variation contributed to lower phytate intake in urban populations compared with the variation in phytate intake of rural vegetarian Indian children, adolescents, and adults. On the other hand, Indian vegetarian diets consisted mainly of cereal-based cooked foods (unleavened chapaties) and cooked beans that are known to contain high amounts of phytate [215]. Bindra et al. [217] indicated that lacto-ovo-vegetarian diets of Punjabi immigrants in Canada contained higher levels of phytate than the diets reported for both omnivorous and vegetarian American diets. The phytate intake for Punjabi immigrants was estimated to be 1487 mg/day. The Punjabi diets consisted mainly of chapaties (prepared from 100% extraction wheat flour) and beans. Preparing whole meal chapaties does not involve leavening; therefore, the phytate is not destroyed or hydrolyzed before consumption. An average Nigerian may consume as much as 2000 to 2200 mg of phytate per day using a typical menu of kidney bean balls, rice, plantains, yams, gari, and pudding [119]. This is three times the estimated intake of phytate in the North American population. Middle Eastern inhabitants also consume very high amounts of phytate in their diets.

The high intake of phytate may have unfavorable effects on the availability of minerals [4]. For proper evaluation of the effect of phytate on mineral status, one needs to know the phytate content of food and the intakes of minerals and fiber in the dicts [201]. Harland and Oberleas [3] presented the phytate content for various foods and drinks in three sets of units: mg phytate per serving, mg phytate per 100 g edible portion, and mg phytate per 100 g dry weight of the material. The phytate intake in a mixed or complex diet can be calculated by using these three sets of phytate units. The role of phytate in absorption of minerals and in cancer and disease prevention are discussed in Chapters 12 and 13.

5. REFERENCES

1. Rose, A.R. 1912. A resume of the literature on inosite phosphoric acid with special reference to the relation of that substance to plants. *Biochem. Bull.* 2:21–49.
2. Averill, H.P. and King, C.G. 1926. The phytin content of foodstuffs. *J. Amer. Chem. Soc.* 48:724–728.
3. Harland, B.F. and Oberleas, D. 1987. Phytate in foods. *World Rev. Nutr. Diet.* 52:235–259.
4. Reddy, N.R., Pierson, M.D., Sathe, S.K., and Salunkhe, D.K. 1989. *Phytates in Cereals and Legumes*. CRC Press, Inc., Boca Raton, FL.
5. Wolters, M.G.E., Diepenmaat, H.B., Hermus, R.J.J., and Voragen, A.G.J. 1993. Relation between *in vitro* availability of minerals and food composition: A mathematical model. *J. Food Sci.* 58:1349–1355.
6. Ravindran, V., Ravindran, G., and Sivalogan, S. 1994. Total and phytate phosphorus contents of various foods and feedstuffs of plant origin. *Food Chem.* 50:133–136.
7. Jackson, J.F., Jones, G., and Linskens, H.F. 1982. Phytic acid in pollen. *Phytochem.* 21:1255–1258.
8. Scott, J.J. and Loewus, F.A. 1986. Phytate metabolism in plants. In *Phytic Acid: Chemistry and Applications*, Graf, E. (Ed.), Pilatus Press, Minneapolis, MN, pp. 23–42.
9. Baldi, B.G., Franceschi, V.R., and Loewus, F.A. 1987. Localization of phosphorus and cation reserves in *Lilium longiflorum* pollen. *Plant Physiol.* 83:1018–1021.
10. Dyer, W.J., Wrenshall, C.L., and Smith, G.R. 1940. The isolation of phytin from soil. *Science* 91:319–321.
11. Caldwell, A.G. and Black, C.A. 1958. Inositol hexaphosphate. III. Content in soils. *Proc. Soil. Sci. Soc. Amer.* 22:290–293.
12. Cosgrove, D.J. 1966. The chemistry and biochemistry of inositol polyphosphates. *Rev. Pure Appl. Chem.* 16:209–224.
13. Rapaport, S. 1940. Phytic acid in avian erythrocytes. *J. Biol. Chem.* 135:403–406.
14. Rapaport, S. and Guest, G.M. 1941. Distribution of acid-soluble phosphorus in the blood cells of various vertebrates. *J. Biol. Chem.* 138:269–275.
15. Oshima, M., Taylor, T.G., and Williams, A. 1964. Variations in the concentration of phytic acid in the blood of the domestic fowl. *Biochem. J.* 92:42–46.
16. Johnson, L.F. and Tate, M.E. 1969. Structure of phytic acids. *Can. J. Chem.* 47:63–73.
17. Isaacks, R.E., Kim, H.D., Bartlett, G.R., and Harkness, D.R. 1977. Inositol pentaphosphate in erythrocytes of a freshwater fish, peraracu (*Arapaima gigas*). *Life Sci.* 20:987–990.
18. Tanaka, K. and Kasai, Z. 1981. Phytic acid in rice grains. In *Antinutrients and Natural Toxicants in Foods*, Ory, R.L. (Ed.), Food and Nutrition Press, Westport, CT, pp. 239–260.
19. Lott, J.N.A. and Ockenden, I. 1986. The fine structure of phytate-rich particles. In *Phytic Acid: Chemistry and Applications*, Graf, E. (Ed.), Pilatus Press, Minneapolis, MN, pp. 43–55.
20. Lott, J.N.A. 1984. Accumulation of seed reserves of phosphorus and other minerals. In *Seed Physiology*, Murray, D.R. (Ed.), Academic Press, New York, NY, pp. 139–166.

21. Raboy, V. and Dickinson, D.B. 1987. The timing and rate of phytic acid accumulation in developing soybean seeds. *Plant Physiol.* 85:841–844.

22. Honke, J., Kozlowska, H., Vidal-Valverde, C., Frias, J., and Gorecki, R. 1998. Changes in quantities of inositol phosphates during maturation and germination of legume seeds. *Z. Lebensm Unters Forsch* A 206:279–283.

23. Mebrahtu, T., Mohamed, A., and Elmi, A. 1997. Accumulation of phytate in vegetable-type soybean genotypes harvested at four developmental stages. *Plant Foods Human Nutr.* 50:179–187.

24. Ogawa, M., Tanaka, K., and Kasai, Z. 1979a. Phytic acid formation in dissected ripening rice grains. *Agric. Biol. Chem.* 43:2211–2213.

25. Ogawa, M., Tanaka, K., and Kasai, Z. 1979b. Energy-dispersive x-ray analysis of phytin globoids in aleurone particles of developing rice grains. *Soil Sci. Plant Nutr.* 25:437–448.

26. Tanaka, K., Yoshida, T., and Kasai, Z. 1974. Radio autographic demonstration of the accumulation site of phytic acid in rice and wheat grains. *Plant Cell Physiol.* 15:147–151.

27. Lott, J.N.A. and Buttrose, M.S. 1978. Globoids in protein bodies of legume seed cotyledons. *Aust. Plant Physiol.* 5:89–111.

28. Ogawa, M., Tanaka, K., and Kasai, Z. 1975. Isolation of high phytin containing particles from rice grains using an aqueous polymer two phase system. *Agric. Biol. Chem.* 39:695–700.

29. Ogawa, M., Tanaka, K., and Kasai, Z. 1977. Note on the phytin-containing particles isolated from rice scutellum. *Cereal Chem.* 54:1029–1034.

30. Wada, T. and Lott, J.N.A. 1997. Light and electron microscopic and energy dispersive x-ray micro analysis studies of globoids in protein bodies of embryo tissues and the aleurone layer of rice (*Oryza sativa* L.) grains. *Can. J. Bot.* 75:1137–1147.

31. Lott, J.N.A. 1980. Protein bodies. In *Biochemistry of Plants*, Volume I, Tolbert, N.E. (Ed.), Academic Press, New York, pp. 589–623.

32. Lott, J.N.A., Goodchild, D.J., and Craig, S. 1984. Studies of mineral reserves in pea (*Pisum sativum*) cotyledons using low-water content procedures. *Aust. J. Plant Physiol.* 11:459–469.

33. Lott, J.N.A., Randall, P.J., Goodchild, D.J., and Craig, S. 1985. Occurrence of globoid crystals in cotyledonary protein bodies of *Pisum sativum* as influenced by experimentally induced changes in Mg, Ca, and K contents of seeds. *Aust. J. Plant Physiol.* 12:341–353.

34. Pratley, C.A. and Stanley, D.W. 1982. Protein-phytate interactions in soybeans. I. Localization of phytate in protein bodies and globoids. *J. Food Biochem.* 6:243–253.

35. Buttrose, M.S. 1978. Manganese and iron in globoid crystals of protein bodies from *Avena* and *Casuarina*. *Aust. J. Plant Physiol.* 5:631–639.

36. Jacobsen, J.V., Knox, R.B., and Pyliotis, N.A. 1971. The structure and composition of aleurone grains in the barley aleurone layer. *Planta* 101:189–192.

37. Liu, D.J. and Pomeranz, Y. 1975. Distribution of minerals in barley at the cellular level by x-ray analysis. *Cereal Chem.* 52:620–623.

38. Batten, G.D. and Lott, J.N.A. 1986. The influence of phosphorus nutrition on the appearance and composition of globoid crystals in wheat aleurone cells. *Cereal Chem.* 63:14–18.

39. Adams, C.A. and Novellie, L. 1975. Composition and structure of protein bodies and spherosomes isolated from ungerminated seeds of *Sorghum bicolor* (Linn) Moench. *Plant Physiol.* 55:1–6.

40. Mikus, M., Bobak, M., and Lux, A. 1992. Structure of protein bodies and elemental composition of phytin from dry germ of maize (*Zea mays* L.). *Bot. Acta.* 105:26–33.

41. Dieckert, J.W., Snowden, J.E., Moore, A.T., Heinzelman, D.C., and Altschul, A.M. 1962. Composition of some subcellular fractions from seed of *Arachis hypogaea*. *J. Food Sci.* 27:321–325.

42. Sharma, C.B. and Dieckert, J.W. 1975. Isolation and partial characterization of aleurone grains of *Arachis hypogaea* seed. *Physiol. Plant.* 33:1–7.

43. Morris, G.F.I., Thurman, D.A., and Boulter, D. 1970. The extraction and chemical composition of aleurone grains (protein bodies) isolated from seeds of *Vicia faba. Phytochem.* 9: 1707–1714.
44. Sobolev, A.M., Suvorou, V.I., and Buzulukova, N.P. 1977. Isolation of aleurone grains from seeds of several plants. *Soviet Plant Physiol.* 24:546–551.
45. Sobolev, A.M. 1966. On the state of phytin in the aleurone grains of mature and germinating seeds. *Soviet Plant Physiol.* 13:177–183.
46. Lui, N.S.T. and Altschul, A.M. 1967. Isolation of globoids from cottonseed aleurone grain. *Arch. Biochem. Biophys.* 121:678–684.
47. Reddy, N.R. and Pierson, M.D. 1987. Isolation and partial characterization of phytic acid-rich particles from Great Northern beans (*Phaseolus vulgaris* L.). *J. Food Sci.* 52:109–112.
48. O'Dell, B.L. and deBoland, A. 1976. Complexation of phytate with proteins and cations in corn germ and oilseed meals. *J. Agric. Food Chem.* 24:804–808.
49. Reddy, N.R., Sathe, S.K., and Pierson, M.D. 1988. Removal of phytate from Great Northern beans (*Phaseolus vulgaris* L.) and its combined density fraction. *J. Food Sci.* 53:107–110.
50. Yao, J.J., Wei, L.S., and Steinberg, M.P. 1983. Effect of maturity on chemical composition and storage stability of soybeans. *J. Amer. Oil Chem. Soc.* 60:1245–1249.
51. Kadam, S.S., Kute, L.S., Lawande, K.M., and Salunkhe, D.K. 1982. Changes in chemical composition of winged bean (*Psophocarpus tetragonolobus* L.) during seed development. *J. Food Sci.* 47:2051–2053.
52. Welch, R.M., House, W.A., and Allaway, W.H. 1974. Availability of zinc from pea seeds to rats. *J. Nutr.* 104:733–740.
53. O'Dell, B.L., deBoland, A., and Koirtyohann, R. 1972. Distribution of phytate and nutritionally important elements among the morphological components of cereal grains. *J. Agric. Food Chem.* 20:718–721.
54. Resurreccion, A.P., Juliano, B.O., and Tanaka, Y. 1979. Nutrient content and distribution in milling fractions of rice grain. *J. Sci. Food Agric.* 30:475–480.
55. Simwemba, C.G., Hoseney, R.C., Variano-Marston, E., and Zeleznak, K. 1984. Certain B-vitamin and phytic acid contents of pearl millets (*Pennisetum americanum* L. Leeke). *J. Agric. Food Chem.* 32:31–34.
56. Ferguson, I.B. and Bollard, E.G. 1976. The movement of calcium in germinating pea seeds. *Ann. Bot.* (London) 40:1047–1051.
57. Beal, L. and Mehta, T. 1985. Zinc and phytate distribution of peas: Influence of heat treatment, germination, pH, substrate, and phosphorus on pea phytate and phytase. *J. Food Sci.* 50: 96–100.
58. Reinhold, J.G. 1975. Phytate destruction by yeast fermentation in whole wheat meals. *J. Amer. Diet. Assoc.* 66:38–41.
59. Singh, B. and Reddy, N.R. 1977. Phytic acid and mineral compositions of triticales. *J. Food Sci.* 42:1077–1083.
60. Reddy, N.R. and Salunkhe, D.K. 1980. Effects of fermentation on phytate phosphorus and minerals of black gram, rice, and black gram-rice blends. *J. Food Sci.* 45:1708–1712.
61. Toma, R.B. and Tabekhia, M.M. 1979. Changes in mineral elements and phytic acid contents during cooking of three California rice varieties. *J. Food Sci.* 44:619–621.
62. Lolas, G.M., Palamids, N., and Markakis, P. 1976. The phytic acid, total phosphorus relationship in barley, oats, soybeans, and wheat. *Cereal Chem.* 53:867–871.
63. Chang, R., Schwimmer, S., and Burr, H.K. 1977. Phytate removal from whole dry beans by enzymatic hydrolysis and diffusion. *J. Food Sci.* 42:1098–1101.
64. Kirby, L.K. and Nelson, T.S. 1988. Total and phytate phosphorus content of some feed ingredients derived from grains. *Nutr. Rep. Intern.* 137:277–280.
65. Eechkhout, W. and Depaepe, M. 1994. Total phosphorus, phytate phosphorus, and phytase activity in plant feed stuffs. *Animal Feed Sci. Technol.* 47:19–29.

66. deBoland, A., Garner, G.B., and O'Dell, B.L. 1975. Identification and properties of phytate in cereal grains and oilseed products. *J. Agric. Food Chem.* 23:1186–1189.

67. Frolich, W. and Nyman, M. 1988. Minerals, phytate, and dietary fibre in different fractions of oat grain. *J. Cereal Sci.* 7:73–80.

68. Doherty, C., Faubion, J.M., and Rooney, L.W. 1982. Semi-automated determination of phytate in sorghum and sorghum products. *Cereal Chem.* 59:373–378.

69. Radhakrishnan, M.R. and Sivaprasad, J. 1980. Tannin content of sorghum varieties and their role in iron availability. *J. Agric. Food Chem.* 28:55–58.

70. Rao, P.U. and Deosthale, Y.G. 1988. *In vitro* availability of iron and zinc in white and colored ragi (*Eleusine coracana*): Role of tannin and phytate. *Plant Foods Hum. Nutr.* 38:35–41.

71. Tongkongchitr, U., Seib, P., and Hoseney, R.C. 1981. Phytic acid: II. Its fate during bread-making. *Cereal Chem.* 58:229–234.

72. Kikunaga, S., Takahashi, M., and Huzisige, H. 1985. Accurate and simple measurement of phytic acid contents in cereal grains. *Plant Cell Physiol.* 26:1323–1330.

73. Harland, B.F. and Prosky, L. 1979. Development of dietary fiber values for foods. *Cereal Foods World* 24:390–394.

74. Franz, K.B., Kennedy, B.M., and Fellers, D.A. 1980. Relative bioavailability of zinc from selected cereals and legumes using rat growth. *J. Nutr.* 110:2272–2283.

75. Morris, E.R. and Ellis, R. 1981. Phytate-zinc molar ratio of breakfast cereals and bioavailability of zinc to rats. *Cereal Chem.* 58:363–366.

76. Kasim, A.B. and Edwards, H.M. 1998. The analysis of inositol phosphate forms in feed ingredients. *J. Sci. Food Agric.* 76:1–9.

77. Matsunaga, A., Yamamoto, A., and Mizukami, E. 1989. Determination of phytic aicd in various foods by indirect photometric ion chromatography. *J. Food Hygienic Soc.* (Japan) 29:408–412.

78. Reddy, N.R. 1976. Milling and biochemical characteristics of triticale. M.S. Thesis, Alabama A & M University, Normal, AL.

79. Tabekhia, M.M. and Donnelly, B.J. 1982. Phytic acid in durum wheat and its milled products. *Cereal Chem.* 59:105–107.

80. Khan, N., Zaman, R., and Elahi, M. 1991. Effect of heat treatments on the phytic acid content of maize products. *J. Sci. Food Agric.* 54:153–156.

81. Harland, B.F. and Oberleas, D. 1986. Anion-exchange method for determination of phytate on foods: Collaborative study. *J. Assoc. Off. Anal. Chem.* 69:667–670.

82. Saastamoinen, M. and Heinonen, T. 1985. Phytic acid content of some oat varieties and its correlation with chemical and agronomical characters. *Ann. Agric. Fenn.* 24:103–105.

83. Larbi, A. and M'barek, E. 1985. Dietary fiber and phytic acid levels in the major items consumed in Morocco. *Nutr. Rep. Intern.* 31:469–476.

84. Ockenden, I., Falk, D.E., and Lott, J.N.A. 1997. Stability of phytate in barley and beans during storage. *J. Agric. Food Chem.* 45:1673–1677.

85. Fretzdorff, B. And Weipert, D. 1986. Phytic acid in cereals. I. Phytic acid and phytase in rye and rye products. *Z. Lebensm. Unters. Forsch.* 182:287–291.

86. Plaami, S. and Kumpulainen, J. 1995. Inositol phosphate content of some cereal-based foods. *J. Food Comp. Analysis* 8:324–335.

87. Cilliers, J.J.L. and Van Niekerk, P.J. 1986. LC determination of phytic acid in foods by post column colorimetric detection. *J. Agric. Food Chem.* 34:680–683.

88. Wheeler, E.L. and Ferrel, R.E. 1971. A method for phytic acid determination in wheat and wheat fractions. *Cereal Chem.* 48:312–315.

89. Ravindran, G. 1991. Studies on millets: Proximate composition, mineral composition and phytate and oxalate contents. *Food Chem.* 39:99–107.

90. Lorenz, K. 1983. Tannins and phytate content in proso millets (*Panicum milliaceum*). *Cereal Chem.* 60:424–426.

91. Khetarpaul, N. and Chauhan, B.M. 1991. Sequential fermentation of pearl millet by yeasts and lactobacilli effect on the antinutrients and *in vitro* digestability. *Plant Foods Human Nutr.* 41:321–327.

92. Kumar, A. and Chauhan, B.M. 1993. Effects of phytic acid on protein digestibility (*in vitro*) and HCl-extractability of minerals in pearl millet sprouts. *Cereal Chem.* 70:504–506.

93. Khetarpaul, N. and Chauhan, B.M. 1990. Effect of germination and pure culture fermentation by yeasts and lactobacilli on phytic acid and polyphenol content of pearl millet. *J. Food Sci.* 55:1180–1181.

94. Abdalla, A.A., El-Tinay, A.H., Mohamed, B.E., and Abdalla, A.H. 1998. Effect of traditional processes on phytate and mineral content of pearl millet. *Food Chem.* 63:79–84.

95. Sripriya, G., Antony, U., and Chandra, T.S. 1997. Changes in carbohydrates, free amino acids, organic acids, phytate, and HCl extractability of minerals during germination and fermentation of finger millet (*Eleusine coracana*). *Food Chem.* 58:345–350.

96. Graf, E. and Dintzis, F.R. 1982. Determination of phytic acid in foods by high performance liquid chromatography. *J. Agric. Food Chem.* 30:1094–1097.

97. Vinh, L.T. and Dworschak, E. 1985. Phytate content of some foods from plant origin from Vietnam and Hungary. *Die Nahrung* 29:161–166.

98. Mameesh, M.S. and Tomar, M. 1993. Phytate content of some popular Kuwaiti foods. *Cereal Chem.* 70:502–503.

99. Latta, M. and Eskin, N.A.M. 1980. A simple and rapid colorimetric method for phytate determination. *J. Agric. Food Chem.* 28:1313–1315.

100. O'Neill, I.K., Sargent, M., and Trimble, M.L. 1980. Determination of phytate in foods by phosphorus[31] flourier transform nuclear magnetic resonance spectrometry. *Anal. Chem.* 52:1288–1291.

101. Swaranjeet, K., Mohinder, K., and Bains, G.S. 1982. Chapaties with leavening and supplements: Changes in texture, residual sugars and phytic acid phosphorus. *Cereal Chem.* 59:367–372.

102. Ellis, R. and Morris, E.R. 1982. Comparison of ion-exchange and iron-precipitation methods for analysis of phytate. *Cereal Chem.* 59:232–233.

103. Lehrfeld, J. and Wu, Y.V. 1991. Distribution of phytic acid in milled fractions of Scout66 hard red winter wheat. *J. Agric. Food Chem.* 39:1820–1824.

104. Gualberto, D.G., Bergman, C.J., Kazemzadeh, M., and Weber, C.W. 1997. Effect of extrusion processing on the soluble and insoluble fiber and phytic acid contents of cereal brans. *Plant Foods Hum. Nutr.* 51:187–198.

105. Wallace, G.W. and Satterlee, L.D. 1977. Calcium binding and its effects on the properties of several food protein sources. *J. Food Sci.* 42:473–476.

106. Nelson, K.J. and Potter, N.N. 1979. Iron binding by wheat gluten, soy isolate, zein, albumin and casein. *J. Food Sci.* 44:104–107.

107. Ranhotra, G.S. 1972. Hydrolysis during breadmaking of phytic acid in wheat protein concentrate. *J. Food Sci.* 37:12–13.

108. Ranhotra, G.S., Loewe, R.J., and Puyat, L.V. 1974. Phytic acid in soy and its hydrolysis during breadmaking. *J. Food Sci.* 39:1023–1025.

109. Yoon, J.H., Thompson, L.U., and Jenkins, D.J.A. 1983. The effect of phytic acid on *in vitro* rate of starch digestibility and blood glucose response. *Amer. J. Clin. Nutr.* 38:835–842.

110. Harland, B.F. and Harland, J. 1980. Fermentative reduction of phytate in rye, white, and whole wheat breads. *Cereal Chem.* 57:226–229.

111. Davies, N.T. and Warrington, S. 1986. The phytic acid, mineral, trace element, protein, and moisture content of UK Asian immigrant foods. *Human Nutr. Appl. Nutr.* 40A:49–59.

112. Lathia, D. and Koch, M. 1989. Comparative study of phytic acid content, *in vitro* protein digestibility and amino acid composition of different types of flat breads. *J. Sci. Food Agric.* 47:353–364.

113. Khan, N., Zaman, R., and Elahi, M. 1986. Effect of processing on the phytic acid content of wheat products. *J. Agric. Food Chem.* 34:1010–1012.

114. Ceruti, G., Finoli, C., and Vecchio, A. 1984. Phytic acid in bran and in natural foods. *Boll. Chim. Farm.* 123:408–410.

115. Ellis, R. and Morris, E.R. 1983. Improved ion-exchange phytate method. *Cereal Chem.* 60:120–124.

116. Camire, A.L. and Clydesdale, F.M. 1982. Analysis of phytic acid in foods by HPLC. *J. Food Sci.* 47:575–578.

117. Poneros, A.G. and Erdman, J.W. 1988. Bioavailability of calcium from tofu, tortillas, non-fat dry milk and mozzarella cheese in rats: Effect of supplemental ascorbic acid. *J. Food Sci.* 53:208–210.

118. Gupta, M. and Khetarpaul, N. 1993. HCl extractability of minerals from rabadi—A wheat flour fermented food. *J. Agric. Food Chem.* 41:125–127.

119. Harland, B.F., Oke, O.L., and Felix-Phipps, R. 1988. Preliminary studies on the phytate content of Nigerian foods. *J. Food Comp. Anal.* 1:202–205.

120. Gahlawat, P. and Sehgal, S. 1993. Antinutritional content of developed weaning foods as affected by domestic processing. *Food Chem.* 47:333–336.

121. Davis, K. 1981. Proximate composition, phytic acid and total phosphorus of selected breakfast cereals. *Cereal Chem.* 58:347–350.

122. Reddy, N.R., Balakrishnan, C.V., and Salunkhe, D.K. 1978. Phytate phosphorus and mineral changes during germination and cooking of black gram (*Phaseolus mungo* L.) seeds. *J. Food Sci.* 43:540–542.

123. Griffiths, D.W. 1982. The phytate content and iron-binding capacity of various field bean (*Vicia faba* L.) preparations and extracts. *J. Sci. Food Agric.* 33:847–851.

124. Deshpande, S.S., Sathe, S.K., Salunkhe, D.K., and Cornforth, D.P. 1982. Effects of dehulling on phytic acid, polyphenols, and enzyme inhibitors of dry beans (*Phaseolus vulgaris* L.). *J. Food Sci.* 47:1846–1850.

125. Ferrando, R. 1983. Natural antinutritional factors present in European plant proteins. *Qual. Plant. Plant Food Hum. Nutr.* 32:455–465.

126. Kumar, K.G., Venkataraman, L.V., Jaya, T.V., and Krishnamurthy, K.S. 1978. Cooking characteristics of some germinated legumes: Changes in phytins, Ca^{++}, Mg^{++}, and pectins. *J. Food Sci.* 43:85–88.

127. Olghobo, A.D. and Fetuga, B.L. 1982. Polyphenols, phytic acid and other phosphorus compounds of lima beans (*Phaseolus lunatus* L.). *Nutr. Rep. Intern.* 26:605–611.

128. Ologhobo, A.D. and Fetuga, B.L. 1983. Investigations on the trypsin inhibitor, hemagglutinin, phytic and tannic acid content of cowpeas (*Vigna unguiculata* L.). *Food Chem.* 12:249–254.

129. Lolas, G.M. and Markakis, P. 1975. Phytic acid and other phosphorus compounds of beans (*Phaseolus vulgaris* L.). *J. Agric. Food Chem.* 23:13–15.

130. Deka, R.K. and Sarkar, C.R. 1990. Nutrient composition and antinutritional factors of *Dolichos lablab* L. seeds. *Food Chem.* 38:239–246.

131. Kotaru, M., Ikeuchi, T., Yoshikawa, H., and Ibuki, F. 1987. Investigations of antinutritional factors of the winged bean (*Psophocarpus tetragonalobus*). *Food Chem.* 24:279–285.

132. Thompson, D.B. and Erdman, J.W., Jr. 1982. Phytic acid determination in soybeans. *J. Food Sci.* 47:513–517.

133. Adewusi, S.R.A. and Falade, O.S. 1996. The effects of cooking on extractable tannin, phytate, sugars, and minerals solubility in some improved Nigerian legume seeds. *Food Sci. Technol. Intern.* 2:231–239.

134. Kantha, S.S., Heittiarachchy, N.S., and Erdman, J.W., Jr. 1986. Nutrient, antinutrient contents and solubility profiles of nitrogen, phytic acid and selected minerals in winged bean flour. *Cereal Chem.* 63:9–13.

135. Tan, N.H., Rahim, Z.H.A., Khor, H.T., and Wong, K.C. 1983. Winged bean (*Psophocarpus tetragonalobus* L.) tannin level, phytate content, and hemagglutinating activity. *J. Agric. Food Chem.* 31:916–917.

136. Kadam, S.S., Smithard, R.R., Eyre, M.D., and Armstrong, D.G. 1987. Effect of heat treatments on antinutritional factors and quality of proteins in winged beans. *J. Sci. Food Agric.* 39:267–274.

137. Sutardi, and Buckle, K.A. 1985. Phytic acid changes in soybeans fermented by traditional inoculum and six strains of *Rhisopus oligosporus*. *J. Appl. Bacteriol.* 58:539–543.

138. Ologhobo, A.D. and Fetuga, B.L. 1984. Distribution of phosphorus and phytate in some Nigerian varieties of legumes and some effects of processing. *J. Food Sci.* 49:199–201.

139. Gad, S.S., Mohamed, M.S., El-Zalaki, M.E., and Mohasseb, S.Z. 1982. Effect of processing on phosphorus and phytic acid contents of some Egyptian varieties of legumes. *Food Chem.* 8:11–19.

140. Eskin, N.A.M. and Wiebe, S. 1983. Changes in phytase activity and phytate during germination of two faba bean cultivars. *J. Food Sci.* 48:270–271.

141. Henderson, H.M. and Ankrah, S.A. 1985. The relationship of endogenous phytase, phytic acid and moisture uptake with cooking time in *Vicia faba* minro cv. Aladin. *Food Chem.* 17:1–11.

142. Kon, S. and Sanshuck, D.W. 1981. Phytate content and its effect on cooking quality of beans. *J. Food Process. Preserv.* 5:169–178.

143. Carnovale, E., Lugaro, E., and Lombardi-Boccia, G. 1988. Phytic acid in faba bean and pea: Effect on protein availability. *Cereal Chem.* 675:114–117.

144. Mazo, F., Aguirre, A., Castiella, M.V., and Alonso, R. 1997. Fertilization effects of phosphorus and sulfur on chemical composition of seeds of *Pisum sativum* L. and relative infestation by *Bruchus pisorhm* L. *J. Agric. Food Chem.* 45:1829–1833.

145. Chen, L.H. and Pan, S.H. 1977. Decrease of phytates during germination of pea seeds (*Pisum sativa*). *Nutr. Rep. Intern.* 46:125–128.

146. Sharma, Y.K., Tiwari, A.S., Rao, K.C., and Misra, A. 1977. Studies on chemical constituents and their influences on cookability in pigeon pea. *J. Food Sci. Technol.* (India) 14:38–40.

147. Rao, P.U. and Deosthale, Y.G. 1983. Effect of germination and cooking on mineral composition of pulses. *J. Food Sci. Technol.* (India) 20:195–197.

148. Igbediah, S.O., Olugbemi, K.T., and Akpapunam, A. 1994. Effects of processing methods on phytic acid level and some constituents in bambara groundnut (*Vigna subterranea*) and pigeon pea (*Cajanus cajan*). *Food Chem.* 50:147–151.

149. Ferguson, E.L., Gibson, R.S., Thompson, L.U., Ounpuu, S., and Berry, M. 1988. Phytate, zinc, and calcium contents of 30 East African foods and their calculated phytate: Zn, Ca: phytate, and [Ca] [phytate]/[Zn] molar ratios. *J. Food Comp. Anal.* 1:316–325.

150. Singh, U., Kherdekar, M.S., Sharma, D., and Saxena, K.B. 1984. Cooking quality and chemical composition of some early, medium, and late maturity cultivars of pigeon pea (*Cajanus cajan* L. Mill). *J. Food Sci. Technol.* (India) 21:367–372.

151. Davis, K. 1981. Effect of processing on consumption and *Tetrahymena* relative nutritive value of green and yellow peas, lentils, and white pea beans. *Cereal Chem.* 58:454–460.

152. Ummadi, P., Chenoweth, W.I., and Uebersax, M. 1995. The influence of extrusion processing on iron dialyzability, phytates, and tannins in legumes. *J. Food Process Preserv.* 19:119–131.

153. Cuadrado, C. Ayet, G., Robredo, L.M., Tabera, J., Villa, R., Pedrosa, M.M., Burbano, C., and Muzquiz, M. 1996. Effect of natural fermentation on the content of inositol phosphates in lentils. *Z. Lebensm. Unters Forsch.* 203:268–271.

154. Ayet, G. Burbano, Cuadrado, C., Pedrosa, M.M., Robredo, L.M., Muzquiz, M., Cuadra, C., Castano, A., and Osagie, A. 1997. Effect of germination under different environmental conditions on saponins, phytic acid and tannins in lentils (*Lens culinaris* L.). *J. Sci. Food Agric.* 74:273–279.

155. Agte, V., Joshi, S., Khot, S., Paknikar, K., and Chiplonkar, S. 1998. Effect of processing on phytate degradation and mineral solubility in pulses. *J. Food Sci. Technol.* (India) 35:330–332.

156. Chitra, U., Singh, U., and Rao, P.U. 1996. Phytic acid, *in vitro* protein digestibility, dietary fiber, and minerals of pulses as influenced by processing methods. *Plant Foods Hum. Nutr.* 49:307–316.

157. Elkowicz, K. and Sosulski, F. 1982. Antinutritive factors in eleven legumes and their air classified protein and starch factions. *J. Food Sci.* 47:1301–1304.

158. Farinu, G.O. and Ingrao, G. 1991. Gross composition, amino acid, phytic acid and trace element contents of thirteen cowpea cultivars and their nutritional significance. *J. Sci. Food Agric.* 55:401–410.

159. Khokar, S. and Chauhan, B.M. 1986. Antinutritional factors in moth bean (*Vigna aconitifolia*): Varietal differences and effects of the methods of domestic processing and cooking. *J. Food Sci.* 51:591–594.

160. Marero, L.M., Payumo, E.M., Aguinaldo, A.R., Matsumoto, I., and Homma, S. 1991. Antinutritional factors in weaning foods prepared from germinated cereals and legumes. *Lebensm. Wiss.U. Technol.* 24:177–181.

161. Yadav, S. and Khetarpaul, N. 1994. Indigenous legume fermentation: Effect on some antinutrients and *in vitro* digestibility of starch and protein. *Food Chem.* 50:403–406.

162. Egbe, I.A. and Akinyele, I.O. 1990. Effect of cooking on the antinutrional factors of lima beans (*Phaseolus lunatus* L.). *Food Chem.* 35:81–87.

163. Trugo, L.C., Donangelo, C.M., Duarte, Y.A., and Tavares, C.L. 1993. Phytic acid and selected mineral composition of seeds from wild species and cultivated varieties of lupine. *Food Chem.* 47:391–394.

164. Knuckles, B.E., Kuzumicky, D.D., and Betschart, A.A. 1982. HPLC analysis of phytic acid in selected foods and biological samples. *J. Food Sci.* 42:1257–1259.

165. Iyer, V.G., Salunkhe, D.K., Sathe, S.K., and Rockland, L.B. 1980. Quick-cooking beans (*Phaseolus vulgaris* L.): II. Phytates, oligosaccharides and antienzymes. *Qual. Plant. Plant Food Hum. Nutr.* 30:45–52.

166. Deshpande, S.S. and Cheryan, M., 1983. Changes in phytic acid, tannins and trypsin inhibitory activity on soaking of dry beans (*Phaseolus vulgaris* L.). *Nutr. Rep. Intern.* 27:371–377.

167. Sathe, S.K., Deshpande, S.S., Reddy, N.R., Goll, D.E., and Salunkhe, D.K. 1983. Effects of germination on proteins, raffinose oligosaccharides and antinutritional factors in the Great Northern beans (*Phaseolus vulgaris* L.). *J. Food Sci.* 48:1796–1800.

168. Siddhuraju, P., Vijayakumari, K., and Janardhan, K. 1996. Chemical composition and protein quality of the little-known legume, Velvet bean (*Mucuna pruriens* L. DC). *J. Agric. Food Chem.* 44:2636–2641.

169. Chau, C.F. and Cheung, C.K. 1997. Effect of various processing methods on antinutrients and *in vitro* digestibility of protein and starch of two Chinese indigenous legume seeds. *J. Agric. Food Chem.* 45:4773–4776.

170. Vijayakumari, K., Siddhuraju, P., and Janardhan, K. 1997. Chemical composition, amino acid content and protein quality of the little-known legume *Bauhinia purpurea* L. *J. Sci. Food Agric.* 73:279–286.

171. Vijayakumari, K., Siddhuraju, P., and Janardhan, K. 1996. Effect of soaking, cooking, and autoclaving on phytic acid and oligosaccharides content of the tribal pulse, *Mucuna monosperma* DC. ex. Wight. *Food Chem.* 55:173–177.

172. Pucciano, M.F., Weingartner, K.E., and Erdman, J.W., Jr. 1984. Relative bioavailability of dietary iron from three processed soy products. *J. Food Sci.* 49:1558–1561.

173. Naczk, M., Rubin, L.J., and Shahidi, F. 1986. Functional properties and phytate content of pea protein preparations. *J. Food Sci.* 51:1235–1237.

174. Winters, D.D., Handel, A.P., and Lohrberg, J.D. 1984. Phytic acid content of crude, degummed and retail soybean oils and its effect on stability. *J. Food Sci.* 49:113–114.

175. Madhusudhan, K.T. and Singh, N. 1983. Studies on linseed proteins. *J. Agric. Food Chem.* 31:959–962.

176. Arntfield, S.D., Ismond, M.A.H., and Murray, E.D. 1985. The fate of antinutritional factors during the preparation of a faba bean protein isolate using a micellization technique. *Can. Inst. Food Sci. Technol.* 18:174–180.

177. Anno, T., Nakanishi, K., Matsuno, R., and Kamikubo, T. 1985. Enzymatic elimination of phytate in soybean milk. *Nippon Shokuhin Kogyo Gakkai-Shi* 32:174–180.

178. Sutardi and Buckle, K.A. 1985. Reduction in phytic acid levels in soybeans during tempeh production, storage, and drying. *J. Food Sci.* 50:261–263.

179. Forbes, R.M., Parker, H.M., Kondo, H., and Erdman, J.W., Jr. 1983. Availability to rats of zinc in green and mature soybeans. *Nutr. Res.* 3:699–704.

180. Rajalakshmi, R. and Vanaja, K.V. 1967. Chemical and biological evaluation of effects of fermentation on the nutritive value of foods prepared from rice and grams. *Brit. J. Nutr.* 21:467–473.

181. Ramakrishnan, C.V. 1979. Studies on Indian fermented foods. *Baroda J. Nutr.* (India) 6:1–25.

182. Fardiaz, D. and Markakis, P. 1981. Degradation of phytic acid in oncom (fermented peanut presscake). *J. Food Sci.* 46:523–525.

183. Ogun, P.O., Markakis, P., and Chenoweth, W. 1989. Effect of processing on certain antinutrients in cowpeas (*Vigna unguiculata*). *J. Food Sci.* 54:1084–1085.

184. Harland, B.F. and Oberleas, D. 1977. A modified method for phytate analysis using an ion-exchange procedure: Application to textured vegetable proteins. *Cereal Chem.* 54:827–832.

185. Weaver, C.M., Heaney, R.P., Proulx, W.R., Hinders, S.M., and Packard, P.T. 1993. Absorbability of calcium from common beans. *J. Food Sci.* 58:1401–1403.

186. Hussein, L., Ghanem, K., Khalil, S., Nassibi, A., and Ezilarabi, A. 1989. The effect of phytate and fiber content on cooking quality of faba beans. *J. Food Qual.* 12:331–340.

187. Gupta, K., Thakral, K.K., Arora, S.K., and Wagle, D.S. 1991. Studies on growth, structural carbohydrate and phytate in corainder (*Coriandrum sativum*) during seed development. *J. Sci. Food Agric.* 54:43–46.

188. Bassiri, A. and Nahapetian, A. 1977. Differences in concentrations and interrelationships of phytate, phosphorus, magnesium, calcium, zinc, and iron in wheat varieties grown under dry land and irrigated conditions. *J. Agric. Food Chem.* 25:1118–1122.

189. Nahapetian, A. and Bassiri, A. 1976. Variations in concentrations and interrelationships of phytate, phosphorus, magnesium, calcium, zinc, and iron in wheat varieties during two years. *J. Agric. Food Chem.* 24:947–950.

190. Miller, G.A., Youngs, V.L., and Oplinger, E.S. 1980. Environmental and cultivar effects on oat phytic acid concentration. *Cereal Chem.* 57:189–191.

191. Miller, G.A., Youngs, V.L., and Oplinger, E.S. 1980. Effect of available soil phosphorus and environment on th phytic acid concentration in oats. *Cereal Chem.* 57:192–196.

192. Feil, B. and Fossati, D. 1997. Phytic acid in triticale grains as affected by cultivar and environment. *Crop Sci.* 37:916–921.

193. Proctor, J.P. and Watts, B.M. 1987. Effect of cultivar, growing location, moisture, and phytate content on the cooking times of freshly harvested navy beans. *Can. J. Plant Sci.* 67:923–930.

194. Griffiths, D.W. and Thomas, T.A. 1981. Phytate and total phosphorus content of field beans (*Vicia faba* L.). *J. Sci. Food Agric.* 32:187–192.

195. Saastmoinen, M. 1987. Effect of nitrogen and phosphorus fertilization on the phytic acid content of oats. *Cereal Res. Commun.* 15:57–62.

196. Raboy, V. and Dickinson, D.B. 1984. Effect of phosphorus and zinc nutrition on soybean seed phytic acid and zinc. *Plant Physiol.* 75:1094–1098.

197. Raboy, V. and Dickinson, D.B. 1993. Phytic acid levels in seeds of *Glycine max* and *G. soja* as influenced by phosphorus status. *Crop Sci.* 33:1300–1305.

198. Michael, B., Zin, F., and Lantzsch, H. 1980. Effect of phosphate application on phytin phosphorus and phosphate fractions in developing wheat grains. *Z. Pflanzenernaehr. Bodenkd.* 143:369–372.

199. Harland, B.F. and Peterson, M. 1978. Nutritional status of lacto-ovo-vegetarian Trappist monks. *J. Amer. Diet. Assoc.* 72:259–264.

200. Ellis, R., Kelsay, J.L., Reynolds, R.D., Morris, E.R., Moser, P.B., and Frazier, C.W. 1987. Phytate:zinc and phytate x calcium:zinc millimolar ratios in self-selected diets of Americans, Asian Indians, and Nepalese. *J. Amer. Diet. Assoc.* 87:1043–1047.

201. Murphy, S.P. and Calloway, D.H. 1986. Nutrient intakes of women in NHANES II, emphasizing trace minerals, fiber, and phytate. *J. Amer. Diet. Assoc.* 86:1366–1372.

202. Harland, B.F., Smith, S.A., Howard, M.P., Ellis, R., and Smith, J.C. 1988. Nutritional status and phytate:zinc and phytate x calcium:zinc dietary molar ratios of lacto-ovo-vegetarian Trappist monks: 10 years later. *J. Amer. Diet. Assoc.* 88:1562–1566.

203. Held, N.A., Buergel, N., Wilson, C.A., and Monsen, E.R. 1988. Constancy of zinc and copper status in adult women consuming diets varying in ascorbic acid and phytate content. *Nutr. Rep. Intern.* 37: 1307–1317.

204. Davies, N.T. 1982. Effects of phytic acid on mineral availability. In *Dietary Fiber in Health and Disease*, Vahoung, G.V. and Kritchevsky, D. (Eds.), Plenum Press, New York, pp. 105–116.

205. Wise, A., Lockie, G.M., and Liddell, J. 1987. Daily intakes of phytate and its meal distribution pattern amongst staff and students in an institution of higher education. *Brit. J. Nutr.* 58:337–346.

206. Torelm, I. and Bruce, A. 1982. Phytic acid in foods. *Var. Foda* 34:79–96.

207. Carnovale, E., Lombardi-Boccia, V.A., and Lugaro, E. 1987. Phytate and zinc content of Italian diets. *Hum. Nutr. Appl. Nutr.* 41A:180–186.

208. Plaami, S. 1997. Myoinositol phosphates: Analysis, content in foods and effects in nutrition. *Lebensm. Wiss. U-Technol.* 30:633–647.

209. Gibson, R.S., Vanderkooy, P.D.S., and Thompson, L.U. 1991. Dietary phytate x calcium/zinc millimolar ratios and zinc nutriture in some Ontario preschool children. *Biol. Trace Elements Res.* 30:87–94.

210. Murphy, S.P., Beaton, G.H., and Calloway, D.H. 1991. Estimated mineral intakes of toddlers: Predicted prevalence of inadequacy in village populations in Egypt, Kenya, and Mexico. *Amer. J. Clin. Nutr.* 56:565–572.

211. Ferguson, E.L., Gibson, R.S., Thompson, L.U., and Ounpuu, S. 1989. Dietary calcium, phytate, and zinc intakes and the calcium, phytate, and zinc molar ratios of the diets of a selected group of East African children. *Amer. J. Clin. Nutr.* 50:1450–1456.

212. Gibson, R.S. 1994. Content and bioavailablity of trace elements in vegetarian diets. *Amer. J. Clin. Nutr.* 59(Supplement):1223S–1232S.

213. Fitzgerald, S.L., Gibson, R.S., Serrano, J.A., Portocarrero, L., Vasquez, A., Epeda, E., Lopez-Palacios, C.Y., Thompson, L.U., Stephen, A.M., and Solomons, N.W. 1993. Trace element intakes and dietary phytate/Zn and Ca x phytate/Zn millimolar ratios of periurban Guatemalan women during the third trimester of pregnancy. *Amer. J. Clin. Nutr.* 57:195–201.

214. Nagi, M. and Mann, S.K. 1991. Nutrient intake by Punjabi women with special reference to iron availability. *J. Food Sci. Technol.* (India) 28:230–233.

215. Khokhar, S., Pushpanjali, and Fenwick, G.R. 1994. Phytate content of Indian foods and intakes by vegetarian Indians of Hissar region, Haryana state. *J. Agric. Food Chem.* 42: 2440–2444.

216. Grewal, P.K. and Hira, C.K. 1995. Intake of nutrients, phytin p, polyphenolic compunds, oxalates and dietary fibre by university residents. *Ecol. Food Nutr.* 34:11–17.

217. Bindra, G.S., Gibson, R.S., and Thompson, L.U. 1986. [Phytate] [calcium]/[zinc] ratios in Asian immigrant lacto-ovo vegetarian diets and their relationship to zinc nutriture. *Nutr. Res.* 6:475–483.

218. Wang, C.F., Tsay, S.M., Lee, C.Y., Liu, S.M., and Aras, N.K. 1992. Phytate content of Taiwanse diet determincd by ^{31}P flourier transform nuclear magnetic resonance spectroscopy. *J. Agric. Food Chem.* 40:1030–1033.

219. Paul, A.A., Bates, C.J., Prentice, A., Day, K.C., and Tsuchiya, H. 1998. Zinc and phytate intakes of rural Gambian infants: Contributions from breast milk and weaning foods. *Intern. J. Food Sci. Nutr.* 49:141–155.

220. Phillippy, B.Q. and Wyatt, C.J. 2001. Degradation of phytate in foods by phytases in fruit and vegetable extracts. *J. Agric. Food Chem.* (in press).

221. Almana, H.A. 2000. Extent of phytate degradation in breads and various foods consumed in Saudi Arabia. *Food Chem.* 70:451–456.

222. McKenzie-Parnell, J.M. and Guthrie, B.E. 1986. The phytate and mineral content of some cereals, cereal products, legumes, legume products, snack bars, and nuts available in New Zealand. *Biol. Trace Element Res.* 10:107–121.

223. Frossard, E., Bucher, M., Machler, F., Mozafar, A., and Hurrell, R. 2000. Potential for increasing the content and bioavailability of Fe, Zn, and Ca in plants for human nutrition. *J. Sci. Food Agric.* 80:861–879.

CHAPTER 4

Biosynthesis of Phytate in Food Grains and Seeds

FRANK A. LOEWUS

1. INTRODUCTION

PHYTATE (myo-inositol hexakisphosphate, $InsP_6$) is a common constituent of plants, largely stored as a complex salt of Mg^{2+}, K^+, and proteins within subcellular single-membrane particles (globoids, aleurone grains) in grains and seeds. As much as 60–80% of the phosphorus present in such organs may be $InsP_6$ [35,36,55]. Other cations including Ca^{2+}, Zn^{2+}, Fe^{3+}, and Cu^{2+} are usually present in measurable quantities. More recently, significant amounts of $InsP_6$ have been found to occur in protista and higher animals, including humans wherein this compound may have significant functions involving signal transduction and cellular regulation [57,63]. This chapter on the biosynthesis of phytic acid begins with an introduction to the biosynthesis of myo-inositol, the carbocyclic structure of $InsP_6$. An overview of myo-inositol mono- and polyphosphates follows. Because specific $Ins(n)P_n$s are involved in discrete processes leading to signal-transducing polyphosphates [$Ins(1,4,5)P_3$, $Ins(1,3,4,5)P_4$, etc.], $InsP_6$ biosynthesis, and $InsP_6$ breakdown, each must be dealt with separately because intermediate phosphate esters are often unique. Finally, selected biochemical properties and functional aspects of phytic acid will be discussed.

2. *MYO*-INOSITOL (1,2,3,5/4,6-HEXAHYDROXYCYCLOHEXANE)

2.1. STEREOCHEMISTRY AND NOMENCLATURE

Inositol has nine configurational isomers [54]. Of these, *myo-*, *scyllo-*, *chiro-*, *muco-*, and *neo*-inositols occur in plants. *Chiro*-inositol possesses neither plane nor center of symmetry and is represented by two optical antipodes (+ *chiro-* or − *chiro*-inositol), both present in plants. Inositol ring structures are not planar due to the tetrahedral nature of carbon, and in the case of *myo*-inositol, there is a preferred conformational structure, 1,2,3,5/4,6, where the upper-numbered substituent hydroxyl groups are above the plane of the ring and the lower-numbered hydroxyl groups are below [54].

To identify specific positions on the *myo*-inositol ring, a set of recommendations was established by the International Union of Pure and Applied Chemistry/International Union of Biochemistry. In 1989, these recommendations were relaxed to accommodate numbering carbon atoms in the ring of *myo*-inositol such that structural relationships could be brought out which otherwise might create confusion should original rules be applied [21]. Both sets of recommendations are found on the Internet at: *http://www.chem.qmw.ac.uk/iupac*.

2.2. BIOSYNTHESIS OF *MYO*-INOSITOL

Conversion of D-glucose 1-P to 1L-*myo*-inositol 1-P, the first committed step toward biosynthesis of all inositol-containing compounds, is catalyzed by 1L-*myo*-inositol-1-P synthase [38]. This cyclization of the six carbon chain conserves the "D-gluco" configuration of C2 through C5 of *myo*-inositol [31]. Because the conventional carbohydrate rule of numbering the carbon chain of D-glucose is replaced by the cyclitol rule in the product, *myo*-inositol, sequence of numbering is reversed, and 1L-*myo*-inositol-1-P is numbered *clockwise*. To accommodate rules for *myo*-inositol nomenclature that were adopted in 1989, positions of substituents on *myo*-inositol are now numbered *counterclockwise*, and the product of 1L-*myo*-inositol-1-P synthase becomes 1D-*myo*-inositol 3-P. As a mnemonic, the abbreviation "Ins" is used to signify counterclockwise numbering of the carbocyclic ring. All substituted products of *myo*-inositol now use this convention. Thus, the abbreviated terms $Ins(3)P_1$ = 1L-*myo*-inositol 1-P = 1D-*myo*-inositol 3-P are equivalent [43]. By the same token, the molecule that links receptor-activated phosphoinositide breakdown to calcium mobilization from intracellular stores is designated $Ins(1,4,5)P_3$ = 1D-*myo*-inositol 1,4,5-trisphosphate = 1L-*myo*-inositol 3,5,6-trisphosphate [19]. The practicality of the 1988 decision is quite evident, even though it ignores the biosynthetic origin of *myo*-inositol. For the remainder of this chapter, the term $Ins(a,b,c...x)P_n$, where "x" refers to the position of substitution on the ring and "n" refers to the number of phosphate substitutions around the ring, is used to describe *myo*-inositol mono- and polyphosphates.

To the best of our knowledge, a single biosynthetic enzyme, Ins(3)P_1 synthase (E.C. 5.5.1.4), catalyzes conversion of hexose-P to Ins(3)P_1. This enzyme oxidizes carbon 5 of D-glucose 6-P to 5-keto-D-glucose 6-P, then catalyzes an aldol condensation between carbon 1 and carbon 6, and finally reduces the product, 2-keto-myoinosose 1-P, to Ins(3)P_1. NAD, the cofactor for the oxidation and reduction steps, is enzyme bound. All substeps are stereospecific with regard to NAD and intermediates [29]. The enzyme is found in prokaryotic and eukaryotic organisms and has been cloned from cyanobacteria, yeasts, fungi, algae, protozoa, and a variety of plant and animal sources [34,38].

Removal of phosphate by a magnesium-dependent monophosphatase (E.C. 3.1.3.25) releases free *myo*-inositol. This enzyme hydrolyzes Ins(3)P_1 and Ins(1)P_1 at similar rates and Ins(2)P_1 at a lower rate [29]. It is inhibited by lithium ion [15], a process of considerable interest in biomedical studies [57].

2.3. FREE *MYO*-INOSITOL

This ubiquitous constituent of plant and animal cells participates in a host of metabolic processes [31] including the following:

- Ion uptake—*myo*-inositol-dependent sodium uptake [45]
- Cell wall biogenesis (*myo*-inositol oxidation pathway) [28,30,32]
- Methylation and isomerization [6,53]
- Conjugation [10]
- Galactosyloligosaccharide and galactosyl cyclitol biosynthesis [46,48,49]
- Phosphoinositide biosynthesis and phospholipid signaling [12,42,61]
- Glycosylphosphatidylinositol and membrane anchoring [44,47]
- Formation of *myo*-inositol polyphosphates including phytic acid [11,16]
- Pyrophosphorylated *myo*-inositol polyphosphates [58]

Normally, free *myo*-inositol is present in plant cells at concentrations lower than free sugars such as D-glucose, but occasionally, higher amounts are encountered as reported for early stages of development of kiwifruit, where it is a major "carbohydrate" constituent [25]. Because *myo*-inositol is the sole carbocyclic structure leading to phytic acid biosynthesis, it is worth considering potential sources of free *myo*-inositol released by metabolic recycling as well as newly synthesized Ins(3)P_1 in this regard. Such sources arise during dephosphorylation of *myo*-inositol monophosphates [examples: Ins(1) P_1 or Ins(2)P_1] or polyphosphates including phytic acid, release of *myo*-inositol following transfer of galactosyl units from galactinol to raffinose or higher homologs, demethylation of O-methyl inositols, de-esterification of *myo*-inositol-auxin conjugates, and breakdown of phosphatidylinositols. Salvage of *myo*-inositol from such sources probably has both spatial and temporal restraints. Although little effort has been made to explore possible regulatory processes involved in modulating free *myo*-inositol levels either intra- or inter-cellularly, such mechanisms must be in place. Lack of information also applies to control of free *myo*-inositol biosynthesis from D-glucose 6-P via $Ins(3)P_1$ synthase and *myo*-inositol monophosphatase. Hexose kinase-mediated sugar sensing [18,22,26,41,62,65] may play a role in regulating both $Ins(3)P_1$ and free *myo*-inositol, although experimental evidence in this regard is needed.

3. PHYTIC ACID BIOSYNTHESIS

3.1. INS(3)P₁ AS THE FIRST STEP IN PHYTIC ACID BIOSYNTHESIS

An important contribution regarding the first step in $InsP_6$ biosynthesis was made by Yoshida et al. [64], whose group isolated a cDNA clone, pRINO1, from rice (*Oryza sativa* L.) callus suspension cultures, that was highly homologous to $Ins(3)P_1$ synthase from yeast and plants [38]. Its transcript appeared in the apical region of globular-stage embryos two days after anthesis, and strong signals were detected in the scutellum and aleurone layer after four days. Phytate-containing particles or globoids appeared in the same tissues at four days, coinciding with the RINO1 transcript. This study demonstrated that $Ins(3)P_1$ synthase is probably the first committed step in phytic acid biosynthesis. A complementary process involving phosphorylation of free *myo*-inositol to $Ins(3)P_1$ by *myo*-inositol kinase (EC 2.7.1.64) remains untested. Curiously, little attention has been given to this kinase that is present in plants, animals, and microorganisms [14]. Its product, $Ins(3)P_1$, has the same configurational structure as that produced by $Ins(3)P_1$ synthase [33]. While one might regard this recycling of *myo*-inositol into the pool of $Ins(3)P_1$ as a salvage mechanism, it fails to take into consideration localization of these enzymic activities or temporal demands during development. Together, $Ins(3)P_1$ synthase and *myo*-inositol kinase constitute ways in which $Ins(3)P_1$ is formed from D-glucose-6-P or free *myo*-inositol in plants. The former enzyme is biosynthetic, while the latter must rely on sources that generate free *myo*-inositol from *myo*-inositol monophosphatase or other *myo*-inositol-conjugated forms. Only further research will provide clues regarding relative contributions from hexose-P or salvaged/stored forms of *myo*-inositol to the pool of $Ins(3)P_1$ during growth and development. One possible approach would be to use antagonists such as 2-O,C-methylene-*myo*-inositol to block access to free *myo*-inositol [4,9,37] or, alternatively, D-glucuronate [32] or 2-deoxy-D-glucose [3] to inhibit $Ins(3)P_1$ biosynthesis.

3.2. STEPWISE PHOSPHORYLATION OF INS(3)P₁ TO PHYTIC ACID

Efforts to explore phosphorylative steps beyond $Ins(3)P_1$ involved in $InsP_6$ biosynthesis have been frustrated by intermingling $InsP_n$s arising from phytase-driven hydrolysis of $InsP_6$, phospholipase-catalyzed release of $InsP_n$s during cleavage of phosphatidylinositols, and release of partially degraded $InsP_n$s by nonspecific phosphatases during experimental procedures. Due to an enormous interest in phosphatidylinositol-linked signal transduction in recent years, analytical tools and standards for most, if not all, of the 63 *myo*-inositol phosphate esters are now available [1,20,23], and attempts to sort out those involved in $InsP_6$ biosynthesis are possible. Notable examples are Stephens and Irvine's stepwise delineation of phosphorylation of *myo*-inositol in *Dictyostelium* [60] and Brearley and Hanke's comparable approach with *Spirodela polyrhiza* [7,8]. The proposed pathways shared common tris- and pentakisphosphate intermediates, but bis- and tetrakisphosphate intermediates had phosphates at 6 versus 4 or 1 versus 5 positions, respectively. In both pathways, phosphorylation of position 2 was the final step to $InsP_6$.

- *Dictyostelium* (slime mold):
  $Ins(3)P_1 \rightarrow Ins(3,6)P_2 \rightarrow Ins(3,4,6)P_3 \rightarrow Ins(1,3,4,6)P_4 \rightarrow$
  $Ins(1,3,4,5,6)P_5 \rightarrow InsP_6$
- *Spirodela polyrrhiza* (duckweed):
  $Ins(3)P_1 \rightarrow Ins(3,4)P_2 \rightarrow Ins(3,4,6)P_3 \rightarrow Ins(3,4,5,6)P_4 \rightarrow$
  $Ins(1,3,4,5,6)P_5 \rightarrow InsP_6$

Formation of $InsP_6$ in animals differs from plants and fungi in that its *myo*-inositol structure arises from free *myo*-inositol via lipase-derived cleavage products of phosphatidylinositol phosphate(s) [58] rather than from stepwise phosphorylation of $Ins(3)P_1$.

To date, efforts to characterize the enzymic steps in plants that lead from $Ins(3)P_1$ to $InsP_6$ have been less fruitful than comparable studies in animals, even though discovery of $InsP_6$ in the latter is a relatively recent event [58]. The idea of stepwise phosphorylation found support when Majumder et al. [39] isolated and partially purified a phosphoinositol kinase from germinating mung bean that phosphorylated $InsP_1$, $InsP_2$, $InsP_3$, $InsP_4$, and $InsP_5$ to the next higher homolog. Positional locations of phosphate in these structures were not examined. More recently, Phillippy et al. [52] purified an $Ins(1,3,4,5,6)P_5$ 2-kinase from immature soybean seeds. As the authors point out, $Ins(1,3,4,5,6)P_5$ appears to be a common precursor of $InsP_6$ in plants and animals and has properties similar to an $InsP_6$-ADP phosphotransferase found in mung bean [5]. Phillippy [50] has also identified two intermediate phosphoinositol kinases in immature soybean seeds, an $Ins(1,3,4)P_3$ 5-kinase and an $Ins(1,3,4,5)P_3$ 6-kinase. Whether these kinases are involved in the stepwise phosphorylation of $Ins(3)P_1$ to $InsP_6$ must still be determined, especially because the *Spirodela* study by Brearley and Hanke [7,8] indicated that $Ins(3,4,6)P_3$ rather than $Ins(1,3,4)P_3$ was the trisphosphate involved in $InsP_6$ biosynthesis.

Efforts to find an enzyme capable of phosphorylating $Ins(1,4)P_2$ to $Ins(1,4,5)P_3$ in suspension-cultured *Nicotiana tabacum* cells were unsuccessful [24], an observation not too surprising if $Ins(3)P_1$ is the first step in $InsP_6$ biosynthesis. This study provided additional evidence that products arising from lipase-catalyzed hydrolysis of phosphatidylinositol and its polyphosphates are not involved in $InsP_6$ biosynthesis in plants.

The need for more information regarding stepwise phosphorylation of $Ins(3)P_1$ to $InsP_6$ is quite obvious, but there seems to be growing evidence for such a stepwise process. By the same token, fresh insight is needed into localization of these kinases and the transport of product from site of synthesis to point of accumulation in globoids as well as the nature of deposition with counterions in that storage organelle. The studies of Yoshida and colleagues on $Ins(3)P_1$ synthase and globoid development in rice provide an important starting point [64].

4. ASPECTS OF InsP₆ METABOLISM UNIQUE TO PLANTS

Before leaving this brief overview of $InsP_6$ biosynthesis, attention should be given to certain aspects that need to be considered for future efforts in this regard. Mandal and Biswas [40] reported $InsP_6$ synthesis during germination in cotyledons of mung beans. Recently, during studies of $InsP_6$ metabolism during postgerminative growth of bean seedlings, this observation was again noted. Rapid $InsP_6$ degradation occurred in embryonic axes of snap and pinto bean seedlings during root emergence, but after two to three days, $InsP_6$ synthesis was detected [13]. No detectable $InsP_6$ was found in vacuoles isolated from roots or hypocotyls, implying that this newly synthesized $InsP_6$ was localized in the cytoplasm and was potentially available as a natural chelator, possibly as an Fe^{3+}-$InsP_6$ complex. Such a complex could prevent cellular damage from oxygen radical-induced processes [17]. Apropos of this observation, pollen and seeds contain an alkaline calcium-activated phytase that removes only three phosphates from $InsP_6$ to produce as an end product, $Ins(1,2,3)P_3$, a trisphosphate with the simplest structure able to bind Fe^{3+} and function as an antioxidant [2,51,56,59].

A more detailed discussion of phytases and $InsP_6$ hydrolysis is found in a recent review [27].

5. REFERENCES

1. Barrientos, L.G. and Murthy, P.P.N. 1996. Conformational studies on inositol phosphates, *Carbohydr. Res.* 296:39–54.
2. Barrientos, L.G., Scott, J.J. and Murthy, P.P.N. 1994. Specificity of hydrolysis of phytic acid by alkaline phytase from lily pollen, *Plant Physiol.* 106:1489–1495.
3. Biffen, M. and Hanke, D.E. 1990. Reduction in the level of intracellular *myo*-inositol in cultured soybean (*Glycine max*) cells inhibits cell division, *Biochem. J.* 265:809–814.
4. Biffen, M. and Hanke, D.E. 1991. Metabolic fate of *myo*-inositol in soybean callus cells, *Plant Sci.* 75:203–213.
5. Biswas, S., Maity, I.B., Chakrabarti, S. and Biswas, B.B. 1978. Purification and characterization of myo-inositol hexaphosphate-adenosine diphosphate phosphotransferase from *Phaseolus aureus*, *Arch. Biochem. Biophys.* 185:557–566.
6. Bohnert, H.J. and Jensen, R.G. 1996. Strategies for engineering stress tolerance in plants, *Trends Biotechnol.* 14:89–97.
7. Brearley, C.A. and Hanke, D.E. 1996a. Inositol phosphates in the duckweed *Spirodela polyrhiza* L., *Biochem. J.* 314:215–225.
8. Brearley, C.A. and Hanke, D.E. 1996b. Metabolic evidence for the order of addition of individual phosphate esters to the *myo*-inositol moiety of inositol hexakisphosphate in the duckweed *Spirodela polyrhiza* L., *Biochem. J.* 314:227–233.
9. Chen, M., Loewus, M.W. and Loewus, F.A. 1977. Effect of a *myo*-inositol antagonist, 2-*O*,*C*-methylene-*myo*-inositol, on the metabolism of *myo*-inositol-2-^{3}H and D-glucose-1-^{14}C in *Lilium longiflorum* pollen, *Plant Physiol.* 59:658–663.
10. Cohen, J.D. and Slovin, J.P. 2001. Recent research advances concerning indole-3-acetic acid metabolism, In: *Mechanism of Action of Plant Hormones*, Eds. Palme, K., Walden, R. and Schell J., Berlin: Springer, (in press).
11. Cosgrove, D.J. 1980. *Inositol Phosphates: Their Chemistry, Biochemistry and Physiology*, Amsterdam: Elsevier, 191 pp.
12. Coté, G.G. and Crain, R.C. 1993. Biochemistry of phosphoinositides, *Annu. Rev. Plant Physiol. Mol. Biol.* 44:333–356.
13. Crans, D.C., Mikuš, M. and Friehauf, R.B. 1995. Phytate metabolism in bean seedlings during post-germinative growth, *J. Plant Physiol.* 145:101–107.
14. English, P.D., Deitz, M. and Albersheim, P. 1966. Myoinositol kinase: partial purification and identification of product, *Science* 151:198–199.
15. Gillaspy, G.E., Keddie, J.S., Oda, K. and Gruissem, W. 1995. Plant inositol monophosphatase is a lithium-sensitive enzyme encoded by a multigene family, *Plant Cell* 7:2175–2185.
16. Graf, E. 1986. *Phytic Acid: Chemistry & Applications*, Minneapolis MN: Pilatus Press, 344 pp.
17. Graf, E., Empson, K.L. and Eaton, J.W. 1987. Phytic acid. A natural food antioxidant, *J. Biol. Chem.* 262:11647–11650.
18. Halford, N.G. and Purcell, P.C. 1999. Reply ... The sugar sensing story, *Trends Plant Sci.* 4:251.
19. Hawthorne, J.N. 1996. Phosphoinositides and Synaptic Transmission, In: *Subcellular Biochemistry, Vol. 26, myo-Inositol Phosphates, Phosphoinositides, and Signal Transduction*, Eds. Biswas, B.B. and Biswas, S., New York: Plenum Press, pp. 43–57.
20. Irvine, R.F. 1990. *Methods in Inositide Research*, New York: Raven Press, 279 pp.
21. IUPAC/IUB, 1989. Numbering of atoms in *myo*-inositol, *Biochem. J.* 258:1–2.
22. Jang, J-C., León, P., Shou, L. and Sheen, J. 1997. Hexokinase as a sugar sensor in higher plants, *Plant Cell* 9:5–19.
23. Johnson, K., Barrientos, L.G., Le, L. and Murthy, P. P. N. 1995. Application of 2D TOCSY for structure determination of individual inositol phosphates in a mixture, *Anal. Biochem.* 231:421–431.

24. Joseph, S.K., Esch, T. and Bonner, W.D. 1989. Hydrolysis of inositol phosphates by plant cell extracts, *Biochem. J.* 264:851–856.
25. Klages, K., Donnison, H., Boldingh H. and MacRae, E. 1998. *myo*-inositol is the major sugar in *Actinidia arguta* during early fruit development, *Aust. J. Plant Physiol.* 25:61–67.
26. Koch, K.E. 1996. Carbohydrate-modulated gene expression in plants, *Annu. Rev. Plant Physiol. Plant Mol. Biol.* 47:509–540.
27. Liu, B-L., Rafiq, A., Tzeng, Y-M. and Rob, A. 1998, The induction and characterization of phytase and beyond, *Enz. Microbiol. Tech.* 22:415–424.
28. Loewus, F. 1973. *Biogenesis of Plant Cell Wall Polysaccharides*, New York: Academic Press, 379 pp.
29. Loewus, F.A. 1990. Inositol biosynthesis, In: *Inositol Metabolism in Plants*, Eds. Morré, D.J., Boss, W.F. and Loewus, F.A., New York: Wiley-Liss, pp. 13–19.
30. Loewus, F.A. and Loewus, M.W. 1983. *myo*-inositol: Its biosynthesis and metabolism, *Annu. Rev. Plant Physiol.* 34:137–161.
31. Loewus, F.A. and Murthy, P.P.N. 2000. *myo*-inositol metabolism in plants, *Plant Sci.* 150: 1–19.
32. Loewus, M.W. and Loewus, F.A. 1974. *myo*-inositol 1-phosphate synthase inhibition and control of uridine diphosphate-D-glucuronic acid biosynthesis in plants, *Plant Physiol.* 54: 368–371.
33. Loewus, M.W., Sasaki, K., Leavitt, A.L., Munsell, L., Sherman, W.R. and Loewus, F.A. 1982. Enantiomeric form of *myo*-inositol-1-phosphate produced by *myo*-inositol-1-phosphate synthase and *myo*-inositol kinase in higher plants, *Plant Physiol.* 70:1661–1663.
34. Lohia, A., Hait, N.C. and Majumder, A.L. 1999. 1L-*myo*-Inositol-1-phosphate synthase from *Entamoeba histolytica, Mol. Biochem. Parasitol.*, 98:67–79.
35. Lott, J.N.A., Greenwood, J.S. and Batten, G.D. 1995. Mechanisms and regulation of mineral nutrient storage during seed development, In: *Seed Development and Germination*, Eds. Kigael, J. and Galili, G., New York: Marcel Dekker, Inc., pp. 215–235.
36. Lott, J.N.A., Ockenden, I., Raboy, V. and Batten, G.D. 2000. Phytic acid and phosphorus in crop seeds and fruits: a global estimate, *Seed Sci. Res.* 10:11–33.
37. Maiti, I.B. and Loewus, F.A. 1978. Myo-inositol metabolism in germinating wheat, *Planta* 142:55–60.
38. Majumder, A.L., Johnson, M.D. and Henry, S.A. 1997. 1L-*myo*-Inositol-1-phosphate synthase, *Biochem. Biophys. Acta*, 1348:245–256.
39. Majumder, A.N.L., Mandel, N.C. and Biswas, B.B. 1972. Phosphoinositol kinase from germinating mung bean seeds, *Phytochemistry* 11:503–508.
40. Mandel, N.C. and Biswas, B.B. 1970. Metabolism of inositol phosphates: Part II—Biosynthesis of inositol polyphosphates in germinating seeds of *Phaseolus aureus, Indian J. Biochem.* 7:63–67.
41. Moore, B.D. and Sheen, J. 1999. Plant sugar sensing and signaling—a complex reality, *Trends Plant Sci.* 4:250.
42. Munnik,T., Irvine, R.F. and Musgrave, A.1998. Phospholipid signaling in plants, *Biochim. Biophys. Acta* 1389:222–272.
43. Murthy, P.P.N. 1996. Inositol phosphates and their metabolism in plants, In: *Subcellular Biochemistry, Vol. 26, myo-Inositol Phosphates, Phosphoinositides, and Signal Transduction*, Eds. Biswas, B.B. and Biswas, S., New York: Plenum Press, pp. 227–255.
44. Nakazato, H., Okamoto, T., Nishikoori, M., Washio, K., Morita, N., Haraguchi, K., Thompson, G.A., Jr. and Okuyama, H. 1998. The glycosylphosphatidylinositol-anchored phosphatase from *Spirodela oligorrhiza* is a purple acid phosphatase, *Plant Physiol.* 118:1015–1020.
45. Nelson, D.E., Koukoumanos, M. and Bohnert, H.J. 1999. *myo*-Inositol-dependent sodium uptake in ice plant, *Plant Physiol.* 119:165–172.

46. Obendorf, R.L. 1997. Oligosaccharides and glactosyl cyclitols in seed desiccation tolerance, *Seed Sci. Res.* 7:63–74.

47. Perotto, S., Donovan, N., Drøbak, B.K. and Brewin, N.J. 1995. Differential expression of a glycosyl inositol phospholipid antigen on the peribacteroid membrane during pea nodule development, *Mol. Plant-Microbe Interact.* 8:560–568.

48. Peterbauer, T. and Richter, A. 1998. Galactosylononitol and stachyose synthesis in seeds of Adzuki bean: Purification and characterization of stachyose synthase, *Plant Physiol.* 117:165–172.

49. Peterbauer, T., Puschenreiter, M. and Richter, A. 1998. Metabolism of galactosylononitol in seeds of *Vigna umbellata*, *Plant Cell Physiol.* 39:334–341.

50. Phillippy, B.Q. 1998. Identification of inositol 1,3,4-trisphosphate 5-kinase and inositol 1,3,4,5-tetrakisphosphate in immature soybean seeds, *Plant Physiol.* 116:291–297.

51. Phillippy, B.Q. and Graf, E. 1997. Antioxidant functions of inositol 1,2,3-trisphosphate and inositol 1,2,3,6-tetraphosphate, *Free Radical Biol. & Med.* 22:939–946.

52. Phillippy, B.Q., Ullah, A.H.J. and Ehrlich, K.C. 1994. Purification and some properties of Ins(1,3,4,5,6)P$_5$ 2-kinase from immature soybean seeds, *J. Biol. Chem.* 269:28393–28399.

53. Popp, M., Lied, W., Bierbaum, U., Gross, M., Grosse-Schulte, T., Hams, S., Oldenettel, J., Schüler, S. and Wiese, J. 1997. Cyclitols—stable osmotica in trees, In: *Trees—Contributions to Modern Tree Physiology*, Eds. Rennenberg, H., Eschrich, W. and Ziegler, H. The Hague: Backhuys Publ. 270 pp.

54. Posternak, T. 1965. *The Cyclitols*, San Francisco, CA: Holden-Day, Inc., pp. 7–24.

55. Reddy, N.R., Pierson, M.D., Sathe, S.K. and Salunkle, D.K. 1989. *Phytates in Cereals and Legumes*, Boca Raton, FL: CRC Press, Inc. pp. 39–56.

56. Scott, J.J. and Loewus, F.A. 1986. A calcium-activated phytase from pollen of *Lilium longiflorum*, *Plant Physiol.* 82:333–335.

57. Shears, S.B. 1996. Inositol pentakis- and hexakisphosphate metabolism adds versatility to the actions of inositol polyphosphates: Novel effects on ion channels, In: *Subcellular Biochemistry, Vol. 26, myo-Inositol Phosphates, Phosphoinositides, and Signal Transduction*, Eds. Biswas, B.B. and Biswas, S., New York: Plenum Press, pp. 187–226.

58. Shears, S.B. 1998. The versatility of inositol phosphates as cellular signals, *Biochim. Biophys. Acta* 1436:49–67.

59. Spiers, I.D., Freeman, S., Poyner, D.R. and Schwable, C.H., 1995. The first synthesis and iron binding studies of the natural product, *myo*-inositol 1,2,3-trisphosphate, *Tetrahedron Lett.* 36:2125–2128.

60. Stephens, L.R. and Irvine, R.F. 1990. Stepwise phosphorylation of *myo*-inositol leading to *myo*-inositol hexakisphosphate in *Dictyostelium*, *Nature* 346:580–581.

61. Stevenson J.M., Perera, I.Y., Heilmann, I., Persson, S. and Boss, W.F. 2000. Inositol signaling and plant growth, *Trends Plant Sci.* 5:252–258.

62. Taylor, C.B. 1997. Sweet Sensations, *Plant Cell* 9:1–4.

63. York, J.D., Odom, A.R., Murphy, R., Ives, F.R. and Wente, S.R. 1999. A phospholipase C-dependent inositol polyphosphate kinase pathway required for efficient messenger RNA export, *Science* 285:96–100.

64. Yoshida, K.T., Wada, T., Koyama, H., Mizobuchi-Fukuoka, R. and Naito, S. 1999. Temporal and spatial patterns of accumulation of the transcript of *myo*-inositol-1-phosphate synthase and phytin-containing particles during seed development in rice, *Plant Physiol.* 119:65–72.

65. Yu, S-M. 1999. Cellular and genetic responses of plants to sugar starvation, *Plant Physiol.* 121:687–693.

CHAPTER 5

Genetics of Phytic Acid Synthesis and Accumulation

VICTOR RABOY
KEVIN A. YOUNG
STEVEN R. LARSON
ALLEN COOK

1. INTRODUCTION

G RAIN and legume crops have two main uses: in human foods and in animal feeds. To be effective, estimations of the impacts of grain and legume phytic acid, and the development of strategies to deal with these impacts, probably need to take into consideration both of these major uses of staple crops. In the case of livestock production, the concerns over grain and legume phytic acid are relatively straightforward [1,2]. Nonruminants such as poultry, swine, and fish, excrete essentially all feed phytic acid they consume [3–5]. Excretion of this large fraction of grain total P can contribute to phosphorus (P) pollution and to the resulting eutrophication of surface waters. This is an important contemporary problem in Europe and the United States [6]. New, more stringent standards for P management in agricultural production are currently being put in place. The magnitude of phytic acid's role in this context is illustrated by the recent estimation that the amount of P synthesized into phytic acid annually worldwide by seed crops may represent more than 50% of the annual fertilizer P application worldwide [7].

There are two main contemporary approaches to the "phytic acid problem" in livestock production. The first involves the production and use of "phytase" enzymes as dietary supplements [5,8–9]. This approach is currently in use in Europe and the United States. The second approach is to reduce the phytic acid content of crops using genetics, such as in the isolation of *low phytic acid* mutants and their use in breeding "low phytate" or "high available P" crops [10].

The issue of dietary phytic acid in human nutrition and health is considerably more complex [11]. The primary historical concern over seed-derived dietary phytic acid has been its role in mineral depletion and deficiency [12–14]. Human populations that subsist on whole grain and/or legume staple foods consume substantial amounts of phytic acid, and this may contribute to their risk for mineral depletion and deficiency. However, dietary phytic acid may also have important positive roles, for example, as an antioxidant and an anticancer agent [15–16]. The putative benefits of dietary phytic acid may be a more important consideration in certain populations than concerns over mineral deficiency. The question of seed-derived dietary phytic acid in human nutrition and health is further complicated by the fact in the cereal grain, phytic acid is deposited in the aleurone and germ, which is also the site for the grain's main mineral stores [17–18]. Removal of these tissues during milling or polishing removes most phytic acid and most of the grain's mineral deposits. The impact of dietary phytic acid in human health must, therefore, be evaluated on a case-by-case basis, considering positive and negative roles in a given population consuming a given diet.

Crop genetics can contribute a powerful experimental approach to studying the question of phytic acid in human nutrition and health. In the long-term, crop genetics may provide sustainable solutions to problems associated with seed phytic acid. This chapter presents the genetics of seed phytic acid synthesis and accumulation, the isolation of *low phytic acid* mutants, and their use in crop breeding. The potential value of the crop genetics approach as an experimental tool in addressing questions of grain- and legume-derived phytic acid in human nutrition and health was also discussed.

2. CONSIDERATIONS PRIOR TO SCREENING FOR "PHYTIC ACID MUTANTS"

Until the early 1990s, there was little or no Mendelian genetics of seed phytic acid metabolism [19]. There were no reports of single-gene mutants that had a proximal effect on seed phytic acid content. The molecular genetics relevant to this field was in its infancy. A first step was to attempt to isolate mutations that greatly reduce seed phytic acid content but that have little or no other effects, such as on seed total P. These then could be used in basic studies of phytic acid metabolism and molecular genetics to study phytic acid's role in nutrition and possibly to breed "low phytic acid" crops.

A first step in any mutant screening project is to predict the phenotype associated with the desired mutant. This prediction would be based on the current state of knowledge concerning the physiology, biochemistry, and genetics of a particular pathway or process. This prediction is necessary to design effective screening methods and to predict and recognize "false positives." This latter issue is very important in mutant screening. If, for some reason, the rate of isolation of false positives is sufficiently great compared to the rate of isolation of the desired mutant, the screening may not work. For example, in the maize kernel, $\geq 80\%$ of the phytic acid is localized in the germ [18]. "Germless" mutants occur frequently in mutagenized maize populations, and a preliminary study [20] showed that germless kernels contain greatly reduced amounts of phytic acid as compared with normal kernels. This is a "pleiotropic" or distal effect of the mutation that results in abortion of the germ and is of little immediate interest or agronomic value.

In the context of our research objective, another type of "false positive" would be the identification of lines that have reduced phytic acid P, where this reduction is due to reduced kernel total P. The long-term interest in the dietary role of phytic acid as antinutrient has resulted in numerous reports of surveys of seed or grain phytic acid content among "wild-type" or "nonmutant" cultivars or lines of various crop species [2,19]. If one analyzes a number of lines of a given species grown in one or more environments, one can observe substantial variation in the phytic acid concentration in the seed or grain of these lines. This variation can reach or surpass 50% of the mean value. However, variation in phytic acid P is highly and positively correlated with variation in seed or grain total P, with correlation coefficients $\geq 95\%$. Therefore, variation in phytic acid P observed in most surveys of crop germplasm is most likely an outcome of variation, due to either environmental or genetic factors, in the uptake and supply of P to the seed or grain. The resulting variation in seed or grain total P is almost completely accounted for by variation in phytic acid P. In normal, nonmutant or "wild-type" seeds or grains, essentially all P over and above a concentration necessary for nominal cellular function accumulates as phytic acid P. The concentration necessary for cellular function, nonphytic acid P, is usually about 15% to 25% of total P in mature grain produced under nominal or standard conditions. If any nonmutagenized population is screened for grain phytic acid P concentration, and lines are identified with "low" or "high" levels, one is most likely selecting for low or high levels of grain total P concentration. A "low phytic acid" line would, therefore, be low total P. In addition, phytic acid P is highly and positively correlated with grain total nitrogen and, to a lesser extent, with several mineral elements [2,21–23]. Selecting "low phytic acid" lines would result in "low protein" and, in certain cases, "low minerals." An approach was needed that could avoid these undesirable "correlated responses," most of which are an outcome of the relationship between phytic acid P and total P. Our objective was to isolate mutants in which this very close relationship between seed total P and phytic acid P is uncoupled, mutants that produce seed with normal levels of total P but greatly reduced levels of phytic acid P. We hypothesized that a mutation in a gene important to phytic acid synthesis or accumulation during seed development may not impact P uptake to the grain, and, thus, accomplish this goal.

The prevailing view at the beginning of this project was that seed phytic acid content is very important to seed function, and therefore, mutations having large effects on phytic acid would be deleterious to the seed. One suggested approach to screen for "phytic acid mutants" was to screen for mutants that perturb germination, a subset of which would be "phytic acid mutants." Subsequent work [10,24] has shown that some *low phytic acid* mutants, including some currently used in breeding, have little effect on germination under nominal conditions. Another physiological consideration is the long-held belief that phytic acid metabolism is a critical component of P homeostasis during seed development and germination [25]. Regulation of cellular P concentration is essential to normal functioning of pathways of major importance in the seed, such as in carbohydrate synthesis. A block in a seed's ability to sequester P as phytic acid might negatively impact the ability of a seed to store carbon as starch, reducing starch accumulation and, therefore, seed weight and yield. The rate-limiting step in starch synthesis is catalyzed by the enzyme ADP-glucose pyrophosphorylase, which is allostearically inhibited by P [26]. Perhaps the most important consideration concerning the physiological roles of seed phytic acid is its putative "storage" function. Both its P and *myo*-inositol contents are recycled during germination. Also, phytic acid is typically deposited in the seed as discrete inclusions called globoids, which contain mixed salts of phytic acid and several mineral cations [17,27–28]. In the cereal grains, these mixed salts primarily contain potassium and magnesium but may also contain several other mineral cations. Recycling of these deposits provides minerals to the germinating seed and seedling. A mutation that blocks phytic acid synthesis may perturb these important storage and redistribution processes.

A classical approach to predicting a given mutant's phenotype would be to consider the metabolic pathways revolving around a given compound and pre-dict the phenotypic outcome of a particular lesion or block in those pathways. Also, each enzymatic step in the phytic acid pathways represents a target gene possibly useful in engineering desired phytic acid levels in grain. A brief review of the metabolic pathways involving phytic acid will illustrate this, using the "D-convention" for numbering of the carbon atoms in phytic acid's *myo*-inositol (Ins) backbone [Figure 5.1(a)]. All pathways to phytic acid begin with the syn-thesis of Ins. The sole synthetic source of Ins is via a simple two-step pathway [29]: (1) the synthesis of D-*myo*-inositol 3-monophosphate [Ins(3)P$_1$] from glu-cose 6-P catalyzed by the enzyme D-*myo*-inositol 3-P$_1$ synthase [MIPS; Figure 5.1(b), Step 1] and (2) the breakdown of the monophosphate to Ins catalyzed by a specific monophosphatase [Figure 5.1(b), Step 2]. The relative contribution of MIPS activity proximal to the accumulation of phytic acid in the seed [30], versus activity of this pathway in the vegetative portion of the parent plant, followed by translocation of Ins to the seed [31], has not yet been determined.

If seed-proximal MIPS activity supplies most Ins for phytic acid synthesis, a second question is whether the product of MIPS is used directly for further phosphorylation to phytic acid, or whether the intermediate steps of dephos-phorylation to yield Ins [Figure 5.1(b), Step 2], followed by rephosphorylation catalyzed by a *myo*-inositol kinase [32] [Figure 5.1(b), Step 3], are required.

a

b

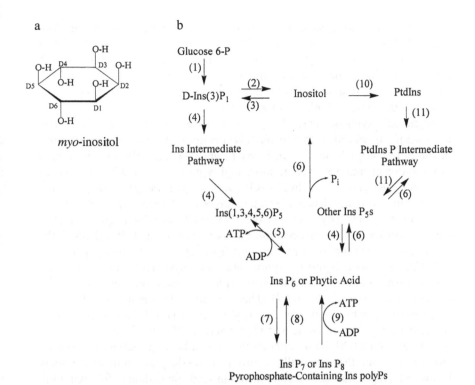

myo-inositol

Figure 5.1 A summary of biochemical pathways involving phytic acid (*myo*-inositol 1,2,3,4,5,6-hexakisphosphate or Ins P_6) in the eukaryotic cell. (a) Structure of *myo*-inositol. The numbering of the carbon atoms in the ring is given following the D-convention. (b) Biochemical pathways. Numbers at arrows indicate the following enzymatic activities: (1) D-*myo*-inositol 3-P_1 synthase (MIPS); (2) D-Ins 3-P_1 phosphatase, or Ins monophosphatase; (3) D-Ins 3-kinase or Ins kinase; (4) Ins P or polyP kinases; (5) Ins 1,3,4,5,6 P_5 2-kinase or phytic acid-ADP phosphotransferase; (6) phytases and phosphatases; (7) Ins P_6 or pyrophosphate-forming kinases; (8) pyrophosphate-specific phosphatases; (9) pyrophosphate-containing Ins PolyP -ADP phosphotransferases; (10) phosphatidylinositol (PtdIns) synthase; and (11) PtdIns and PtdIns P kinases followed by PtdIns P-specific phospholipase C.

The product of *myo*-inositol kinase is the same as MIPS: D-*myo*-inositol(3)P_1 [33]. The pathway to phytic acid then may proceed via stepwise phosphorylation of soluble Ins and Ins Ps to phytic acid, catalyzed by two or more specific kinases [34–35] [Figure 5.1(b), Steps 4 and 5]. In an early proposal [34], one enzyme referred to as "phosphoinositol kinase" first converts Ins(3)P_1 to Ins(1,3,4,5,6)P_5 [Figure 5.1(b), Step 4]. A second enzyme, referred to as "phytic acid-ADP phosphotransferase," then converts Ins(1,3,4,5,6)P_5 to Ins P_6 [36] [Figure 5.1(b), Step 5]. This later enzyme has been more recently studied in the soybean, where it was referred to as Ins(1,3,4,5,6)P_5 2-kinase [37]. In the presence of the right substrate concentrations, this enzyme can regenerate ATP from ADP. More recent studies with the slime mold *Dictyostelium* describe a pathway from Ins to Ins P_6 that involves a series of site-specific kinases that in sequence add individual P esters to the Ins ring, beginning with Ins kinase [35]. Through characterization of Ins Ps in the duckweed *Spirodela* [38], a pathway similar to that observed in *Dictyostelium* was described.

Phytic acid synthesis may also proceed in part via intermediates such as Ins(1,4,5)P_3 produced from phosphatidylinositol phosphate (PtdIns P) intermediates [39–40] [Figure 5.1(b), Steps 10 and 11]. The relative contribution of these two alternative routes to more highly phosphorylated inositols in the developing seed, solely via "free" Ins Ps, versus a pathway proceeding through PtdIns P intermediates, has not been unequivocally determined at present.

In a current view, phytic acid is seen not simply as a P-storage product or end product for Ins phosphorylation but as a pool for both P and Ins P, the latter function possibly of importance to signaling and bond-energy pathways [41–42]. Ins Ps, more highly phosphorylated than phytic acid, such as Ins P_7 and Ins P_8, have been documented to occur widely in eukaryotic cells [41,43–46] [Figure 5.1(b), Steps 7, 8, and 9]. These compounds contain pyrophosphate moieties and may be involved in ATP regeneration [41,47]. This is reminiscent of the decades-old hypothesis that phytic acid itself may serve as a donor in ATP regeneration [48]. It now appears that this early conjecture may have been closer to the truth than originally believed. A classic question in seed biology is how ATP regeneration is accomplished in seeds immediately following imbibition, prior to establishment of membrane integrity. Perhaps pyrophosphate-containing Ins polyPs in fact serve in this role. It is possible that Ins polyP pyrophosphate metabolism plays a key role in phytic acid synthesis and breakdown and in ATP regeneration in early germination. However, there has to date been very little progress in the study of these pyrophosphate-containing compounds in plant systems, with only one report at present [43].

Genomic and/or cDNA sequences encoding Ins 3-P_1 synthases and monophosphatases have been isolated from a number of plant species [30,49–56]. This work is leading to an advanced understanding of Ins metabolism in plant biology. However, to date, no cDNA or genomic sequences have been reported for an Ins kinase in any organism. A full understanding of Ins metabolism as it relates to phytic acid synthesis will require identification of Ins kinase sequences and genes.

Sequences encoding a number of Ins P and polyP kinases and phosphatases, PtdIns and PtdIns phosphate kinases and phosphatases, and phytases, have been reported from a number of plant and animal species. Specific points of relevance are that at present no sequences encoding Ins P kinase, Ins polyP kinase, or PtdIns P kinase have been shown to be specifically involved in phytic acid synthesis in the plant cell. Progress at the molecular level has been made in studies of Ins P_6 metabolism in nonplant systems. For example, a yeast Ins

1,3,4,5,6-P_5 2-kinase gene was recently isolated [40]. A gene was isolated [57] that encodes an Ins polyP kinase similar in some ways to the "phosphoinositol kinase" of Biswas et al. [34], in that it encodes an enzyme that acts on a series of Ins polyP substrates of increasing phosphorylation to yield more highly phosphorylated derivatives. The genetic resources and methods developed in these studies will probably lead to breakthroughs in the isolation of plant genes encoding similar functions. In the near future, the genetics of both Ins and Ins P_6 will begin to reach maturity. A clearer understanding of phytic acid metabolism and its biology will then follow.

At present, however, a great deal remains to be accomplished concerning the phytic acid pathways in seeds. Little is definitively known concerning these pathways. In addition to the questions concerning the central biosynthetic pathway discussed above, other aspects of phytic acid metabolism in seeds, such as transport, localization and compartmentalization functions, and pathway regulation, remain obscure at the molecular genetic level. Therefore, estimates of the number of target genes of potential use in engineering seed phytic acid levels may range from a minimum of two or three loci, encoding MIPS, and at least one or two Ins P and polyP kinases [Figure 5.1(b), Steps 1, 4, and 5], up to seven or more (the sum of all steps leading to phytic acid in Figure 5.1(b), possibly plus one or more regulatory or other functions).

Because Ins is an important component of numerous pathways in plant tissues, a mutation that completely blocks its synthesis, throughout the plant and seed, would undoubtedly be lethal. However, a mutation in a specific copy of a MIPS gene with an expression that is seed-specific and serves mostly to supply Ins for phytic acid synthesis may not be lethal. A mutation that reduces MIPS activity but does not eliminate it may also not be lethal and may reduce phytic acid accumulation. In the early stages of this work, we thought that a mutation early in the pathway to phytic acid, such as a MIPS null, may have a more deleterious effect on the plant and seed function than a mutation in the later Ins P or Ptd Ins P pathways. At that time, the importance of the Ins P and PtdIns P pathways to all cells were not well understood, and these latter pathways were thought to be more specific to phytic acid synthesis in the seed. Subsequent work has shown that both types (early versus late pathway mutations) can have varying effects on plant and seed functions, in an allele-specific manner.

3. THE ISOLATION OF CEREAL LOW PHYTIC ACID MUTANTS

Based on the above considerations, a search was started for phytic acid pathway mutants using maize as the model system. Maize is particularly well-suited for this work. An excellent method of maize mutagenesis, the "pollen-treatment" method [58], produces high rates of mutation, and the materials produced are easy to analyze in genetic terms. Maize is well suited for screening for seed-specific traits. The maize seed is large enough that an experienced researcher can quickly and easily recognize, by visual inspection, segregation for mutations that perturb the germ, aleurone, or other aspects of kernel morphology. As a class, mutants of this type are referred to as defective kernel mutants (DEKs). A preliminary study [20] indicated that DEKs in which the germ and/or aleurone tissues are perturbed, the tissues containing phytic acid, have greatly reduced phytic acid levels. Also, reduced phytic acid P levels were accompanied by "molar-equivalent" increases in inorganic P. However, these mutants are lethal, and the reduction in phytic acid is probably a pleiotropic effect. We [24] sought mutants that had greatly reduced phytic acid but that had morphologically normal seeds and that were viable as homozygotes.

For screening, kernels from the M2 generation were sampled that were "to the unaided eye" phenotypically wild-type (having normal-looking germs and aleurone layers). For initial screening work, we [24] used a "brute force" method employing chromatographic separation and detection of inorganic P and Ins Ps including phytic acid, in single-seed extracts. The high-voltage paper electrophoresis (HVPE) method simultaneously assays for reductions in kernel phytic acid (as compared with nonmutant controls), increases in Ins Ps with less than six P esters, or increases in inorganic P [20,24]. Five "wild-type-looking" kernels were sampled from each M2 progeny ear. These were individually crushed and extracted in 10 μL 0.4 M HCl per mg kernel dry weight (overnight at 4°C). Ten μL of extract was fractionated using HVPE. This method could test a relatively large number of kernels inexpensively and fairly quickly. The levels of phytic acid typical of a nonmutant kernel are high enough to be readily assayed using HVPE. In addition, a test that clearly assays for phytic acid, other Ins Ps (Ins P_2 through Ins P_5s), and inorganic P provides a wide "window" that is preferable for initial work.

Using the above approach, two mutants were identified [24,59] in the first ~1000 M2s and were termed *Zm low phytic acid 1-1* (indicating that this is the first recessive allele at the *Zea mays lpa1* locus) and *Zm low phytic acid 2-1* (Figure 5.2). Quantitative analyses of P and Ins P fractions in kernels homozygous for these mutants indicated that in both mutants, phytic acid P was reduced by at least 50% (Figure 5.2). In the case of *Zm lpa1-1*, the reduction in kernel phytic acid P was accompanied by a molar-equivalent (in terms of P) increase in inorganic P. No unusual increases in other, nonphytic acid Ins Ps were observed. In the case of *Zm lpa2-1*, the decrease in phytic acid P is accompanied by an increase in inorganic P and increased levels of at least two to three Ins P_3s, Ins P_4s, and Ins P_5s, based on paper electrophoresis, HPLC and NMR. Based on these phenotypes, we [24] hypothesized that *Zm lpa1-1* is a mutation in a gene early in the pathway to phytic acid, perhaps a MIPS gene, and that *Zm lpa2-1* is a mutation in a gene later in the pathway, perhaps an Ins P kinase-encoding gene.

Genetic analyses [24] indicated that these two mutants are single-gene mutants, nonlethal and nonallelic, and that they map to two sites on the short

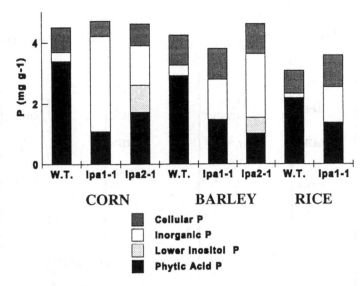

Figure 5.2 Seed phosphorus fractions in wild-type (nonmutant) and *low phytic acid 1* and *low phytic acid 2* lines of maize and barley and wild-type and *low phytic acid 1* lines of rice. All fractions are expressed as their phosphorus (P, atomic weight 32) contents. W.T. = wild-type; *lpa1-1* = *low phytic acid 1-1*; *lpa2-1* = *low phytic acid 2-1*. "Cellular P" represents all seed P-containing compounds other than inorganic P, phytic acid P, and "lower inositol P." For analytical methods, see References [20,24]. The "lower inositol P" primarily consists of Ins tri-, tetrakis- and pentakisphosphates.

arm of maize chromosome 1 (chromosome 1S, Figure 5.3). MIPS-homologous sequences were isolated from maize, and mapping experiments indicated that these sequences mapped to up to seven sites in the maize genome [53]. One MIPS sequence mapped to the same vicinity of maize chromosome 1S, as does *Zm lpa1*. This represents the first evidence in support of the "candidate gene" hypothesis that *Zm lpa1-1* is a MIPS mutant.

Further genetic analyses [24] indicated that for both *Zm lpa1-1* and *Zm lpa2-1*, the seed phenotype is seed-specific and not due to any effect of these mutations on the maternal plant. These studies also indicated that there is a seed-specific reduction in dry weight in seeds homozygous for either mutant. In the case of *Zm lpa1-1*, the dry weight reduction typically ranges from 5% to 15% and appears to be endosperm-specific. In some progenies, no dry weight difference is observed between sibling wild-type and mutant kernels. In the case of *Zm lpa2-1*, observed dry weight reductions were less than those observed with *Zm lpa1-1*, ranging from 5% to 10%.

Because one can readily obtain viable plants homozygous for either *Zm lpa1-1* or *Zm lpa2-1* and can then maintain these mutants as homozygotes, neither mutation is conditionally lethal. While preliminary observations indicated

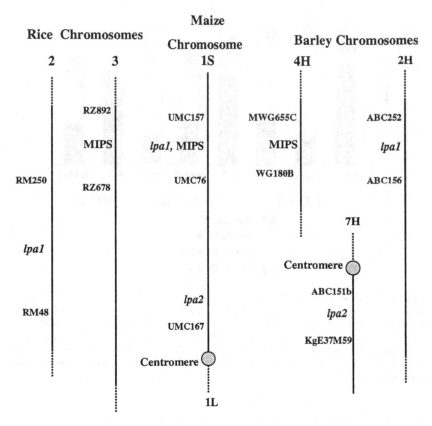

Figure 5.3 Diagrammatic map of chromosomal regions containing the *lpa1* and *lpa2* loci and *myo*-inositol 1-phosphate synthase (MIPS) genes of barley, maize, and rice. This map summarizes results of three earlier studies that mapped the position of *lpa* and MIPS loci [24,53,54,61], and was also constructed using data from a number of additional mapping studies that are cited in these three papers. No attempt was made to represent on this map actual physical or genetic distances between the indicated loci. Rather, two flanking markers are given for each *lpa* and MIPS locus to provide reference points for interested readers.

that homozygosity for *Zm lpa1-1* has relatively little effect on plant growth habit or vigor [20,24], initial unpublished results indicate that while *Zm lpa2-1* homozygotes are viable, this mutation has a consistently negative impact on plant and seed function. These results, therefore, indicate that it may be possible to produce vigorous "low phytate" types by modulating Ins synthesis or supply (the early part of the pathway to phytic acid), but that mutations in the later Ins P pathway have a more consistently severe plant and seed phenotype. A third maize low phytic acid locus has not yet been identified.

In a "wild-type" seed of nearly any species, inorganic P concentration is tightly maintained throughout development and germination. Inorganic P levels in the mature dry seed typically represent ≤5% of seed total P. Homozygosity for either of the first two maize mutants results in several-fold increases in inorganic P (Figure 5.2)—from the ≤5% of total P in wild-type to ∼25% (a fivefold increase) in the case of *Zm lpa2-1* and ∼50% (a 10-fold increase) in the case of *Zm lpa1-1*. If one inspects individual maize kernels for viable germs and the wild-type morphology of nonmutant kernels (removing the most frequent type of false-positives in terms of kernel phytic acid P), then a test for the "high inorganic P" (HIP) phenotype provides a highly reliable, sensitive, and quick assay for homozygosity of these mutants. A simple and inexpensive micro-assay for inorganic or "free" P, conducted in micro-titre plates, has been used for numerous genetic analyses of these mutants [24] and for screening for new mutants.

Populations of mutagenized barley M2s, obtained following treatment of seed with sodium azide [60], were screened for *lpa* mutants using the HIP assay. The findings to date were remarkably similar to those found for maize. At present [61], we have found recessive alleles at only two loci, termed *Hordeum vulgare low phytic acid 1* (*Hv lpa1*) and *Hv lpa2*. The seed P phenotypes also parallel closely those observed with their respective maize counterparts (Figure 5.2). In both cases, homozygosity for the mutant allele does not greatly alter kernel total P. In barley, *lpa1-1* phytic acid is reduced by ≥50%, and this is accompanied by a molar-equivalent increase in inorganic P, with no unusual increases in other Ins Ps. In *Hv lpa2-1*, phytic acid P is reduced by ≥50%, and this is accompanied both by an increase in inorganic P and increased levels of a series of Ins Ps similar to that observed in *Zm lpa2-1*.

The *Hv lpa1* and *Hv lpa2* loci map to barley chromosomes 2H and 7H, respectively [61] (Figure 5.3). The phenotypic similarity suggests that the maize *lpa1* and *lpa2* loci are orthologous (ancestrally related) to barley *lpa1* and *lpa2*, respectively. Chromosomal mapping experiments indicate that there may be a conservation in marker order (synteny) between the maize chromosome 1s and barley 7H segments containing their respective *lpa2* loci. While further studies are required, it is, therefore, possible that the maize and barley *lpa2* loci are orthologous. However, the segment of maize chromosome 1s containing the *Zm lpa1* locus may not correspond to barley 2H segment containing *Hv lpa1* (Figure 5.3). The single barley MIPS identified to date maps to chromosome 4H [53]. Therefore, at present, *Hv lpa1* does not appear to be orthologous to *Zm lpa1*. *Hv lpa1* may be a mutation in a gene encoding a third function, not that mutated in either *Zm lpa1*, *Zm lpa2*, and *Hv lpa2*, or a MIPS gene. Definitive proof will require cloning of each of these loci.

We [54] have completed the first round of screening for rice (*Oriza sativa*) *lpa* mutants obtained following the gamma irradiation of seeds. We have isolated a number of putative mutants, but at present, have only identified one that is nonlethal. Kernels homozygous for *Oz lpa1-1* are phenotypically similar to the maize and barley *lpa1* mutants (Figure 5.2) [54]. We have not yet identified a mutant in rice that is phenotypically similar to the *lpa2* mutants—one that accumulates nonphytic acid Ins Ps in the mature grain. The immediate objective of the rice research is to provide genetic resources of value in rice improvement. A more basic rationale is to contribute to a research program that can take advantage of the strengths of several related species as genetic models. Comparative biology of these three cereal species also may provide insight not obtainable by conducting studies with just one. As in barley, rice *lpa1* and the rice genome's single-copy MIPS gene map to different chromosomes, chromosomes 2 and 3, respectively (Figure 5.3). The chromosomal segment containing the rice MIPS gene is syntenic with those containing the barley genome's single-copy MIPS and the maize genome's chromosome 1s MIPS/*lpa1* segment. However, no clear relationship between the maize, barley, and rice genome's chromosomal segments containing their respective *lpa1* loci has been elucidated. Taken together, these results indicate that there are at least three loci recessive alleles of which can have large effects on seed phytic acid but little other effect on plant or seed function: MIPS/*Zm lpa1*, *Hv lpa1*, and *Zm/Hv lpa2*.

4. BREEDING "LOW PHYTIC ACID" CROPS USING *lpa* MUTANTS

The maize *lpa* mutations represented the first-generation technology useful in breeding "low phytic acid" maize hybrids. Initial agronomic studies [10] compared the performance of 14 pairs of hybrids, consisting of the nonmutant version of the hybrid and an *lpa1-1* version (essentially near-isogenic pairs, "isolines"). This study indicated that under the production conditions used, there was little or no effect of *lpa1-1* on germination or stand establishment. There were also no yield differences in six of the 14 hybrid pairs. In eight of the 14, the *lpa1-1* hybrid yielded less than the nonmutant hybrid, and overall, the yield reduction observed in *lpa1-1* hybrids was about 6% as compared with the yields observed in the wild-type hybrids. These results appear to parallel the observation that homozygosity for *Zm lpa1-1* causes a variable kernel dry weight reduction. Perhaps the yield loss is due to a direct effect of the mutation on kernel dry weight.

The first multi-hectare-scale production of grain produced by a maize hybrid that is homozygous for *lpa1-1* was conducted in 1996 [10]. Two neighboring fields (approximately 10 to 15 hectares each) were planted with a test hybrid that was either "wild-type," or nonmutant, or homozygous *lpa1-1*. This grain was used for the first-generation animal trials [10] that tested the level of grain "Available P" (the percentage of grain total P that is available to the animal). An extensive literature indicates that, in most cases, grain phytic acid P is essentially nonavailable to nonruminants. In theory, Available P is, therefore, equal to grain nonphytic acid. In a simple yet fairly accurate model, in nonmutant maize, Available P represents about 25%, and nonavailable P 75%, of grain total P. Based on the quantitative analyses, this should be almost exactly reversed in *lpa1-1* grain.

Results of the first poultry (chicken) trial supported this theoretical estimate [10]. Several measures of "Available P" ranged from 30% to 48% in wild-type grain and from 70% to 90% in *lpa1-1* grain, and fecal P was reduced from 9% up to 40%. As a result of this type of study's confirmation that *lpa1-1* maize is also "high available P" in terms of animal nutrition, some refer to such maize as HAP maize. A second interesting finding of this first study was the positive effect on calcium nutrition observed. Bone calcium was 11% to 13% higher, and blood calcium was 29% to 36% higher, in birds consuming *lpa1-1* versus wild-type feeds containing the same level of total calcium. This improvement in calcium use was observed despite the fact that all diets in this study contained ~20% by weight soy meal, which contributes significant amounts of phytic acid to the feed. A number of additional poultry, swine, and fish trials have been conducted or are currently underway to further investigate the level of Available P in *lpa1-1* grain and to look at other aspects of the nutritional impacts of this change in grain P chemistry [62–63]. The most interesting question concerns the potential benefits of reduced phytic acid consumption distinct from benefits due to enhanced Available P, such as improvements in calcium retention. Reduced dietary phytic acid may enhance mineral nutritional health in general and may also positively affect other dietary processes like protein utilization.

The first study of "low phytic acid" maize in human nutrition compared iron retention from a meal consisting of tortillas prepared using either normal (wild-type) maize or *lpa1-1* maize [64]. These tortillas were extrinsically labeled with a stable isotope of iron, and iron retention was measured as the incorporation of labeled iron into the red blood cells of 14 non-anemic men two weeks after intake. Iron absorption was 49% higher from *lpa1-1* tortillas than from wild-type tortillas (8.2% of intake versus 5.5% of intake). While these fractional absorption levels are low, these results support the potential value of reduced phytic acid consumption in improving iron nutrition.

A more recent study measured zinc retention following consumption of "polentas" prepared using either normal maize or *lpa1-1* [65]. During the cooking process, ^{67}Zn and ^{70}Zn stable isotopes were incorporated into the *lpa1-1* and wild-type polentas, respectively. Five healthy adults (ages 23 to 39) consumed *lpa1-1* polentas for one day and wild-type polentas on a second day (the order was randomly assigned). Mean fractional absorption of the ^{67}Zn in the *lpa1-1* food was 0.30, whereas mean fractional absorption of the ^{70}Zn in the wild-type food was 0.17. This represents a 78% improvement in zinc retention. These results also support the potential for reduced dietary phytic acid intake, via the use of "low phytate" food grains, for achieving improvements in mineral nutrition of those who depend on such grains as staple foods.

The performance of the first generation of "low phytic acid" hybrids was adequate to serve as "proof of principle," that at a minimum, hybrids with "adequate" yields could be developed in which the chemistry of grain P was greatly altered. Studies are underway to determine if the "low phytic acid" or "high inorganic P" trait is specifically responsible for yield reductions and to see if further breeding or bioengineering can overcome this loss in yield. In the context of the hybrid corn industry, it would be preferable for such a trait to be dominant, not recessive as are these mutants, and targeted only to the seed. Such targeting might reduce the impact of a lesion or alteration in Ins or Ins P metabolism on plant productivity, yield, or performance, such as disease or pest resistance, and stress tolerance. Cloning and sequencing of *lpa* loci would be a first step toward these goals. If *lpa1* is a mutation in chromosome 1s MIPS gene, this initial step has already largely been achieved.

Initial breeding efforts are currently underway with the first barley mutant, *Hv lpa1-1*. As in the case of the first maize mutant, the initial results are encouraging. Studies that compared the cultivar "Harrington," a "two-rowed" malting barley, and the same cultivar homozygous *Hv lpa1-1*, indicated that there were no large effects on germination, plant growth habit, yield, and grain "test weight." Current barley research focuses on studies of a selected group of mutants, three of which have increasing reductions in phytic acid P, each accompanied by molar-equivalent increases in inorganic P (*Hv lpa1-1* with a ~50% reduction, M2 635 with a ~75% reduction, and M2 955 with a ~95% reduction). A set of near-isogenic lines representing each of the mutants in the Harrington background, and the nonmutant Harrington as control, has been prepared. The use of such a collection as an experimental tool and model will be discussed below.

5. DISCUSSION

These studies indicate that one can isolate mutations in cereal genes that as homozygotes reduce grain phytic acid P levels substantially, while having relatively little effect on other grain constituents such as total P, minerals, protein, oil, or starch. For some mutants, large effects on viability, germination, and yield are not observed. However, other alleles at the same *lpa* loci can have noticeable effects on viability, germination, and yield. Several other private and public sector [66] research programs are also involved in isolating and working with *lpa* mutants in various cereal grain and legume species. These mutants can be used to breed "low phytic acid" crops. Molecular genetic approaches to achieve the desired *lpa* phenotype with a minimal impact on other plant processes are in development. The extent to which such cultivars and hybrids find use in the "real world" of agricultural production depends on a number of technical and cultural factors. Additional research is required to answer two outstanding questions: can we develop such hybrids or cultivars that yield as well across environments as the most elite contemporary germplasm? and what is the actual "end-use" value or benefit of such germplasm as a component of animal feeds and human foods? Additional studies are required to address the disease susceptibility, pest and stress tolerance, and storage and handling problems possibly associated with "low phytic acid" crops. If these studies reveal that "low phytic acid" hybrids or cultivars are feasible in technical terms and competitive in yield, then their subsequent use will depend on the economics and politics of a given culture or production environment.

The most progress to date has been made with maize, and commercialization of "low phytic acid" maize hybrids began in 2001. These hybrids produce grain with phytic acid reductions ranging from 66% to 80%. The "low phytate" or "high available P" trait is very desirable in poultry and swine production. A large fraction of all maize produced in the United States is used in nonruminant feeds, explaining the interest in this trait. The question of grain and legume phytic acid in nonruminant production is more straightforward than the human nutrition question, because it is largely quantitative and concerns a part of the phytic acid molecule itself, its P content. It is a "primary" effect. In the context of poultry and swine production, any substantial increase in the availability or use of phytic acid P, and corresponding reduction in waste P, is desirable. Thus, a "low phytic acid" maize that has the same level of grain total P as normal maize, but 50% to 75% less phytic acid P, would have 37% to 50% more "available P" than normal maize. A 50% increase in grain available P, if it comes at the expense of grain nonavailable P (phytic acid P), would contribute significantly to meeting new standards for P management in agricultural production.

Evaluating the impact of dietary phytic acid in human nutrition and health is far more complex. Any consideration of grain or legume phytic acid in human nutrition must be made on a case-by-case basis. The extreme cases range from children growing up in communities truly dependent on whole-grain and legume crops as bulk foods, where dietary phytic acid may contribute significantly to less than optimal mineral nutrition, to adult and aging individuals in the developed world, who perhaps may be more interested in dietary phytate's positive roles. Therefore, there probably is not one type of crop, in terms of phytic acid content, that is suitable for all purposes. Perhaps *lpa* or HAP crops will find a niche, in the marketplace and in production, along other "specialty-use" versions of crops. Developing new types of crops that are improved in terms of a specific end use is an important current strategy in plant breeding. Diversification in agriculture is healthy, and these new lines provide enhanced-value alternatives to standard commodity crop production.

There is one circumstance where it may be wise to recommend *lpa* crops for immediate use, if preliminary studies continue to indicate that they are safe as foods. When environmental or political upheavals result in famine, and food aid, in the form of basic grain crops, is required for prolonged periods of time, perhaps *lpa* crops should be used. In this context, the P and mineral nutrition of children, women, and men may take precedence over concerns about long-term issues like aging and cancer incidence.

The impact of dietary phytic acid on human mineral nutrition primarily concerns the interaction of phytic acid and minerals, a "secondary" effect. In the case of iron and zinc, this effect may appear more qualitative than quantitative. Seeds may contain 15 μmoles of phytic acid for each μmole of zinc or iron. Even in the presence of higher concentrations of other mineral "counter" cations, this amount of phytic acid would probably bind a substantial fraction of the zinc and iron, given basic principles of ion exchange chemistry. Grain may typically contain 50 ppm iron and 50 ppm zinc but may contain about 10,000 ppm phytic acid. Put in terms of charge, grain may contain 5 μmoles of total iron and zinc positive charge per gm but 100 μmoles of phytic acid P negative charge per gm. In this context, if grain phytic acid were reduced by 50% to 75%, the 25% to 50% remaining would represent from 25–50 μmoles negative charge, or five to ten times the counterion charge necessary to chelate zinc or iron. Phytic acid binds iron so efficiently that even greatly reduced levels in a diet may prove to be problematic in this context.

Efforts are underway to develop maize lines in which grain phytic acid is reduced by 90% or more. A nonlethal barley mutant, M2 955, produces grain with a ∼99% reduction. There may be a physiological limit to the level of phytic acid reduction possible in adequately yielding lines, but it has not yet been determined. Evidence is accumulating that reductions of up to 66% to 80%, obtained using classical genetic methods, may be possible in adequately yielding lines. Going beyond this may require use of molecular genetic approaches. The acceptability of such approaches, involving sequence manipulation and some form of transformation, will depend on the outcome of current worldwide debate.

The near-isogenic lines developed in this work may provide a powerful tool in evaluating the positive versus negative roles of dietary phytic acid in human health and nutrition. It may be that the crop that has seed that contains 65% to 80% less phytic acid than normal crops would in fact be highly beneficial as staple foods for certain communities, in terms of zinc and iron nutrition, and mineral nutrition in a general or global sense. The first two studies with human subjects consuming maize *lpa1-1* indicated that simply substituting maize grain with a 66% reduction in phytic acid for normal grain resulted in small to medium improvements in iron and zinc retention. Sets of near-isogenic lines can be used to better quantitate the impacts of dietary phytic acid. For example, sets of lines that are similar in most ways, and have seed that contains similar levels of minerals, fiber, protein, oil and starch, but have either normal or 50%, 66%, 80%, and 95% reductions in phytic acid, are currently being produced in barley, and are in development in maize. The set of barley near-isogenic lines are currently being used in nutritional evaluation studies with poultry, swine, and fish.

Perhaps the most important type of study made possible by the development of normal and *lpa1-1* near-isolines would be to provide these types of grain to communities in which maize serves as the traditional staple food. Families in these communities would be assigned either normal or *lpa1-1* maize, and sufficient grain would be provided to supply these families for periods of two months or longer. Large-scale production (hundred-hectare plot sizes or larger) of a variety of maize hybrids and their *lpa1-1* isolines is underway in the United States, and sufficient grain would be available for these studies in the future. The mineral nutritional status of individuals within these families can be followed over the long term. Safety of *lpa* crops is a primary issue in this research. A growing number of animal trials have revealed no obvious safety issues, and the two first trials with human subjects also showed no safety issues.

These studies would permit quantitation of any "steady state" differences in mineral nutrition between similar individuals consuming one or the other of the two types of maize. It would be the first type of study to use a crop genetics approach to quantitate "chronic" phenomena or effects of dietary phytic acid in human health versus the acute phenomena assayed in most previous studies. Depending on the results of a first study, future studies may include a wider variety of "low phytic acid" maize lines, hopefully including those with grain that has 80% and 95% reductions in phytic acid, and when they become available, "low phytic acid" legumes.

Phytic acid metabolism does not appear to be essential for P homeostasis during seed development and germination, as originally thought. It may play a role, but the role is probably relatively minor. Seeds homozygous for most *lpa* mutants, containing such high levels of inorganic P, appear normal in terms of other major grain constituents, such as total protein, starch, and oil. If P homeostasis is critical to nominal seed development, and to nominal synthesis and accumulation of protein, starch and oil, then these results indicate that P homeostasis may be a physical, rather than chemical, process. Perhaps the elevated levels of inorganic P are being sequestered in micro-vacuoles as a salt, and it is not necessary to sequester the P as phytic acid P. Clearly, the phytic acid levels typical of nonmutant grains do not appear to be essential for seed function and viability under the standard or nominal cultural practices.

While kernel phytic acid P metabolism may not be essential for P homeostasis, it may play an important role in P and mineral storage processes. Grains homozygous for *lpa* mutants may leach P and minerals to a greater extent than do wild-type grains. Such a difference may not be significant for cultivated species but may be very important for the survival of noncultivated species "in the wild." However, there are many additional possible effects of these mutants, such as on mineral accumulation and on other processes such as cell wall polysaccharide synthesis, that need to be investigated. For example, while total protein or starch does not appear to be greatly affected by reduced phytic acid P in barley or maize, perhaps the types of proteins or starches, or types of polysaccharides that accumulate in the endosperm, not the total amount, are altered in these mutants.

In the research that helped to pioneer the study of phytic acid metabolism at the biochemical level [34], a minimum of three enzymes were implicated in phytic acid synthesis: either MIPS or *myo*-inositol kinase, a "phosphoinositol kinase," and an ADP-phytic acid phosphotransferase. More recent research indicates that there may be a series of at least five distinct kinases required for phytic acid synthesis. In addition to the "structural" pathway to phytic acid (the pathway of enzymatic steps that synthesize it), there most likely are also genes encoding "regulatory" functions that determine which tissue and under which conditions the pathway is expressed. It is, therefore, surprising that up to this point, we have found mutants at only two loci in maize and barley [24,54]. Perhaps further research will identify additional loci. Perhaps in these genomes, additional functions in the structural pathway to phytic acid are encoded by duplicated genes and, therefore, would not be identified as single-gene mutants with large effects. However, the most parsimonious theory at present is that the pathway to phytic acid is a simple one, catalyzed by a very few functions encoded by a very few genes [34,36].

6. REFERENCES

1. Cosgrove, D.J. 1980. *Inositol Phosphates: Their Chemistry, Biochemistry and Physiology.* Amsterdam, Netherlands: Elsevier Scientific.
2. Raboy, V. 1997. Accumulation and storage of phosphate and minerals, in B.A. Larkins and I.K. Vasil, (eds.), *Cellular and Molecular Biology of Plant Seed Development.* Dordrecht, Netherlands: Kluwer Academic Publishers, pp. 441–477.
3. Cromwell G.L. and Coffey, R.D. 1991. Phosphorus—a key essential nutrient, yet a possible major pollutant—its central role in animal nutrition, in T.P. Lyons, (ed.), *Biotechnology in the Feed Industry.* Nicholasville, KY: Alltech Tech Publ., pp. 133–145.
4. Jongbloed, A.W. and Lenis, N.P. 1992. Alteration of nutrition as a means to reduce environmental pollution by pigs. *Livestock Production Sci.* 31:75–94.
5. Nelson, T.S., Shieh, T.R., Wodzinski, R.J. and Ware, J.H. 1968. The availability of phytate phosphorus in soybean meal before and after treatment with a mold phytase. *Poultry Sci.* 47:1842–1848.
6. Sharpley, A.N., Charpa, S.C., Wedepohl, R., Sims, J.Y., Daniel, T.C. and Reddy, K.R. 1994. Managing agricultural phosphorus for protection of surface waters: Issues and options. *J Environ Qual.* 23:437–451.
7. Lott, J.N.A., Ockenden, I., Raboy, V. and Batten, G.D. 2000. Phytic acid and phosphorus in crop seeds and fruits: A global estimate. *Seed Sci. Res.* 10:11–33.
8. Gibson, D.M. and Ullah, A.B.J. 1990. Phytases and their action on phytic acid, in D.J. Morre, W.F. Boss, and F.A. Loewus (eds.), *Inositol Metabolism in Plants.* New York: Wiley-Liss, pp. 77–92.
9. Pen, J., Verwoerd, T.C., van Paridon, P.A., Beudeker, R.F., van den Elzen, P.J.M., Geerse, K., van der Klis, J.D, Versteegh, H.A.J., van Ooyen, A.J.J. and Hoekema, A. 1993. Phytase-containing transgenic seeds as a novel feed additive for improved phosphorus utilization. *Bio/Tech* 11:811–814.

10. Ertl, D., Young, K.A. and Raboy, V. 1998. Plant genetic approaches to phosphorus management in agricultural production. *J. Environ. Qual.* 27:299–304.

11. Harland B.F. and Morris, E.R. 1995. Phytate: A good or a bad food component. *Nutr. Res.* 15:733–754.

12. Brown, K.H. and Solomons, N.W. 1991. Nutritional problems of developing countries. *Infectious Disease Clinics of North America* 5:297–317.

13. Erdman, J.W. 1981. Bioavailability of trace minerals from cereals and legumes. *Cereal Chem.* 58:21–26.

14. McCance, R.A. and Widdowson, E.M. 1935. Phytic acid in human nutrition. *Biochem. J.* 29:42694–42699.

15. Graf, E. and Eaton, J.W. 1993. Suppression of colonic cancer by dietary phytic acid. *Nutr Cancer* 19:11–19.

16. Graf, E., Epson, L.L. and Eaton, J.W. 1987. Phytic acid: a natural anitoxidant. *J. Biol. Chem.* 262:11647–11650.

17. Lott, J.N.A. 1984. Accumulation of seed reserves of phosphorus and other minerals, in D.R. Murray (ed.), *Seed Physiology*. New York: Academic Press, pp. 139–166.

18. O'Dell, B.L., de Boland, A.R. and Koirtyohann, S.R. 1972. Distribution of phytate and nutritionally important elements among the morphological components of cereal grains. *J. Agr. Food Chem.* 20:718–721.

19. Raboy, V. 1990. The biochemistry and genetics of phytic acid synthesis, in D.J. Morre, W. Boss, and F. Loewus (eds.), *Inositol Metabolism in Plants*. New York: Alan R Liss, pp. 52–73.

20. Raboy, V., Dickinson, D.B. and Neuffer, M.J. 1990. A survey of maize mutants for variation in phytic acid. *Maydica* 35:383–390.

21. Raboy, V., Below, F.E. and Dickinson, D.B. 1989. Recurrent selection for maize kernel protein and oil has altered phytic acid levels. *J. Heredity* 80:311–315.

22. Raboy, V., Dickinson, D.B. and Below, F.E. 1984. Variation in seed total phosphorus, phytic acid, zinc, calcium, magnesium, and protein among lines of *Glycine max*, and *G. soja*. *Crop Sci.* 24:431–434.

23. Raboy, V., Noaman, M.M., Taylor, G.A. and Pickett, S.G. 1991. Grain phytic acid and protein are highly correlated in winter wheat. *Crop Sci.* 31:631–635.

24. Raboy, V., Gerbasi, P.F., Young, K.A., Stoneberg, S.D., Pickett, S.G., Bauman, A.T., Murthy, P.P.N., Sheridan, W.F. and Ertl, D.S. 2000. Origin and seed phenotype of maize *low phytic acid 1-1* and *low phytic acid 2-1*. *Plant Physiol.* 124:355–368.

25. Strother, S. 1980. Homeostasis in germinating seeds. *Ann. Bot.* 45:217–218.

26. Plaxton, W.C. and Preiss, J. 1987. Purification and properties of non-proteolytic degraded ADP-glucose pyrophosphorylase from maize endosperm. *Plant Physiol.* 83:105–112.

27. Lott, J.N.A., Greenwood, J.S. and Batten, G.D. 1995. Mechanisms and regulation of mineral nutrient storage during seed development, in J. Kigel and G. Galili (eds.), *Seed Development and Germination*. New York: Marcel Dekker Inc., pp. 215–235.

28. Pernollet J-C. 1978. Protein bodies of seeds: ultrastructure, biochemistry, biosynthesis and degradation. *Phytochem.* 17:1473–1480.

29. Loewus, F.A. 1990. Inositol biosynthesis, in D.J. Morre, W.F. Boss and F.A. Loewus (eds.), *Inositol Metabolism in Plants*. New York: Wiley-Liss, pp. 13–19.

30. Yoshida, K.T., Wada, T., Koyama H., Mizobuchi-Fukuoka, R. and Naito, S. 1999. Temporal and spatial patterns of accumulation of the transcript of *myo*-inositol-1-phosphate synthase and phytin-containing particles during seed development in rice. *Plant Physiol.* 119:65–72.

31. Sasaki, K. and Loewus, F.A. 1990. Metabolism of *myo*-[2-^{3}H]Inositol and *scyllo*-[R-^{3}H]Inositol in ripening wheat kernels. *Plant Physiol.* 66:740–745.

32. English, P.D., Dietz, M. and Albersheim, P. 1966. Myoinositol kinase: Partial purification and identification of product. *Science* 151:198–199.

33. Loewus, M.W., Sasaki, D., Leavitt, A.L., Munsell, L., Sherman, W.R. and Loewus, F.A. 1982. The enantiomeric form of myo-inositol-1-phosphate produced by myo-inositol 1-phosphate synthase and myo-inositol kinase in higher plants. *Plant Physiol.* 70:1661–1663.

34. Biswas, B.B., Biswas, S., Chakrabarti, S. and De, B.P. 1978. A novel metabolic cycle involving myo-inositol phosphates during formation and germination of seeds, in W.W. Wells and F. Eisenberg Jr. (eds.), *Cyclitols and Phosphoinositides*. New York: Academic Press, pp. 57–68.

35. Stephens, L.R. and Irvine, R.F. 1990. Stepwise phosphorylation of myo-inositol leading to myo-inositol hexakisphosphate in *Dictyostellium*. *Nature* 346:580–583.

36. Biswas, S., Maity, I.B., Chakrabarti, S. and Biswas, B.B. 1978. Purification and characterization of myo-inositol hexaphosphate-adenosine diphosphate phosphotransferase from *Phaseolus aureus*. *Arch. Biochem. Biophys.* 185:557–566.

37. Phillippy, B.Q., Jaffor Ullah, A.H. and Ehrlich, K.C. 1994. Purification and some properties of inositol 1,3,4,5,6-pentakisphosphate 2-kinase from immature soybean seeds. *J. Biol. Chem.* 269:28393–28399.

38. Brearley, C.A. and Hanke, D.E. 1996. Metabolic evidence for the order of addition of individual phosphate esters to the myo-inositol moiety of inositol hexakisphosphate in the duckweed *Spirodela polyrhiza* L. *Biochem. J.* 314:227–233.

39. Van der Kaay, J., Wesseling, J. and Van Haastert, P.J.M. 1995. Nucleus-associated phosphorylation of Ins(1,4,5)P_3 to InsP$_6$ in *Dictyostelium*. *Biochem J.* 312:911–917.

40. York, J.D., Odom, A.R., Murphy, R., Ives, E.B. and Wente, S.R. 1999. A phospholipase C-dependent inositol polyphosphate kinase pathway required for efficient messenger RNA export. *Science* 285:96–100.

41. Safrany, S.T., Caffrey, J.J, Yang, X. and Shears, S.B. 1999. Diphosphoinositol polyphosphates:the final frontier for inositide research? *Biol Chem.* 380:945–951.

42. Van Dijken, P., de Haas, J-R., Craxton, A., Erneux, C., Shears, S.B. and Van Haastert, P.J.M. 1995. A novel, phospholipase C-independent pathway of inositol 1,4,5-triphosphate formation in *Dictyostelium* and rat liver. *J. Biol. Chem.* 270:29724–29731.

43. Brearley, C.A. and Hanke, D.E. 1996. Inositol phosphates in barley (*Hordeum vulgare* L.) aleurone tissue are stereochemically similar to the products of breakdown of Ins P$_6$ *in vitro* by wheat bran phytase. *Biochem. J.* 318:279–286.

44. Mayr, G.W., Radenberg, T., Thiel, U., Vogel, G. and Stephens, L.R. 1992. Phosphoinositol diphosphates: non-enzymic formation *in vitro* and occurrence *in vivo* in the cellular slime mold *Dictyostelium*. *Carbohydrate Res.* 234:247–262.

45. Menniti, F.S., Miller, R.N., Putney, J.W. Jr. and Shears, S.B. 1993. Turnover of inositol polyphosphate pyrophosphates in pancreatoma cells. *J. Biol. Chem.* 268:3850–3856.

46. Stephens, L., Radenberg, T., Thiel, U., Vogel, G., Khoo, K.-H., Dell, A., Jackson, T.R., Hawkins, P.T. and Mayr, G.W. 1993. The detection, purification, structural characterization, and metabolism of diphosphoinositol pentakisphosphate(s) and bisdiphosphoinositol tetrakisphosphates. *J. Biol. Chem.* 268:4009–4015.

47. Voglmaier, S.M., Bembenek, M.E., Kaplin, A.I., Dormán, G., Olszewski, J.D., Prestwich, G.D. and Snyder, S.H. 1996. Purified inositol hexakisphosphate kinase is an ATP synthase: Diphosphoinositol pentakisphosphate as a high-energy phosphate donor. *Proc. Natl. Acad. Sci.* (USA) 93:4305–4310.

48. Morton, R.K. and Raison, J.K. 1963. A complete intracellular unit for incorporation of amino-acid into storage protein utilizing adenosine triphosphate generated from phytate. *Nature* 200:429–433.

49. Dean Johnson, M. 1994. The *Arabidopsis thaliana* myo-inositol 1-phosphate synthase (EC 5.5.1.4). *Plant Physiol.* 105:1023–1024.

50. Dean Johnson, M. and Burk, D. 1995. Isozyme of 1L-myo-inositol-3-phosphate synthase from Arabidopsis. (Accession No. U30250)(PGR95-067). *Plant Physiol.* 109:721.

51. Gillaspy, G.E., Keddie, J.S., Oda, K. and Gruissem, W. 1995. Plant inositol monophosphatase is a lithium-sensitive enzyme encoded by a multigene family. *Plant Cell* 7:2175–2185.
52. Ishitani, M., Majumder, A.L., Bornhouser, A., Michalowski, C.B., Jensen, R.G. and Bohnert, H.J. 1996. Coordinate transcriptional induction of *myo-inositol* metabolism during environmental stress. *Plant J.* 9:537–548.
53. Larson, S.R. and Raboy, V. 1999. Linkage mapping of maize and barley *myo-*inositol 1-phosphate synthase DNA sequences: correspondence with a *low phytic acid* mutation. *Theor. Appl. Genet.* 99:27–36.
54. Larson, S.R., Rutger, J.N., Young, K.A. and Raboy, V. 2000. Isolation and genetic mapping of a non-lethal rice *low phytic acid* mutation. *Crop Sci.* 40:1397–1405.
55. Smart, C.C. and Fleming, A.J. 1993. A plant gene with homology to D-myo-inositol-3-phosphate synthase is rapidly and spatially up-regulated during an abscisic acid induced morphogenic response in *Spirodela polyrrhiza*. *Plant J.* 4:279–293.
56. Wang, X. and Dean Johnson, M. 1995. An isoform of 1L-myo-Inositol 1-Phosphate Synthase (EC 5.5.1.4) from *Phaseolus vulgaris* (Accession No. U38920)(PGR95-121). *Plant Physiol.* 110:336.
57. Saiardi, A., Erdjument-Bromage, H., Snowman, A.M., Tempst, P. and Snyder, S.H. 1999. Synthesis of diphosphoinositol pentakisphosphate by a newly identified family of higher inositol polyphosphate kinases. *Current Biology* 9:1323–1326.
58. Neuffer, M.G. and Coe, E.H. 1978. Paraffin oil technique for treating mature corn pollen with chemical mutagens. *Maydica* 23:21–28.
59. Raboy, V. and Gerbasi, P. 1996. Genetics of *myo*-inositol phosphate synthesis and accumulation, in B.B Biswas and S. Biswas (eds.), *Subcellular Biochemistry, Vol. 26:myo-Inositol Phosphates, Phosphoinositides, and Signal Transduction*. New York: Plenum Press, pp. 257–285.
60. Nilan, R.A., Kleinhofs, A. and Sander, C. 1975. Azide mutagenesis in barley. *Barley Genetics* III:113–132.
61. Larson, S.R., Young, K.A., Cook, A., Blake, T.K. and Raboy, V. 1998. Linkage mapping two mutations that reduce phytic acid content of barley grain. *Theor. Appl. Genet.* 97:141–146.
62. Huff, W.E., Moore, P.A., Waldroup, P.W., Waldroup, A.L., Balog, J.M., Huff, G.R., Rath, N.C., Daniel, T.C. and Raboy, V. 1998. Effect of dietary phytase and high available phosphorus corn on broiler chicken performance. *Poultry Sci.* 77:1899–1904.
63. Sugiura, S.H., Raboy, V., Young, K.A., Dong, F.M. and Hardy, R.W. 1998. Availability of phosphorus and trace elements in low-phytate varieties of barley and corn for rainbow trout (*Oncorhynchus mykiss*). *Aquaculture.* 170:285–286.
64. Mendoza, C., Viteri, V.E., Lönnerdal, B., Young, K.A., Raboy, V. and Brown, K.H. 1998. Effect of genetically modified, low-phytic acid maize on absorption of iron from tortillas. *Am. J. Clin. Nutr.* 68:1123–1128.
65. Adams, C., Raboy, V., Krebs, N., Westcott, J., Lei, S. and Hambidge, M. 2000. The effect of low-phytic acid corn mutants on zinc absorption. *FASEB J.* 14:A359.11.
66. Rasmussen, S.K. and Hatzack, F. 1998. Identification of two low-phytate barley (*Hordeum vulgare* L.) grain mutants by TLC and genetic analysis. *Hereditas* 129:107–112.

CHAPTER 6

Phytase Expression in Transgenic Plants

ELIZABETH A. GRABAU

1. INTRODUCTION

PHOSPHORUS is a critical nutrient for all living organisms as a component of membrane phospholipids, nucleic acids, ATP, and many other biological molecules. A recent editorial addressing a potential crisis in phosphorus availability highlighted a variety of areas in which loss of this resource is a common concern [1]. Phosphorus availability plays a key role in issues of soil fertility and crop production, animal health and nutrition, as well as waste management and water quality. While many areas of the world lack sufficient phosphorus to sustain good crop yields, areas of intensive animal production experience an accumulation of excess phosphorus in soil as a result of repeated applications of phosphorus-rich manure. Runoff from pastures and croplands with elevated soil phosphorus levels can contaminate surface water and lead to environmental phosphorus pollution and eutrophication.

Phytate (*myo*-inositol hexa*kis*phosphate) is the major storage form of phosphorus in plant seeds and a principal factor limiting phosphorus availability in the diets of many animals, including humans. Phytate cannot be digested efficiently by nonruminants and is excreted as waste. Concomitant loss of complexed minerals exacerbates the antinutritional impact of phytate. Strategies to improve phosphorus and mineral availability include the addition of phytase to feed, the use of feed components genetically modified to express a phytase gene, or reduction in the synthesis or accumulation of phytate in seeds.

As food and feed components, cereals and legumes provide obvious targets for genetic modification to improve nutrient availability. This review summarizes the results of studies on the introduction of a phytase gene from *Aspergillus niger* into several different plants including tobacco, soybean, canola, and alfalfa. Synthesis, processing, and localization of recombinant phytase have been characterized in plant cells. In addition, poultry feeding studies have been conducted to determine whether plant-derived recombinant phytase can serve as an effective dietary supplement.

2. SEED PHYTATE AND PHYTASE IN ANIMAL DIETS

Plant seeds such as corn and soybean are major components of livestock feed, including diets of nonruminant animals such as swine and poultry. Phytate phosphorus is utilized inefficiently by monogastric animals, which can result in serious nutritional and environmental consequences. Diets of nonruminants must be supplemented with inorganic phosphate to meet animal growth requirements. Undigested phytate is excreted in manure, which typically is applied as fertilizer to agricultural fields. This practice can lead to elevated soil phosphorus levels in areas of intensive animal production and the potential for phosphorus runoff into lakes and streams. High phosphorus levels can decrease water quality due to eutrophication, because phosphorus is the limiting nutrient for aquatic plant growth [2].

Phytate is also considered an anti-nutrient because it chelates essential minerals such as calcium, iron, zinc, and magnesium, lowering their bioavailability in animals [3,4]. Phytate in plant seeds is present predominantly as the calcium-magnesium-potassium salt called phytin [5]. Phytate also interacts with seed proteins to form protein-cation-phytate complexes that have been reported to lower protein digestibility [6].

Supplementation of animal diets with enzymes is an increasingly popular method for improving digestibility and nutrient utilization. The enzyme phytase catalyzes the removal of orthophosphate from phytate and other *myo*-inositol phosphates [3]. Phytase as a feed supplement is available commercially as Natuphos® (BASF). This supplement is derived from the fungus *Aspergillus niger* (NRRL 3135) which produces high levels of the enzyme as an extracellular glycoprotein [7,8]. Almost 30 years ago, supplementation of animal diets with phytase (E.C. 3.1.3.8) from *Aspergillus niger* was shown to improve phosphorus availability [9]. Numerous other supplementation studies have demonstrated the efficacy of this method for improving animal diets and lowering phosphorus excretion in recent years [5,10–14].

One limitation of using phytase as a feed supplement is the cost associated with production and application of the enzyme. In countries where regulations controlling nutrient management have been implemented, such as the Netherlands, avoiding fines for excess phosphorus output can offset the expense associated with enzyme supplementation. However, in areas where animal waste management remains less heavily regulated, enzyme costs have prevented widespread use. Another factor limiting use has been the inability of the commercially available phytase supplements to withstand the elevated temperatures associated with feed pelleting. To have an impact on nutrient utilization, an enzyme present in the plant seed components of animal feed must retain activity until it is consumed and can act to release phytate phosphorus in the animal digestive tract. Improving the thermal and protease stability of phytases is an active area of research (see Chapter 7).

The synthesis and accumulation of active, stable enzymes in transgenic plants could provide a unique strategy for producing and packaging enzymes for feed supplementation. Plant seeds already make up a major portion of animal diets and may provide a convenient delivery system for inclusion of supplemental enzymes. Crop plants supply a simple and inexpensive source of biomass and should compete well with other methods of enzyme production. To test the feasibility of this novel approach, recent efforts have been made to introduce a fungal phytase gene into the genomes of different plant species.

3. FUNGAL PHYTASE EXPRESSION IN TOBACCO

Scientists at Mogen International (Leiden, Netherlands) first reported the introduction and expression of a recombinant phytase gene in plants [15]. Their studies included a characterization of phytase-expressing tobacco seeds to assess the molecular weight, glycosylation patterns, and enzymatic stability of the recombinant phytase, and efficacy of plant-derived phytase as a feed supplement. The gene for the fungal phytase (*phy*A) had been previously cloned and sequenced [16–18]. The mature *phy*A coding region was fused to the DNA sequence for the signal peptide from the tobacco PR-S protein. Included in the Mogen binary vector was the alfalfa mosaic virus RNA4 leader sequence to improve messenger stability, the cauliflower mosaic virus (CaMV) 35S promoter to drive transcription, and the nopaline synthase polyadenylation signal for termination. Following *Agrobacterium tumefaciens*-mediated transformation and selection for kanamycin-resistant plants, seeds from transgenic tobacco lines were analyzed for the presence of the phytase by immunodetection and phytase activity assays. The recombinant phytase accounted for up to 1% of total soluble protein in tobacco seed extracts, and the molecular mass was estimated at 67 kD, less than the 85 kD reported for the phytase from *A. niger* [19].

The PR-S signal peptide should direct the recombinant phytase to the endomembrane system for glycosylation and extracellular secretion of the enzyme via the default pathway. To verify that the lower molecular mass observed for the recombinant phytase was due to differences in glycosylation patterns, the *A. niger* and recombinant enzymes were treated with trifluoromethansulfonic acid (TFMS) or Endo-β-*N*-acetylglycosaminidase H (EndoH) and examined by immunoblot analysis. TFMS treatment, which removes all glycans, resulted in proteins of identical molecular mass for both enzymes (60 kD). EndoH treatment of the fungal phytase also resulted in a polypeptide of 60 kD. However, EndoH digestion, which cleaves only high mannose structures, reduced the apparent molecular mass of the recombinant phytase to approximately 64 kD, indicating that not all glycans were present as high mannose structures on the plant-derived enzyme. Phytase activity remained high after one year in whole seeds maintained at 4°C or at room temperature and in milled seeds stored at room temperature, demonstrating the stability of the enzyme.

The potential for using the plant-derived recombinant phytase as a feed supplement was initially tested by *in vitro* assays under simulated digestive conditions and in poultry feeding trials [15]. Samples of standard poultry rations supplemented with milled transgenic tobacco seeds or commercial phytase preparations were incubated in reactions mimicking poultry crop and stomach conditions. This included a 60-minute incubation in distilled water at 39°C, followed by acidification and the addition of pepsin to the samples for a further 90-minute incubation. The reactions were assayed at 30-minute intervals, and the results indicated comparable effects on the release of phosphorus from plant phytate in feed supplemented with either phytase source. In a poultry feeding study conducted over a four-week time course, an increase in growth was observed for broilers whose diets were supplemented with milled transgenic tobacco seeds compared to birds fed basal diets alone or basal diets supplemented with nontransgenic seed. Increases in body weight in response to recombinant phytase supplementation were comparable to weight gains obtained for supplementation with inorganic phosphate or with commercial phytase preparations, demonstrating that transgenic seeds could substitute for supplemental phosphorus in poultry diets.

In a second report by the same group, accumulation of the recombinant phytase was examined in other tissues of the transgenic tobacco plants [20]. The authors found that the recombinant phytase exhibited different glycosylation patterns depending upon the plant tissue analyzed. Recombinant phytase expressed in tobacco leaves had an apparent molecular mass of approximately 70 kD, intermediate between the *A. niger* and seed-derived enzyme. Again, EndoH treatment removed some, but not all, of the carbohydrates, demonstrating the presence of both high mannose and complex glycans in the leaf-derived phytase. The secretion of the recombinant phytase was confirmed by assaying extracellular fluid for enzyme activity. Accumulation of phytase in the extracellular space was also observed by immunogold localization in electron micrographs. Three transgenic tobacco lines were examined for phytase RNA and protein expression. Phytase mRNA was highly expressed in transgenic tobacco plants with levels up to 0.02% of total mRNA, which was similar to results observed for transgenic plants expressing human serum albumin [21]. Phytase protein accumulation in transgenic plants showed a 20-fold increase over seven weeks with a maximum accumulation of 14.4% of soluble protein. Thus, high levels of protein expression and good enzyme stability, two essential features for industrial enzyme production, have been demonstrated for recombinant phytase in transgenic plants using tobacco as a model system.

4. RECOMBINANT FUNGAL PHYTASE EXPRESSION IN SOYBEAN

4.1. EXPRESSION OF *phy*A CONSTRUCTS IN SOYBEAN SUSPENSION CULTURES

To address the feasibility of expressing phytase in a plant that is used as an animal feed component, we examined *phy*A transgene expression in soybean. A number of constructs were generated for microprojectile bombardment into suspension culture cells to test whether phytase could be properly expressed and processed by soybean cells [22]. These constructs were designed to include different promoters and to test the requirement for a plant signal sequence (Fig. 6.1). Plasmids contained the mature coding sequence of the fungal *phy*A gene, both with and without a patatin signal sequence [23]. In constructs without

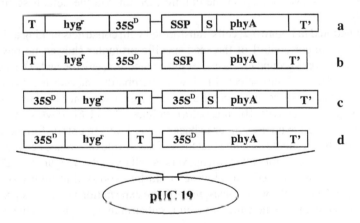

Figure 6.1 Phytase constructs used to transform soybean suspension culture cells. Vectors containing the *phy*A coding region were generated with (a) and (c) and without (b) and (d) a signal sequence (indicated by the S). Expression of the mature *phy*A coding sequence was controlled by the soybean β-conglycinin α' subunit seed-specific promoter (SSP) or constitutive dual-enhanced CaMV 35S promoter ($35S^D$). Constructs containing the seed-specific promoter were created by insertion into the multiple cloning site of a soybean expression vector that also contains the terminator for the β-conglycinin gene (T'). A *Hin*dIII-*Kpn*I fragment containing the dual-enhanced CaMV 35S promoter was used to replace the seed-specific promoter to generate the constitutive constructs in (c) and (d). The hygromycin resistance gene (hyg^r) expression cassette was inserted into the vector as a 2.0 kb *Hin*dIII fragment. T indicates the CaMV 35S polyadenylation signal.

a signal sequence, an ATG codon was generated during DNA amplification of the *phy*A coding sequence. Two different promoters were utilized to control phytase expression: the constitutive, dual-enhanced CaMV 35S promoter [24–26] or a seed-specific promoter from the soybean β-conglycinin α' subunit gene [27]. The gene for hygromycin phosphotransferase (HYG) was used as a marker for selection of transformed cells. Cells were transformed by microprojectile bombardment by the methods of Finer et al. [28,29].

To test for expression of active phytase in transformed soybean suspension culture cells, culture media and cell extracts were assayed for activity by measuring phosphorus released during incubation of samples with phytate [19,30]. A total of two control cultures and eleven transformed, hygromycin-resistant cultures containing the different phytase constructs were examined. The two controls were untransformed cells (UT) and transformants containing the vector with the hygromycin resistance gene alone (HYG). The 11 *phy*A cultures were derived from four different constructs containing the A. *niger* phytase gene under control of the constitutive dual-enhanced CaMV 35S or seed-specific β-conglycinin promoter. The dual-enhanced CaMV 35S promoter directs constitutive phytase expression in tissue culture cells. However, the seed-specific promoter should remain inactive in tissue culture, leading to an absence of detectable activity. The presence of a signal sequence should result in the appearance of phytase activity in the culture medium via the default secretory pathway.

As illustrated in Figure 6.2, cells transformed with constructs lacking a signal sequence or under control of the seed-specific promoter showed no phytase activity above background levels in the culture medium, as expected. The media from two of the cultures selected following bombardment with the construct containing the constitutive promoter and plant signal sequence contained high levels of phytase activity [Figure 6.2(a), cultures 3 and 6]. Phytase activity was low in all cell extracts, similar to background levels in untransformed cells [Figure 6.2(b)]. It should be noted that there is a 250-fold difference between the enzyme activity scale for culture media versus cell extracts in Figure 6.2. These experiments demonstrated that phytase could be expressed at high levels in soybean, and the results were consistent with the expectation that a glycoprotein would be secreted into the culture medium rather than be retained in the cells.

4.2. GROWTH OF TRANSFORMED SOYBEAN SUSPENSION CULTURES IN PHYTATE-MODIFIED MEDIUM

Expression of recombinant phytase should confer on transgenic soybean cultures the ability to utilize phytate as a phosphorus source. To test this, transformed cells were assayed for growth in MS medium [31] that was modified by replacement of the potassium phosphate (monobasic) with dipotassium phytate (Sigma Chemical Co.). This amount of dipotassium phytate maintained a level of potassium equivalent to regular MS medium (1.25 mM), but the amount of

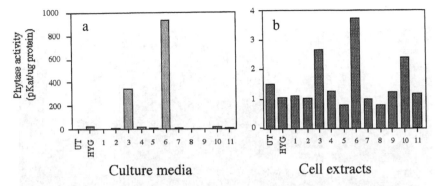

Figure 6.2 Phytase activity in soybean suspension cultures. Phytase activity was assayed in media (a) and cell extracts (b) from transgenic cultures. UT: untransformed control Williams 82 cells, HYG: cells containing the control vector with hygromycin resistance gene alone, 1 and 2: cells containing constructs under control of the CaMV 35S promoter without a signal sequence, 3–6: cells containing constructs under control of the CaMV 35S promoter with a signal sequence, 7–9: cells containing constructs under control of the seed-specific promoter without a signal sequence, and 10 and 11: cells containing constructs under control of the seed-specific promoter with a signal sequence. Each sample value is the mean of three independent assays. Reprinted with permission of the American Society of Plant Physiologists, copyright 1997. (*Source*: Reference [22].)

phosphate was three-fold higher (3.75 mM) due to the increased number of phosphate groups in phytate (assuming complete hydrolysis of phytate by the transgenic soybean cells). Phytase-expressing soybean suspension culture cells were compared to control cultures transformed with the hygromycin resistance marker alone. Following several washes to remove the remaining phosphorus, equal volumes of packed cells were used to initiate the cultures. After the first week, visual inspection revealed no obvious difference in growth between the *phy*A-containing cells and the control cells containing only the hygromycin resistance gene (data not shown). This result may be attributed to residual pools of inorganic phosphate in the cells. After three weeks, growth of the *phy*A-containing cells was two-fold higher than the control cells, as measured by packed cell volume. In addition, the control cells appeared brown [Figure 6.3(a)]. The results indicated that transgenic soybean cells producing phytase could utilize phytate as the sole phosphorus source. In contrast, cells lacking the recombinant phytase activity failed to thrive, presumably due to a phosphorus deficiency.

To compare growth curves for the two soybean cell lines, cultures were initiated in the phytate-modified MS medium, and growth was monitored by measuring cell fresh weights over a two-week period [Figure 6.3(b)]. Data were expressed as an increase in fresh weight compared to the day zero value. Both cultures increased in fresh weight up to nine days, but the control cells showed a slower growth rate, and the final fresh weight for *phy*A-expressing cells was 20% greater than the control culture without the *phy*A transgene.

(a)

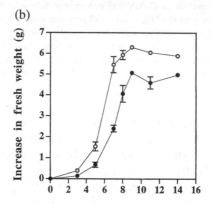

*phy*A construct   *hyg* control

(b)

Days after transfer to MS-phytate medium

Figure 6.3 Growth comparison of *phy*A-containing and control soybean cultures in phytate-modified MS medium. Appearance of cultures after 20 days of growth in phytate-modified MS medium (a). Dipotassium phytate was used as the sole phosphorus source. Growth curves of cell cultures transformed with *phy*A-containing or control constructs in phytate-modified MS medium (b). Results for control cells containing the vector with the hygromycin resistance gene alone (•) and the cells transformed with the constitutive *phy*A construct (○) are indicated. Each point represents the mean from three cultures. Standard errors were calculated for all points and indicated by error bars.

4.3. COMPARISON OF FUNGAL AND RECOMBINANT PHYTASE

An analysis of the activity of the recombinant phytase from soybean cells [22] revealed temperature and pH optima similar to those reported for the phytase from *Aspergillus niger* [19]. The activities of the recombinant phytase and a commercially available fungal phytase (Sigma Chemical Co.) were compared as illustrated in Figure 6.4. The dual pH optima were identical for both phytase

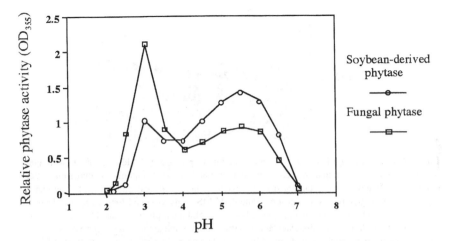

Figure 6.4 Comparison of pH optima of fungal and recombinant phytase. Assays were performed at 58°C by the method of Ullah and Gibson [19]. Recombinant phytase was assayed from the culture medium of transformed cells. Commercial fungal phytase was dissolved in medium from untransformed cells at a concentration of 0.1 mg/mL. For data presentation, the activity of the recombinant phytase was normalized to an equivalent total value to the fungal phytase activity. All data points for fungal and recombinant enzyme represent means of three assays. Reprinted with permission of the American Society of Plant Physiologists, copyright 1997. (*Source:* Reference [22].)

sources (approximately pH 3 and 5.5), however, the relative distribution of the amount of activity between the two optima differed, with a greater proportion of the total activity found at the lower pH for the commercial phytase preparation. A possible explanation for this difference may be the presence of an additional phytase activity that is known to exist in *Aspergillus niger* [17,32]. This second phytase, the product of the *phy*B gene, has a single optimum at the lower pH and is believed to be the same enzyme previously identified as a pH 2.5 acid phosphatase [33,34]. The fungal phytase preparation from Sigma may contain both activities (encoded by *phy*A and *phy*B genes) leading to a higher relative proportion of activity at the lower pH. The pH profile for phytase activity measured in transformed soybean cells can be attributed to the action of the *phy*A gene product because the fungal *phy*B gene was not included in the constructs.

The molecular masses of the recombinant and *A. niger* phytases were compared to examine potential differences in posttranslational modification. Both phytase samples were treated by enzymatic digestion with endoglycosidase F/N-glycosidase or chemical deglycosylation with TFMS [35] to verify glycosylation of the proteins. Protein samples were separated by SDS polyacrylamide gel electrophoresis and subjected to immunoblot analysis. Bands were detected using polyclonal antibody to *A. niger* phytase (provided by

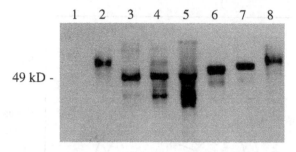

Figure 6.5 Western blotting of fungal and recombinant phytase. Polyclonal antibody to fungal phytase was used for immunodetection of phytase bands. Lane 1: culture medium from HYG control cells, lanes 2 and 8: untreated fungal phytase, lane 3: fungal phytase treated by endo-glycosidase F/N-glycosidase F digestion, lane 4: TFMS-deglycosylated fungal phytase, lane 5: TFMS-deglycosylated recombinant phytase, lane 6: recombinant phytase treated by endoglycosi-dase F/N-glycosidase F digestion, and lane 7: untreated recombinant phytase. Reprinted with per-mission of the American Society of Plant Physiologists, copyright 1997. (*Source*: Reference [22].)

J. Ullah, USDA-ARS, New Orleans, LA) and a chemiluminescent detection kit (Amersham Pharmcia Biotech, Piscataway, NJ). Recombinant phytase from the culture medium was observed as a doublet at approximately 69–71 kD, compared to 85–100 kD observed for fungal phytase (Figure 6.5). Complete deglycosylation with TFMS revealed a protein with an apparent molecular mass of 49 kD for both phytase sources (lanes 4 and 5), which is the predicted size of the mature phytase protein. Enzymatic digestion of the fungal phytase also re-sulted in a 49 kD protein, while enzymatic digestion of the recombinant phytase resulted in incomplete removal of glycans. This is consistent with the speci-ficity of the glycosidases, which are unable to cleave unique glycosidic linkages found in plant glycans [36,37].

4.4. POULTRY FEEDING TRIAL

To determine the efficacy of feed supplementation with phytase produced in transgenic soybeans, a three-week poultry feeding trial was conducted [38]. Graded levels of phytase were added to a basal poultry diet as either milled transgenic soybeans or as a commercially available enzyme supplement (Natuphos®). To generate transgenic soybean plants, Asgrow A5403 soybeans were transformed with expression vector pWR2787 by the method of McCabe et al. [39] at Agracetus (Middleton, WI). Plasmid pWR2787 was generated by amplifying the *phy*A gene via the polymerase chain reaction to generate a product that contained two amino acid substitutions compared to the reported sequence [16–18]. The plasmid construct contained a cauliflower mosaic virus 35S promoter to drive expression, the sequence for leader and signal peptide from tobacco extensin (*Nicotiana plumbaginafolia*) fused in-frame with the mature *phy*A coding sequence, and a nopaline synthase polyadenylation sequence to provide a termination signal. Additionally, the plasmid contained a chimeric β-glucuronidase (GUS) marker cassette for screening of transgenic soybean plants. Soybean seeds used in the feeding studies expressed recombinant phytase at 11.2 units per gram of seed in a bulked heterozygous population.

Basal poultry diets contained a suboptimal level of nonphytate phosphorus (0.2%) to which graded levels of phytase were added (at 400, 800, and 1200 units per kilogram of diet). Test diets were fed to broilers from age 1 to 3 weeks. Each dietary treatment was fed to four pens of birds (eight birds per pen), except for the basal diet, which was fed to eight pens of birds. Growth was measured as body weight gain in the second and third weeks. Phosphorus availability was determined using toe ash as an indicator of bone mineralization. Middle toes were collected at the termination of the experiment, dried, weighed, and ashed. Toe ash measurements were calculated as a percentage of dry weights. Phosphorus excretion was measured as the phosphorus concentration in excreta per kg of diet consumed (days 18 through 20). All data are presented as a percentage of results obtained for control birds (birds fed basal diets without supplementation), as illustrated in Figure 6.6. Supplementing the diets with either source of phytase resulted in increased growth (measured as body weight gain) and increased phosphorus availability (measured as toe ash). These results indicated that transgenic soybeans expressing fungal phytase enhanced growth performance when used as a feed supplement. In addition, phosphorus excretion in manure was reduced as a result of supplementation with transgenic soybeans or with Natuphos®.

Even with the promising results obtained for recombinant phytase expression in soybean, the use of transgenic soybeans as a source for phytase supplementation has several limitations. As components of animal feed, soybeans are generally added in the form of meal, which has been roasted to inactivate anti-nutrients such as trypsin inhibitors, and solvent-extracted to remove oil, which is an economically valuable commodity. Soybean meal prepared by these standard methods should not retain activity of the recombinant phytase. In addition, the expression level in seeds (11.2 units per gram of seed) was not high enough to be competitive with production of the extracellular enzyme in *A. niger.* An alternative approach to addressing problems associated with the high phytate content of plant seeds used in food and feed is the targeted expression of the recombinant phytase in developing seeds. In contrast to the expression of extracellular phytase for use as a supplement, seed-specific expression of a phytase targeted to the site of phytate accumulation may allow the reduction of phytate levels in seed prior to harvest. A possible candidate for this approach would be a soybean phytase that should contain the signals for proper subcellular protein localization to the site of substrate accumulation.

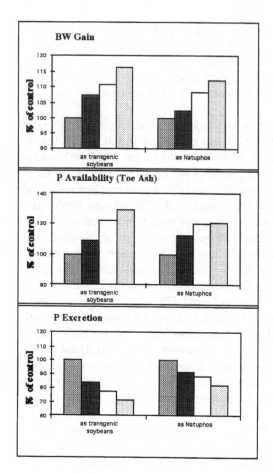

Figure 6.6 Effects of phytase supplementation on growth, phosphorus availability, and phosphorus excretion. Broiler diets were supplemented with phytase supplied as transgenic soybean seeds or a commercial enzyme preparation (Natuphos®). Basal diets contained 0.2% nonphytate phosphorus, which is below optimal requirements for broilers 0–3 weeks of age. Growth was measured as body weight (BW) gain in weeks 2 and 3. P availability was measured as percentage toe ash. P excretion was measured as P concentration in total excreta from days 18–20. Results are presented as a percentage of the values obtained for birds fed the control basal diet. (*Source*: Reference [38].)

5. CANOLA SEEDS AS A SOURCE OF FUNGAL PHYTASE

Transgenic canola seeds may provide a suitable alternative to tobacco for the expression, packaging, and delivery of recombinant phytase as a feed supplement. The efficacy of tobacco-derived phytase in poultry feeding studies was previously documented [15], but tobacco seeds are not normally used in animal diets. The potential for canola to serve as a component of animal feed and the ease of transformation provided the rationale for scientists at Mogen International and Gist-Brocades (Delft, Netherlands) to adopt this plant as a "biofactory" for recombinant phytase production. Phytase constructs were generated using the promoter from the endogenous gene for a storage protein (cruciferin) in oilseed rape (A. Ponstein, personal communication). The constructs also included the signal peptide for the seed-specific cruciferin gene. Primary transformants obtained from *Agrobacterium tumefaciens*-mediated transformation of *Brassica napus* cv. Westar showed variable phytase expression levels that differed with the site and number of insertions. Following selection for enzyme expression and copy number, and propagation through several generations, average phytase levels were found to range from 1.4–2.4% of soluble seed protein for homozygous, single-copy plants. Selection for homozygous offspring from multi-copy transformants resulted in phytase levels up to 13% of total soluble protein. The recombinant phytase produced in canola had a lower molecular mass than the fungal enzyme as determined by polyacylamide gel electrophoresis, similar to observations for tobacco and soybean [15,20,22]. This result can most likely be attributed to different glycosylation patterns of the recombinant enzyme.

Recombinant phytase from canola supplied as Phytaseed® has been used in feeding studies to evaluate its efficacy and safety as a feed supplement [40,41]. Phytaseed® phytase was found to be as effective as Natuphos® phytase in improving phosphorus availability when supplemented in the diets fed to turkey poults. Mortality was not influenced by the different diets, but measurements of feed intake, body weight gain, feed conversion, and toe and tibia ash increased with increasing levels of phytase from either phytase source [40]. Another comparison of Natuphos® and Phytaseed® for enhancing the utilization of phytate phosphorus from corn-soybean meal diets was conducted in a five-week study with young broilers [41]. In this study, body weight gain, feed intake, gain-to-feed ratios, apparent digestibility of dry matter, calcium, and phosphorus increased similarly for both sources of phytase. In addition, gross necropsy and histologic examination determined that there were no adverse effects from Natuphos® or Phytaseed® phytase at supplementation rates of 500 or 2500 units/kg of feed.

6. FUNGAL PHYTASE DERIVED FROM TRANSGENIC ALFALFA

An alternative source of phytase and other industrial enzymes is the juice of transgenic plants. The ability to synthesize recombinant proteins in plant parts other than seeds takes advantage of the tremendous biomass produced by green plants. A forage crop such as alfalfa offers additional advantages in this role because it is a hardy, perennial, leguminous crop that requires no nitrogen fertilizer and has the potential for several harvests per year. A feasibility study for the commercial production of industrially important enzymes in the soluble protein fraction of alfalfa was conducted by Austin et al. [42] using a reporter gene for β-glucuronidase. More recently, two patents have been issued that describe the application of this approach to expression of phytase [43,44].

U.S. Patents 5,824,779 and 5,900,525 describe a method for deriving phytase, dietary protein, and xanthophylls from alfalfa juice [43] and an application of this technique in the formulation of animal feed [44]. Transgenic alfalfa plants were produced by *Agrobacterium tumefaciens*-mediated transformation with binary vectors containing the *phy*A coding sequence that was fused to the coding region for a signal peptide of plant origin and placed under regulatory control of a series of different promoters. The three promoters described include the CaMV 35S promoter, an *Arabidopsis thaliana* Rubisco small submit promoter, or a hybrid promoter containing distal elements from the CaMV 35S promoter and proximal elements from the *Agrobacterium* mannopine synthase promoter. All three constructs achieved significant accumulation of phytase that was differentially glycosylated compared to the fungal enzyme. This is consistent with the observations of other plants. Constitutive expression of phytase activity throughout the plant resulted in phytase concentrations ranging from 0.1–2% of total soluble protein in juice. As an indicator of enzyme stability, fresh alfalfa juice expressing the recombinant phytase activity could be assayed after treatment for 100 minutes at 0, 30, or 50°C with no detectable loss of activity. The juice retained 88% of initial phytase activity after one month of storage at room temperature. The stability of phytase from alfalfa juice in formulated poultry feed showed good stability for three weeks when stored at room temperature, and a poultry feeding experiment confirmed retention of phytase activity.

Six groups of 25 chicks each were fed different diets: diets 1 to 3 were supplemented with three different levels of inorganic phosphate (0.053, 1.06, and 1.60% monocalcium phosphate), diet 4 was supplemented with juice from nontransformed alfalfa, diet 5 was supplemented with juice from phytase-expressing alfalfa, and diet 6 was unsupplemented. Transgenic alfalfa juice was added to poultry diets at a level that should have provided 400 units/kg of feed based on activity assay of the juice. However, the authors reported that the actual measurements of activity in feed samples remained at approximately 250 units/kg throughout the experiment. Chicks fed the unsupplemented diets or diets supplemented with nontransformed alfalfa juice showed severe malnutrition after five days and were euthanized. The weight gain for birds fed the other diets indicated that transgenic alfalfa juice was able to improve phosphorus availability in poultry diets. Based on body weight gain measurements, supplementation with phytase in the form of alfalfa juice was equivalent to the addition of 0.55% monocalcium phosphate. It should be noted that the level of phytase provided by supplementation with alfalfa juice in these experiments resulted in phosphorus levels that were suboptimal for growth. Supplementation with alfalfa-derived phytase gave a growth response equivalent to phosphorus levels at approximately 33% of the recommended guidelines of the National Research Council (NRC). This was observed as a lower body weight gain (approximately 20% lower) for birds supplemented with transgenic alfalfa juice compared to birds supplemented with monocalcium phosphate at 100% of NRC recommended amounts. Even at the suboptimal phytase levels used in these experiments, the results suggest that alfalfa juice can provide an alternative source of phytase and other required dietary components (protein and xanthophylls) for poultry rations.

7. GENE CLONING OF PLANT PHYTASES

All examples of recombinant phytase expression in plants reported to date have involved the introduction and expression of a fungal phytase gene (*phy*A). Plants also express endogenous phytases during germination that function in the degradation of phytic acid reserves. Plant phytase activity has been demonstrated for a number of plant species (for phytase review see Reference [34]). The amounts of plant phytase synthesized are insufficient to provide an economical source for commercial purposes. Maize is the only plant from which the cloning and characterization of a phytase gene have been reported [45,46]. The maize cloning strategy relied on a previously described procedure for purifying phytase from germinating seedlings [47]. The maize seedling phytase is a homodimer with a subunit molecular mass of 38 kD. The purified protein was used to generate a rabbit polyclonal antibody that was used to screen a λgt11 expression library, generated from poly(A)+ RNA from three- to four-day-old maize seedlings. The resulting maize phytase cDNA sequence showed little overall sequence similarity to the fungal phytases but contained one region of 33 amino acids with marked homology to other histidine acid phosphatases, including a conserved RHGXRXP motif [45]. The cDNA was used to screen a maize genomic library and resulted in the isolation of two phytase genes. The presence of two genomic copies of the phytase gene confirmed the results obtained by Southern blot analysis, where two hybridizing bands were observed with the labeled cDNA probe. The two phytase genes showed 99.2% and 96.7% nucleotide identity to the cDNA and were tightly linked (1 cM) on the long arm of chromosome 3 [46].

The maize phytase cDNA sequence was inserted into a vector for expression in *E. coli*. The resulting recombinant phytase appeared to undergo cleavage of the N-terminus to yield a polypeptide that migrated in SDS-polyacylamide gels at the same position as the phytase from maize seedling extracts. Although the recombinant phytase was capable of dimer formation as determined by electrophoresis on native protein gels, no phosphatase activity was associated with the maize phytase produced in *E. coli*. Staining with periodic acid-Schiff reagent indicated the absence of carbohydrate on the phytase from maize seedlings.

This suggested that the lack of activity from the enzyme produced in *E. coli* could not be attributed to a lack of glycosylation but may have been due to the inability of the bacterial system to carry out some other posttranslational modification. Future studies on the phytases from maize and other plants will be required to assess the utility of overexpressing plant phytase genes in heterologous systems.

8. DISCUSSION AND FUTURE OUTLOOK

Successful introduction of a fungal phytase gene into several plant species has been reported. Phytase-expressing plants were obtained using constructs containing the *phy*A gene under control of constitutive and seed-specific promoters and a signal sequence to direct the protein to the plant endomembrane system for secretion. Transgenic soybean suspension culture cells secreted active phytase into the culture medium which allowed the hydrolysis of phytate as the sole phosphorus source for culture growth. Transgenic tobacco showed extracellular localization of the enyzme by immunocytochemistry. Glycosylation patterns for the plant-derived enzyme differed from the *A. niger* phytase depending upon plant species and specific tissue analyzed. Biochemical characteristics of the recombinant enzyme in soybean cultures were similar to those reported for *A. niger* phytase [19]. Thus, plant cells are capable of expressing active phytase, but all plant systems studied to date resulted in differential glycosylation compared to expression of the enzyme in *Aspergillus niger*.

The stability of phytases as a function of glycosylation has been addressed in other systems [48,49] and is reviewed in Chapter 7 of this book. A fungal phytase with superior thermal stability has also been reported [50] and may represent an important advance in the utilization of phytase for food and feed applications. A significant preliminary report on the introduction of a gene for a thermostable phytase into plants was reviewed in a summary of the 16th International Botanical Congress [51]. The review highlighted efforts underway to improve rice through introduction of genes to enhance the availability of iron and other nutrients. Iron and other mineral deficiencies are common in populations consuming diets rich in seeds and grains due to the chelating effects of phytate. Phytase is among the genes that are being introduced to improve the nutritional value of rice. The expression of a thermostable phytase to lower phytate content in rice and other crops would represent a critical nutritional enhancement.

The use of recombinant phytase in animal feeding studies has addressed some of the nutritional aspects of phytase supplementation. Poultry feeding experiments have been performed to compare the results of supplementation with phytase from different sources. Performance based on a variety of parameters such as body weight gain, phosphorus availability, and phosphorus excretion was equivalent for recombinant phytase derived from plants and the commercial fungal phytase, Natuphos®. These results indicate successful accumulation of active recombinant phytase in transgenic plants. While transgenic canola seeds and alfalfa juice may provide economically viable approaches to phytase production for feed supplementation, transgenic soybeans are not ideally suited as a phytase source. Soybeans are normally roasted and processed to produce meal [52,53], which would lead to inactivation of the recombinant phytase. It has been demonstrated that recombinant phytase produced in soybean does not exhibit sufficient thermal stability to withstand production of meal [22]. Enzyme stability during the manufacture of feed remains another hurdle to wider use of phytase supplements in animal production. The feed pelleting process requires elevated temperatures [10] that inactivate the phytase. This necessitates the development of new methods for adding the phytase, such as application by post-pelleting sprays, but that also involve additional expense for the producer.

An alternative strategy for improved phosphorus utilization in food and feed that avoids the limitations of thermal stability and expense of supplementation is the production of seeds with reduced phytate levels. Low phytic acid (*lpa*) maize has been developed by a mutagenesis approach, and a patent describing this process has been issued [54]. This genetic approach to improving phosphorus availability relies on a reduction in the synthesis of phytate in maize. There are several key steps in the biosynthesis of phytic acid that are potential targets for this promising strategy (see Chapter 4 for the genetics of seed phytate biosynthesis). Low phytic acid corn, alternatively referred to as "high available phosphate" corn, has been used in feeding studies to demonstrate the effectiveness of this approach [55–58]. The phytate biosynthetic pathway is the focus of several other patent applications describing methods for the alteration of seed phytic acid [59–61].

Another potential route to improved phosphorus availability is the targeted expression of phytase to modify phosphorus reserves in mature seeds. Subcellular localization of a phytase to the site of phytate accumulation during seed development may allow a reduction in seed phytate levels with an accompanying increase in available phosphorus. To achieve proper targeting, a better understanding of the site of phytate biosynthesis and accumulation as well as signals for protein localization will be required. Two areas of research underway in my laboratory to address the modification of seed phytate levels include the alteration of the fungal *phy*A gene to include protein targeting signals and the isolation of a soybean phytase gene. The goal of our efforts to isolate the soybean phytase gene is the eventual reintroduction of a modified phytase gene into soybean under control of a seed-specific promoter. The soybean phytase is effective at releasing phosphorus from phytic acid during seed germination and may be able to perform that function when expressed during seed development. Lower phytic acid levels in harvested seed should enhance phosphorus bioavailability and reduce phosphorus and mineral loss.

9. ACKNOWLEDGEMENTS

Special thanks go to Anne Ponstein for making the information on transgenic canola available prior to publication.

10. REFERENCES

1. Abelson, P.H. 1999. A potential phosphate crisis, *Science* 283:2015.
2. Sharpley, A.N., Chapra, S.C., Wedepohl, R., Sims, J.T., Daniel, T.C. and Reddy, K.R. 1994. Managing agricultural phosphorus for protection of surface waters: issues and options, *J. Environ. Qual.* 23:437–451.
3. Reddy, N.R., Pierson, M.D., Sathe, S.K. and Salunkhe, D.K. 1989. *Phytates in Cereals and Legumes*, Boca Raton, Florida, CRC Press.
4. Ravindran, V., Bryden, W.L. and Kornegay, E.T. 1995. Phytates: occurrence, bioavailability and implications in poultry nutrition, *Poultry and Avian Biology Rev.* 6:125–143.
5. Swick, R.A. and Ivey, R.J. 1992. The value of improving phosphorus retention, *Feed Manag.* 43:8–17.
6. Prattley, C.A. and Stanley, D.W. 1982. Protein-phytate interactions in soybean. I. Localization of phytate in protein bodies and globoids, *J. Food Biochem.* 6:243–253.
7. Shieh, T.R. and Ware, J.H. 1968. Survey of microorganisms for the production of extracellular phytase, *Appl. Microbiol.* 16:1348–1351.
8. Howson, S.J. and Davis, R.P. 1983. Production of phytate-hydrolysing enzyme by some fungi, *Enzyme Microb. Technol.* 5:377–382.
9. Nelson, T.S., Shieh, T.R., Wodzinski, R.J. and Ware, J.H. 1971. Effect of supplemental phytase on the utilization of phytate phosphorus by chicks, *J. Nutr.* 101:1289–1294.
10. Simons, P.C.M., Versteegh, H.A.J., Jongbloed, A.W., Kemme, P.A., Slump, P., Bos, K.D., Wolters, M.G.E., Beudeker, R.F. and Verschoor, G.J. 1990. Improvement of phosphorus availability by microbial phytase in broilers and pigs, *Brit. J. Nutrition* 64:525–540.
11. Cromwell, G.L., Coffey, R.D., Monegue, H.J. and Randolph, J.H. 1995. Efficacy of low-activity, microbial phytase in improving the bioavailability of phosphorus in corn-soybean meal diets for pigs, *J. Anim. Sci.* 73:449–456.
12. Denbow, D.M., Ravindran, V., Kornegay, E.T., Yi, Z. and Hulet, R.M. 1995. Improving phosphorus availability in soybean meal for broilers by supplemental phytase, *Poultry Sci.* 74:1831–1842.
13. Ravindran, V., Kornegay, E.T., Denbow, D.M., Yi, Z. and Hulet, R.M. 1995. Response of turkey poults to tiered levels of Natuphos® phytase added to soybean meal-based semi-purified diets containing three levels of nonphytate phosphorus, *Poultry Sci.* 74:1843–1854.
14. Yi, Z., Kornegay, E.T., Ravindran, V. and Denbow, D.M. 1996. Improving phytate phosphorus availability in corn and soybean meal for broilers using microbial phytase and calculation of phosphorus equivalency values for phytase, *Poultry Sci.* 75:240–249.
15. Pen, J., Verwoerd, T.C., van Paridon, P.A., Beudeker, R.F., van den Elzen, P.J.M., Geerse, K., van der Klis, J.D., Versteegh, H.A.J., van Ooyen, A.J.J. and Hoekema, A. 1993. Phytase-containing transgenic seeds as a novel feed additive for improved phosphorus utilization, *Bio/Technology* 11:811–814.

16. Mullaney, E.J., Gibson, D.M. and Ullah, A.H.J. 1991. Positive identification of a lambda gt11 clone containing a region of fungal phytase gene by immunoprobe and sequence verification, *Appl. Microbiol. Biotechnol.* 55:611–614.
17. Piddington, C.S., Houston, C.S., Paloheimo, M., Cantrell, M., Miettinen-Oinonen, A., Nevalainen, H. and Rambosek, J. 1993. The cloning and sequencing of the genes encoding phytase (*phy*) and pH 2.5-optimum acid phosphatase (*aph*) from *Aspergillus niger* var. *awamori, Gene* 133:55–62.
18. van Hartingsveldt, W., van Zeijl, C.M.J., Harteveld, G.M., Gouka, R.J., Suykerbuyk, M.E.G., Luiten, R.G.M., van Paridon, P.A., Selten, G.C.M., Veenstra, A.E., van Gorcom, R.F.M. and van den Hondel, C.A.M.J.J. 1993. Cloning, characterization and overexpression of the phytase-encoding gene (*phyA*) of *Aspergillus niger, Gene* 127:87–94.
19. Ullah, A.H.J. and Gibson, D.M. 1987. Extracellular phytase (E.C. 3.1.3.8) from *Aspergillus ficuum* NRRL 3135: purification and characterization, *Prep. Biochem.* 17:63–91.
20. Verwoerd, T.C., van Paridon, P.T., van Ooyen, A.J.J., van Lent, J.W.M., Hoekema, A. and Pen, J. 1995. Stable accumulation of *Aspergillus niger* phytase in transgenic tobacco leaves, *Plant Physiol.* 109:1199–1205.
21. Sijmons, P.C., Dekker, B.M.M., Schrammeijer, B., Verwoerd, T.C., van den Elzen, P.J.M. and Hoekema, A. 1990. Production of correctly processed human serum albumin in transgenic plants, *Bio/Technology* 8:217–221.
22. Li, J., Hegeman, C.E., Hanlon, R.W., Lacy G.H., Denbow, D.M. and Grabau, E.A. 1997. Secretion of active recombinant phytase from soybean cell-suspension cultures, *Plant Physiol.* 114:1103–1111.
23. Iturriaga, G., Jefferson, R.A. and Bevan, M.W. 1989. Endoplasmic reticulum targeting and glycosylation of hybrid proteins in tobacco, *Plant Cell* 1:381–390.
24. Fang, R.X., Nagy, R., Sivasubramanian, S. and Chua, N.H. 1989. Multiple cis regulatory elements for maximal expression of the cauliflower mosaic virus 35S promoter in transgenic plants, *Plant Cell* 1:141–150.
25. Kay, R., Chan, A., Daly, M. and McPherson, J. 1987. Duplication of CaMV 35S promoter sequences creates a strong enhancer for plant genes, *Science* 236:1299–1302.
26. Carrington, J.C. and Freed, D.D. 1990. Cap-independent enhancement of translation by a plant potyvirus 5' nontranslated region, *J. Virol.* 64:1590–1597.
27. Chen, Z-L, Schuler, M.A. and Beachy, R.N. 1986. Functional analysis of regulatory elements in a plant embryo-specific gene, *Proc. Natl. Acad. Sci. USA* 83:8560–8564.
28. Finer, J.J. and McMullen, M.D. 1991. Transformation of soybean via particle bombardment of embryogenic suspension culture tissue, *In Vitro Cell. Dev. Biol.* 27P:175–182.
29. Finer, J.J., Vain, P., Jones, M.W. and McMullen, M.D. 1992. Development of the particle inflow gun for DNA delivery to plant cells, *Plant Cell Rep.* 11:323–328.
30. Heinonen, J.K. and Lahti, R.J. 1981. A new and convenient colorimetric determination of inorganic orthophosphate and its application to the assay of inorganic pyrophosphatase, *Anal. Biochem.* 113:313–317.
31. Murashige, T. and Skoog, F. 1962. A revised medium for rapid growth and bioassays with tobacco tissue cultures, *Physiol. Plant* 15:473–497.
32. Ehrlich, K.C., Montalbano, B.G., Mullaney, E.J., Dischinger, H.C. Jr. and Ullah, A.H.J. 1993. Identification and cloning of a second phytase gene (*phyB*) from *Aspergillus niger* (*ficuum*), *Biochem. Biophys. Res. Comm.* 195:53–57.
33. Ullah, A.H.J. and Cummins, B.J. 1987. Purification, *N*-terminal amino acid sequence and characterization of pH 2.5 optimum acid phosphatase (E.C. 3.1.3.2) from *Aspergillus ficuum, Prep. Biochem.* 17:397–422.
34. Wodzinski, R.J. and Ullah, A.H.J. 1996. Phytase, *Adv. Appl. Micro.* 42:263–302.
35. Edge, A.S.B., Faltynek, C.R., Hof, L., Reichert, L.E, Jr. and Weber, P. 1981. Deglycosylation of glycoproteins by trifluoromethanesulfonic acid, *Anal. Biochem.* 118:131–137.

36. Tretter, V., Altmann, F. and Maerz, L. 1991. Peptide-N^4-(N-acetyl-β-glucosaminyl)asparagine amidase F cannot release glycans with fucose attached $\alpha 1 \rightarrow 3$ to the asparagine-linked N-acetylglucosamine residue, *Eur. J. Biochem.* 199:647–652.
37. Chrispeels, M.J. and Faye, L. 1996. The production of recombinant glycoproteins with defined non-immunogenic glycans, in M.R. Owen and J. Pen (eds.), *Transgenic Plants. A Production System for Industrial and Pharmaceutical Proteins*, London, Wiley, pp. 99–114.
38. Denbow, D.M., Grabau, E.A., Lacy, G.H., Kornegay, E.T., Russell, D.R. and Umbeck, P.F. 1998. Soybeans transformed with a fungal phytase gene improve phosphorus availability for broilers, *Poultry Sci.* 77:878–881.
39. McCabe, D.E., Swain, W.F., Martinell, B.J. and Christou, P. 1988. Stable transformation of soybean (*Glycine max*) by particle acceleration, *BioTechnology* 6:923–926.
40. Ledoux, D.R., Broomhead, J.N., Firman, J.D. and Bermudez, A.J. 1998. Efficacy of Phytaseed®, a phytase containing canola, to improve phytate phosphorus utilization from corn-soybean meal diets fed to turkey poults from day 1 to 35, *Poultry Sci.* (supplement) 77:54.
41. Zhang, Z.B., Kornegay, E.T., Denbow, D.M., Larsen, C.T. and Veit, H.P. 1998. Comparison of genetically engineered microbial and plant phytase for young broilers, *Poult. Sci.* (supplement) 77:71.
42. Austin, S., Bingham, E.T., Koegel, R.G., Mathews, D.E., Shahan, M.N., Straub, R.J. and Burgess, R.R. 1994. An overview of a feasibility study for the production of industrial enzymes in transgenic alfalfa, *Annals NY Acad. Sci.* 721:234–244.
43. Koegel, R.G., Straub, R.J. and Austin-Phillips, S. 1998. Phytase-protein-pigmenting concentrate derived from green plant juice, *United States Patent* 5,824,779.
44. Austin-Phillips, S., Koegel, R.G., Straub, R.J. and Cook, M. 1999. Animal feed compositions containing phytase derived from transgenic alfalfa and methods of use thereof. *United States Patent* 5,900,525.
45. Maugenest, S., Martinez, I. and Lescure, A-M. 1997. Cloning and characterization of a cDNA encoding a maize seedling phytase, *Biochem. J.* 322:511–517.
46. Maugenest, S., Martinez, I., Godin, B., Perez, P. and Lescure, A-M. 1999. Structure of two maize phytase genes and their spatio-temporal expression during seedling development, *Plant Mol. Biol.* 39:503–514.
47. Laboure, A-M., Gagnon, J. and Lescure, A-M. 1993. Purification and characterization of a phytase (*myo*-inositol-hexakisphosphate phosphohydrolase) accumulated in *maize* (*Zea mays*) seedlings during germination, *Biochem. J.* 295:413–419.
48. Wyss, M., Pasamontes, L., Friedlein, A., Rémy, R., Tessier, M., Kronenberger, A., Middendorf, A., Lehmann, M., Schnoebelen, L., Röthlisberger, U., Kusznir, E., Wahl, G., Müller, F., Lahm, H-W., Vogel, K. and van Loon, A.P.G.M. 1999. Biophysical characterization of fungal phytases (*myo*-inositol hexakisphosphate phosphohydrolases): molecular size, glycosylation pattern, and engineering of proteolytic resistance, *Appl. Environ. Micro.* 65: 359–366.
49. Han, Y. and Lei, X.G. 1999. Role of glycosylation in the functional expression of an *Aspergillus niger* phytase (*phyA*) in *Pichia pastoris*, *Arch. Biochem Biophys.* 364:83–90.
50. Pasamontes, L., Haiker, M., Wyss, M., Tessier, M. and van Loon, A.P.G.M. 1997. Gene cloning, purification, and characterization of a heat-stable phytase from the fungus *Aspergillus fumigatus*, *Appl. Environ. Micro.* 63:1696–1700.
51. Gura, T. 1999. New genes boost rice nutrients, *Science* 285:994–995.
52. Cowan, J.C. 1973. Processing and products, in B.E. Caldwell (ed.), *Soybeans: Improvement, Production and Uses*, Madison, WI, American Society of Agronomy, pp. 619–664.
53. Mounts, T.L., Wolf, W.J. and Martinez, W.H. 1987. Processing and utilization, in J.R. Wilcox (ed.) *Soybeans: Improvement, Production and Uses*, Madison, WI, American Society of Agronomy, pp. 819–866.

54. Raboy, V. 1997. Low phytic acid mutants and selection thereof, *United States Patent* 5,689,054.
55. Ertl, D.S., Young, K.A. and Raboy, V. 1998. Plant genetic approaches to phosphorus management in agricultural production, *J. Environ. Qual.* 27:299–304.
56. Huff, W.E., Moore, P.A. Jr., Waldroup, P.W., Waldroup, A.L., Balog, J.M., Huff, G.R., Rath, N.C., Daniel, T.C. and Raboy, V. 1998. Effect of dietary phytase and high available phosphorus corn on broiler chicken performance, *Poultry Sci.* 77:1899–1904.
57. Kersey, J.H., Saleh, E.A., Stilborn, H.L., Crum, R.C. Jr., Raboy, V. and Waldroup, P.W. 1998. Effects of dietary phosphorus level, high available phosphorus corn, and microbial phytase on performance and fecal phosphorus content 1. Broilers grown to 21 d. in battery pens, *Poultry Sci.* (supplement) 77:71.
58. Yan, F., Kersey, J.H., Stilborn, H.L., Crum, R.C. Jr., Rice, D.W., Raboy, V. and Waldroup, P.W. 1998. Effects of dietary phosphorus level, high available phosphorus corn, and microbial phytase on performance and fecal phosphorus content. 2. Broilers grown to market weights in litter floor pens, *Poultry Sci.* (supplement) 77:71.
59. Hitz, W.D. and Sebastian, S.A. 1998. Soybean plant producing seeds with reduced levels of raffinose saccharides and phytic acid, *International Patent Application* WO 98/45448.
60. Martino-Catt, S.J., Wang, H., Beach, L.R., Bowen, B.A. and Wang, X. 1999. Genes controlling phytate metabolism in plants and uses thereof, *International Patent Application* WO 99/05298.
61. Keeling, P.L., Chang, M-T., Guan, H. and Wilhelm, E.P. 1999. Controlled germination using inducible phytate gene, *International Patent Application* WO 99/07211.

CHAPTER 7

Stability of Plant and Microbial Phytases

BRIAN Q. PHILLIPPY

1. INTRODUCTION

PHYTASES are phosphatases that can utilize phytate (*myo*-inositol hexakis phosphate; Ins(1,2,3,4,5,6)P_6 as a substrate. Phytate and phytases appear to be ubiquitous in nature and likely exist in all types of cells [1,2,3]. Because they exhibit considerably greater phytate-cleaving activity than the phytases from animal cells, plant and microbial phytases are utilized industrially to degrade the phytate in foods and feeds. The specificity difference in the initial phosphate removed by the best studied plant and fungal phytases is shown in Figure 7.1. *Triticum aestivum* (wheat) phytase preferentially hydrolyzes the phosphate at position 4, whereas *Aspergillus niger* phytase first cleaves mainly at position 3 [4,5]. The remaining phosphates are also removed by these enzymes, but inositol 2-monophosphate tends to accumulate at the end of the reaction rather than free *myo*-inositol [4,6]. The functional significance of the differences in product specificity of the various phytases is currently unknown.

The stability of phytases has become an important topic of research in recent years as a result of the expanding interest in the use of phytase in animal feeds. Because insufficient phytase is produced in the guts of pigs, poultry, and fish to break down dietary phytate, inorganic phosphate must be supplemented to their diets. The original work showing the feasibility of adding phytase from *Aspergillus ficuum* (*A. niger*) to chicken feeds [7] was not adopted immediately because the cost of inorganic phosphate was lower than that of phytase. The turning point occurred when environmental concerns over pollution from phytate in farm runoffs reached a critical level [8]. The impetus of expected legal constraints on animal waste disposal to prevent excessive growth

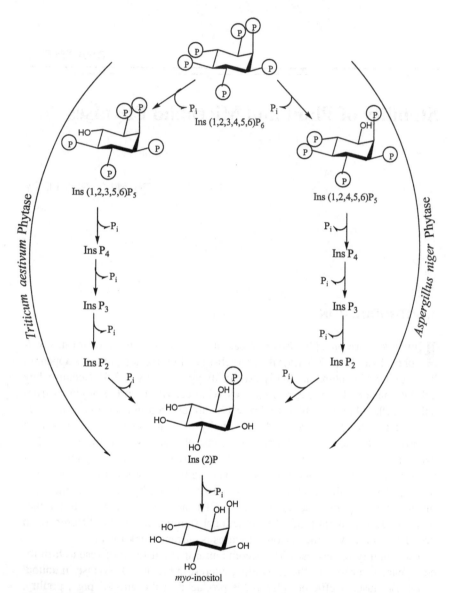

Figure 7.1 Predominant pathways for the hydrolysis of phytate by plant and fungal phytases.

of algae such as the dinoflagellate *Pfiesteria piscicida* has revitalized phytase research.

As phytases from a variety of sources were studied, it became evident that a new emphasis in the commercialization process would be on enzyme stability. The primary initial concern was for thermostability, because many animal feeds are manufactured using pelleting machines that heat the feed to temperatures of about 80°C, which inactivates enzymes such as *A. niger* phytase, which is only stable to 63°C [9]. Although the majority of recent publications on phytase stability have focused on thermostability, additional areas recognized as important include stability during storage and stability in the gut. Phytases are now being produced transgenically and reengineered to optimize their activity in an industrial setting. The use of stabilizers to enhance their performance is another aspect drawing the attention of researchers.

2. THERMOSTABILITY OF PHYTASES

2.1. PLANT PHYTASES

Plant phytases can be categorized according to pH optima as acid phytases or alkaline phytases. Acid phytases predominate in plant tissues and have been studied most extensively. They have similar thermostabilities and generally lose most of their activity below 60°C (Table 7.1).

The thermostability of phytase from wheat grain was determined by Peers [10]. During 10 min incubations, an aqueous extract purified about 20-fold by acetone and ammonium sulfate precipitations retained most of its activity up to 55°C. Two homogeneously purified phytases from wheat bran retained high activity following a 15 min incubation up to 40°C, but higher thermostabilities were observed in wheat bran phytases purified from seeds of other wheat varieties sown in the spring rather than in the fall [11]. Phytase from spelt, which is an old variety of wheat, was purified from kernels after 2 days of germination [12]. Following 8 h preincubation at 55°C, 32% of its activity was recovered, whereas no loss occurred at 45°C.

Although some grain contain considerable amounts of phytase, others appear to require germination in order to yield significant activity. Purified rice bran phytase retained most of its activity during incubation of 30 min at temperatures up to 50°C [13]. After 90 min, the phytase from ungerminated rye retained 85%

TABLE 7.1. Thermostability of Plant Phytases.

Source	Approximate Stability Limit	Reference
Triticum aestivum (wheat) grain	55°C	[10]
Triticum aestivum (wheat) bran	40°C	[11]
Triticum spelta (spelt), germinated	55°C	[12]
Oryza sativa (rice) bran	50°C	[13]
Secale cereale (rye) grain	55°C	[14]
Zea mays (maize) seedling	55°C	[15]
Avena sativa (oat) seedling	40°C	[16]
Hordeum vulgare (barley) seedling	50°C	[17]
Glycine max (soybean) seed	60°C	[18]
Allium fistulosum (scallion) leaf	55°C	[20]

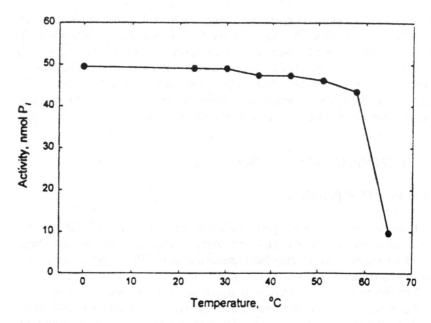

Figure 7.2 Thermostability of purified scallion leaf phytase. Activity was assayed 30 min at 37°C following 10 min preincubation at the indicated temperatures. Reprinted from Reference [20].

of its activity at 50°C but only 20% at 60°C [14]. Maize phytase purified from seedlings germinated 5 or 6 days was completely stable for 10 min at 55°C, but lost 70% of its activity at 65°C [15]. An oat phytase retained 85% of its activity at 40°C but only 29% at 50°C after 90 min incubations [16]. Two phytases purified from barley seedlings and exposed for 90 min to 50°C kept at least 76% of their activity, whereas only 22–53% remained after 90 min at 60°C [17].

It is not clear whether soybean seed phytase may be slightly more thermostable than the other plant phytases. The partially purified enzyme retained most of its activity at 60 and 65°C for 10 min at pH 4.8 [18]. However, a soybean seed phytase purified 22-fold from cotyledons of seedlings germinated 10 days lost most of its activity between 55 and 70°C during 10 min of preincubation at pH 4.8 [19].

Recently, it was discovered that significant levels of phytase are present in a variety of vegetables [20]. A phytase purified from scallion leaves was stable for 10 min at 58°C (Figure 7.2).

2.2. FUNGAL PHYTASES

Aspergillus niger phytase was originally selected for use in animal feeds because the activity yield was greater than that of the other microorganisms tested

TABLE 7.2. Thermostability of Fungal Phytases.

Source	Approximate Stability Limit	Reference
Aspergillus niger (*phy*A)	60°C	[9]
Aspergillus niger (*phy*B, pH 2.5 acid phosphatase)	80°C	[23]
Aspergillus fumigatus	100°C	[25]
Myceliophthora thermophila	50°C	[25]
Aspergillus terreus	45°C	[25]
Thermomyces lanuginosus	60°C	[27]
Peniophora lycii	80°C	[28]

[21]. However, *A. niger* phytase proved unable to survive existing industrial processes to manufacture feed pellets. As a result, the phytases from a number of thermotolerant and thermophilic fungi have been cloned to determine the genetic features of thermostability. Their approximate stability limits are shown in Table 7.2.

A. niger produces two phytases [22]. The enzyme with optima at pH 2.0 and 5.5 is known as *phy*A or simply *A. niger* phytase, and the other has been called *phy*B or pH 2.5 acid phosphatase. Upon purification, the phytase secreted by *A. niger* with optima at pH 2.0 and 5.5 had maximum activity at 58°C but lost most of its activity after 10 min at 68°C [9]. The second phytase from *A. niger* with a single optimum near pH 2.5 retained most of its activity after 20 min at 80°C [23]. The latter enzyme had been cloned and overexpressed in *A. niger* and might be expected to be equivalent to the native phytase. However, when the native enzyme was preincubated 10 min at 75°C, no activity was detected in a subsequent assay using *p*-nitrophenylphosphate as the substrate [24].

Phytases from *Aspergillus fumigatus*, *Myceliophthora thermophila*, and *Aspergillus terreus* 9A were overexpressed in *A. niger* [25]. The latter two enzymes retained most of their activity following 20 min exposure to 45 and 50°C, respectively, but not at higher temperatures. In contrast, *A. fumigatus* phytase retained more than 80% of its activity after 20 min at 100°C, which was the highest temperature used (Figure 7.3). Further investigation revealed that the resistance of this enzyme to high temperatures was primarily due to its ability to refold properly after heat denaturation [23]. Active phytases cloned from *Emericella (Aspergillus) nidulans* and *Talaromyces thermophila* have also been expressed in *A. niger*, but their thermostabilities were not reported [26].

When phytase from *Thermomyces lanuginosus* was expressed in *Fusarium venatum*, about half of its activity was retained after a 20 min incubation at 65°C, but about 30% of its original activity still remained after 20 min at 75°C [27]. This residual activity, attributed to reversible thermal denaturation, was considerably greater than the 5% displayed by *A. niger* phytase [27]. However, *A. niger* T213 or CB phytase, which differs from the widely used *A. niger*

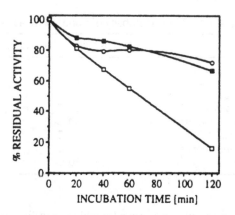

Figure 7.3 Residual enzymatic activity of the purified *A. fumigatus* phytase after exposure for different periods to 60°C (○), 90°C (■) and 100°C (□). Reprinted with permission from Reference [25].

NRRL 3135 phytase in 12 amino acids, displayed 25% of its original activity after 20 min at 90°C [23]. Because the *A. niger* phytases were not compared under identical conditions, it is uncertain whether the amino acid sequence or the experimental procedures accounted for the observed difference.

In contrast to the above fungal phytases from the phyllum *Ascomycota*, the phytase from *Peniophora lycii*, a Basidiomycete, recovered most of its activity after refolding following its thermal denaturation [28]. At pH 5.5, the activity of *P. lycii* phytase expressed in *Aspergillus oryzae* was stable for 60 min at 50°C, whereas 50–60% of the activity remained after 60 min at 60, 70, or 80°C (28).

2.3. YEAST PHYTASES

A few secreted yeast phytases have been evaluated for thermostability. The phytase from *Schwanniomyces castelli* has optimum activity at 77°C and was stable for 1 h at 74°C [29]. The optimal temperature for phytate hydrolysis for the phytase from *Arxula adenovirans* was 75°C, but the enzyme lost 90% of its activity following 20 min at that temperature [30]. During an extensive survey of yeast strains, an additional species, *Pichia spartiae*, was found to secrete a phytase that had optimal activity at 75–80°C and retained a large fraction of its activity at more than 80°C [31].

2.4. BACTERIAL PHYTASES

In contrast to the phytases produced by eukaryotes, bacterial phytases have none of the posttranslational modifications that may provide stability in an

TABLE 7.3. Thermostability of Bacterial Phytases.

Source	Approximate Stability Limit	Reference
Bacillus sp.DS11	90°C	[32]
Enterobacter sp.4	60°C	[33]
Escherichia coli	50°C	[34]
Klebsiella terrigena	50°C	[35]

extracellular environment. Nevertheless, they appear to be effective when fed to animals, and are also of interest for possible food applications. A few of these enzymes have been evaluated, and the phytase purified from *Bacillus* sp.DS11 had very good thermostability under optimal conditions (Table 7.3).

Similar to the plant phytases, the bacterial enzymes could be classified as acid or alkaline phytases according to their pH optima. The phytases from *Bacillus* sp.DS11 [32] and *Enterobacter* sp.4 [33] have alkaline optima of pH 7.0 and pH 7.5, respectively. The *Bacillus* phytase was only stable to about 40°C for 10 min in the absence of calcium (Figure 7.4). However, in the presence of 5 mM CaCl$_2$, 50% of its activity was recovered following 10 min at 90°C. When a crude phytase solution from *Enterobacter* was preincubated 20 h, most of the activity was recovered at 50°C, and about 30% was recovered at 60°C.

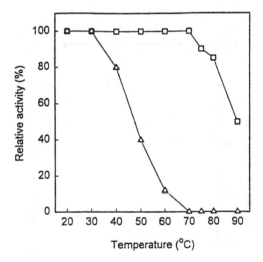

Figure 7.4 Effect of temperature on stability of *Bacillus* sp.DS11 phytase. The enzyme was preincubated at various temperatures for 10 min without (△) and with (□) 5 mM CaCl$_2$, and the remaining activity was measured at 37°C for 30 min. Reprinted from *Enzyme and Microbial Technology*, 22, Kim, Y.-O., Kim, H.-K., Bae, K.-S., Yu, J.-H. and Oh, T.-K., Purification and Properties of a Thermostable Phytase from *Bacillus* sp. DS11, 2–7, Copyright 1998, with permission from Elsevier Science.

The acid phytases from *Escherichia coli* [34] and *Klebsiella terrigena* [35] have optimum activities at pH 4.5 and pH 5.0, respectively. When incubated for 1 h at 50°C, *E. coli* phytase was completely stable, while at 60°C only 24% of the activity was recovered [34]. However, *E. coli* phytase was more stable below its pH optimum, with approximately 50% of the original activity recovered after 30 min at 60°C at pH 1–3 [36]. *K. terrigena* phytase preincubated 1 h retained 80% of its activity at 50°C and was completely inactivated at 60°C. However, in the presence of 20 mg/L bovine serum albumin, the latter enzyme still displayed 30% of its original activity after 5 h at 60°C.

3. STABILIZERS FOR PROCESSING AND STORAGE

There are times when the stability of phytases becomes critical before the enzymes are activated to break down phytate in foods or feeds. For animal feeds the event of most concern is the pelleting step, when the temperature may temporarily reach 60–90°C [23]. During 10 min incubations, wheat phytase in whole meal was stable at 81°C (Figure 7.5), whereas an aqueous extract purified about 20-fold retained most of its activity only up to 55°C [10]. The large decrease in thermostability resulting from purification of this phytase

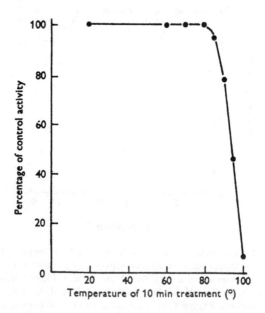

Figure 7.5 Thermal inactivation of wheat phytase in whole meal heated to various temperatures in sealed ampoules for 10 min. Activity was determined in buffered substrate, pH 5.15, at 55°C [10].

demonstrated the protective effects of the endogenous seed components. In fact, more than 60% of the activity remained after heating the whole meal for 90 min at 81°C [10]. Similar findings were reported by Skoglund et al. [37], who observed no decrease in the intrinsic phytase activity of a barley-rapeseed-pea diet for pigs after pelleting at 81 ± 5°C. Three recombinant fungal phytases retained most of their activity after pelleting maize-soybean-fish meal feed at 75°C, even though the purified recombinant *A. niger* phytase in solution lost most of its activity at 70°C [23]. However, after pelleting the feed at 85°C, only the *A. fumigatus* phytase was mostly recovered, while most of the *A. niger* phytase and the *A. niger* phytase also known as pH 2.5 acid phosphatase were inactivated. In a comparison of four recombinant fungal and two bacterial phytases during pelleting of wheat-based feed, *E. coli* and *A niger* phytases were the most stable at 70°C, but only *P. lycii* phytase retained more than half of its activity at 80°C [38].

Although no systematic study has been conducted to determine the relative contributions of individual feed components to phytase thermostability, several stabilizing compounds have been identified. In the presence of soluble starch and sorghum liquor wastes, 67% and 18% of the activity of Natuphos (*A. niger* phytase) survived 10 and 30 min, respectively, at 80°C, whereas no residual activity of unstabilized Natuphos remained after 30 min at 60°C [39]. Phytate has been reported to confer heat resistance to *A. adeninivorans* phytase, which has an optimal temperature of 75°C but was 90% inactivated after 20 min at 75°C in the absence of substrate [30]. In the absence of calcium, most of the activity of *Bacillus* sp.DS11 was lost after 10 min at 50°C [32]. However, in the presence of 5 mM $CaCl_2$, half of the original activity remained after 10 min at 90°C. Similarly, the phytase from *Bacillus subtilis* strain VTT E-68013 required $CaCl_2$ in all purification steps to maintain activity [40]. *K. terrigena* phytase lost 20% and 100% of its activity after 1 h at 50 and 60°C, respectively, but in the presence of 20 mg/L serum albumin, 100% and 30% of the activity remained after 5 h at those temperatures [35]. It is not surprising that dramatic effects of stabilizers have been observed for the nonglycosylated bacterial phytases. A stabilizing effect of glycosylation has been shown for recombinant *A. niger* phytase expressed in *Pichia pastoris* and *Saccharomyces cerevisiae*, where deglycosylation by Endo H_f resulted in losses in thermostability of 34% and 40%, respectively [41,42].

Fungal phytases produced in liquid culture are dried and stored prior to their formulation into edible products. When various salts were tested for their effects on the spray drying of *A. niger* phytase, the smallest losses, which were similar to the control, were obtained for $MgSO_4$, $ZnSO_4$, and $(NH_4)_2SO_4$ at 80 mmol/ 100 mL phytase [43]. Following subsequent storage for 8 weeks at 35°C, 52% and 46% of the activity was lost in the control and with $(NH_4)_2SO_4$, respectively, whereas $MgSO_4$ and $ZnSO_4$ limited the losses to just 15% and 9%, respectively.

4. STABILITY IN THE GUT

4.1. pH STABILITY

The stability of phytases to changes in pH has not received nearly as much attention as their thermostability. The effect of a broad range of pH values on stability per se has been determined mainly for the plant and bacterial phytases, although the fungal phytase of *P. lycii* was reported to be stable for 1 h at 40°C from pH 3–9 [28].

Phytase from rice bran retained most of its activity after 24 h at 4°C from pH 4 to pH 8 [13], and wheat bran phytase was stable for 12 h at 4°C from pH 4–7 [11]. Spelt, rye, oat, and barley phytases were relatively stable for 5–10 days at 4°C between pH 3.0 and pH 7.5 [12,14,16,17]. As a prelude to determining their susceptibility to inactivation by gastrointestinal enzymes, wheat phytase was exposed to pH 2.5, 3.0, or 3.5 for 15 to 60 min at 37°C [44]. Activity was stable at pH 3.5, decreased slightly at pH 3.0, and was completely lost after 15 min at pH 2.5 (Figure 7.6).

The phytase from *Enterobacter* sp.4 retained most of its activity only between pH 7 and pH 8 after 20 h at room temperature [33]. Most of the activity of *Bacillus* sp.DS11 phytase survived a preincubation of 1 h at 37°C between pH 4 and pH 8 [32]. The addition of 5 mM $CaCl_2$ extended the range of stability to pH 3–12 [45]. Neither *E. coli* nor *Klebsiella terrigena* phytase lost any activity

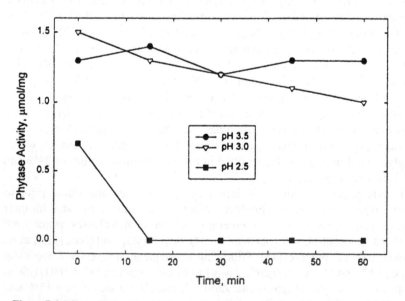

Figure 7.6 Effect of pH on the stability of wheat phytase. Reprinted from Reference [44].

after 10 days at 4°C from pH 3.0 to pH 9.0 [34,35]. *E. coli* phytase also retained 80–90% of its activity for 2 h at 37°C from pH 2–10, and 40% of its activity survived under similar conditions at pH 1 [36]. In a comparison of recombinant *A. niger* phytase expressed in *E. coli* with native *A. niger* phytase, both enzymes were stable at pH 3, but at pH 2, the native enzyme remained stable, whereas the recombinant phytase was completely inactivated following 2 h at 23°C [46].

4.2. PROTEOLYTIC RESISTANCE

The effect of pepsin on the activities of wheat and *A. niger* phytases was determined at 37°C and pH 3.5, because both phytases were completely stable at that pH [44]. At pepsin concentrations that inactivated significant portions of wheat phytase, nearly all of the *A. niger* activity was recovered. Results similar to the above were obtained when pancreatin at pH 6.0 was substituted for the pepsin. Another study compared the proteolytic resistance at 37°C of recombinant *A. niger* phytase overexpressed in *A. niger* with *E. coli* phytase overexpressed in *Pichia pastoris* [47]. Unexpectedly, the recombinant *E. coli* phytase (r-AppA) exhibited increased activity after incubation with pepsin at pH 2.0, whereas recombinant *A. niger* phytase lost most of its activity (Figure 7.7). Trypsin at pH 7.5 caused both phytases to lose activity, with the *E. coli* phytase experiencing the greater loss (Figure 7.7). When four recombinant fungal and two bacterial phytases were compared for susceptibility to inactivation by pepsin or pancreatin during 60 min at 40°C, the fungal enzymes were mostly inactivated [38].

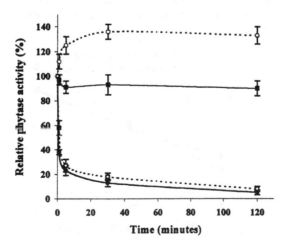

Figure 7.7 Residual phytase activity after trypsin [r-PhyA (■), r-AppA (●)] or pepsin [r-PhyA (□), r-AppA(○)] hydrolysis. The ratios of trypsin/phytase and pepsin/phytase used were 0.01 (w/w) and 0.005 (w/w), respectively. Reprinted with permission from Reference [47].

In contrast, *E. coli* phytase was stable to both proteases, and *B. subtilis* phytase retained most of its activity in the presence of pancreatin. Further testing of the six phytases for stability in various segments of the digestive tract of chickens showed that *E. coli* phytase was also the most stable during 60 min at 40°C in the presence of digesta supernatant fractions from the stomach [38]. In a similar study, 80% of *E. coli* phytase activity was recovered after 2 h at 37°C in the presence of pepsin at pH 2.5, compared to 38% recovery following equivalent exposure to a mixture of pancreatic proteases at pH 7 [36].

In addition to hydrolysis by intestinal enzymes, phytases secreted into culture fluids may also undergo proteolysis during storage and processing. Recombinant *A. fumigatus* and *E. nidulans* phytases were degraded at 50°C by culture supernatant fluids from *A. niger* but not from *Hansenula polymorpha* [48]. Cleavage of the *A. fumigatus* phytase resulted in considerable loss of activity, whereas cleavage of the *E. nidulans* enzyme had no effect. Site-directed mutagenesis of the protease-sensitive sites reduced their sensitivity to proteolysis and decreased the rate of inactivation of the *A. fumigatus* phytase [48].

4.3. EFFECTIVENESS *IN VIVO*

Dozens of reports published in the past decade, especially within the last few years, have documented the effectiveness of recombinant *A. niger* phytase in poultry (e.g., References [49,50]), pigs (e.g., References [51,52]) and fish (e.g., References [53,54]). Satisfactory results have also been obtained with cereal [55] and bacterial [56] phytases. However, in addition to their stability in the gut, other characteristics of these enzymes, such as a low K_m for phytate and a high K_i for inorganic phosphate, may contribute to their effectiveness

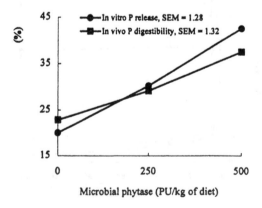

Figure 7.8 Effect of adding microbial phytase on *in vitro* P release and *in vivo* P digestibility of growing pigs fed a corn-soybean meal diet. Reprinted with permission from Reference [58]. Copyright 1997 American Chemical Society.

in vivo [11]. The direct method to determine the stability of phytase within an animal's gut is to collect the digestive contents and analyze them for phytase activity. Alternative methods developed at the University of Missouri utilize *in vitro* digestion to predict the phytase activity in meals fed to poultry [57] and pigs [58]. The *in vitro* phosphorus release corresponded to *in vivo* phosphorus digestibility in pigs with a correlation coefficient of 0.999 [58] (Figure 7.8).

5. STABILITY OF RECOMBINANT PHYTASES

5.1. MOLECULAR MODELING

Initial efforts to increase the thermostability of the phytases used in animal feeds have involved screening a variety of plants and microorganisms to determine the range in properties that evolved in nature. Once the groundwork has been laid, alterations will be made in the structures of these enzymes through mutation and recombination of their amino acid sequences and additionally by controlling the posttranslational modification events with possible emphasis on glycosylation.

In order to identify specific amino acid residues to target for change, the crystal structures of the two phytases from *A. niger* were determined [59,60]. The high sequence homology of the enzymes (Figure 7.9) resulted in an overall structural similarity. The active site of each could be divided into a catalytic center and a substrate specificity site. Although the catalytic centers were almost identical, the substrate specificity site of the dual optima phytase contained four basic amino acids that the phytase, also known as pH 2.5 acid phosphatase, lacked. This helped to explain both the lack of activity of the latter at pH 5, where its negatively charged active site would repel phytate, as well as its lower dependence on the total negative charge of its substrates, because its substrate specificity site was not positively charged. Any modifications designed to increase stability need to preserve the functional integrity of the catalytic center and, to some extent, the substrate specificity region of the active site.

Based on the three-dimensional structure of *A. niger* phytase, the glutamine at position 27 in the amino acid sequence of *A. fumigatus* phytase was suspected of contributing to its lower activity compared to *A. terreus* phytase [61]. This glutamine appeared to form a hydrogen bond with the 6-phosphate of phytate, and *A. terreus* phytase had a leucine at the equivalent position. Site-directed mutagenesis to convert the glutamine at position 27 to a leucine increased the activity of *A. fumigatus* phytase from 26.5 U/mg to 92.1 U/mg [61]. Unfortunately, however, this also probably lowered its thermostability, because the mutation resulted in a decrease from 62.5 to 56°C in the unfolding temperature. Similar attempts to increase the thermostability of *A. terreus* phytase, based on the structure of *A. niger* phytase, by point mutations were unsuccessful,

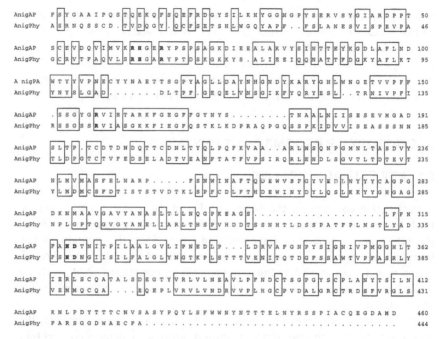

Figure 7.9 Amino acid sequence alignment of *A. niger* pH 2.5 acid phosphatase (AnigAP) with *A. niger* phytase (AnigPhyA). Similar amino acid residues are surrounded by boxes. The catalytically important residues are in bold type. Reprinted with permission from Reference [60].

but replacing a 31 amino acid α-helix gave the desired result, possibly due to improved packing in the protein interior [62].

Using the protein sequences of 13 fungal phytases, a consensus phytase was constructed containing at each position the amino acid occurring most frequently at that position [63]. The consensus phytase had an unfolding temperature of 78°C compared to 55.7–63.3°C for the parent enzymes. Molecular modeling suggested that the increased thermostability was likely due to tighter hydrophobic packing, the formation of additional hydrogen bonds, and stabilization of a loop. The consensus phytase also had an increase in temperature optimum of 16–26°C over its precursors and displayed normal catalytic properties at 37°C. Replacing 24 amino acids in or immediately adjacent to the active site of the consensus phytase with residues of *A. niger* phytase gave a phytase with an unfolding temperature of 70.4°C and a pH-activity profile more similar to that of *A. niger* phytase [64]. Including six additional fungal phytases in the consensus calculations improved the unfolding temperature to 85.4°C, and site-directed mutagenesis to eliminate four destabilizing mutations and introduce one stabilizing mutation resulted in a catalytically uncompromised phytase that unfolded at 90.4°C [65].

The crystal structures of *E. coli* [66] and *Bacillus* sp.DS11 [67] phytases have recently been published. A mutant *E. coli* phytase with more than 40% residual activity following heating 15 min at pH 2.5 and 90°C was created, apparently by eliminating a disulfide bond, thus increasing the number of hydrophobic interactions [68].

5.2. TRANSGENIC PRODUCTION OF PHYTASES

Phytases have been cloned and expressed transgenically in plants, yeast, fungi, and bacteria. These exercises have revealed valuable information on how differences in posttranslational modification *in vivo* and processing for secretion or storage affect phytase stability. The trait most often evaluated has been thermostability, although some reports have focused on stability during storage, processing, or digestion.

The effect of transgenic expression on the thermostability of *A. niger phy*A is shown in Table 7.4. In *Glycine max*, the thermostability was slightly increased despite the fact that the soybean version was significantly less glycosylated as evidenced by a molecular mass of 69,000 to 71,000 compared to 85,000 in *A. niger* [69]. A similar increase in activity was observed in *Pichia pastoris*, which produced a more heavily glycosylated phytase with a molecular mass of 95,000 [41]. Different laboratories reported that *A. niger* phytase expressed in *Saccharomyces cerevisiae* had either a significant increase or no change in thermostability [42,48]. The reason for the apparently conflicting results is not obvious, as similar times and temperatures were used for the incubations. Although not tested specifically for thermostability, *A. niger* phytase expressed in *E. coli* had no activity above 55°C, presumably due to a lack of glycosylation [46]. In related experiments, *A. fumigatus* phytase expressed in *A. niger, H. polymorpha*, and *S. cerevisiae* had identical activities at temperatures from 37–90°C [48].

From an *E. coli* strain isolated from pig colon, Rodriguez et al. [70] cloned a phytase with seven amino acids different from the phytase sequence from *E. coli* BL21(DE3). When expressed in *P. pastoris*, these two *E. coli* enzymes exhibited similar thermostabilities that were slightly higher between 37 and 60°C and

TABLE 7.4. Effect of Transgenic Expression on Thermostability of *Aspergillus niger* Phytase.

Transgenic Host	Effect on Thermostability	Reference
Glycine max	Slight increase	[69]
Pichia pastoris	Slight increase	[41]
Saccharomyces cerevisiae	Increase	[42]
Saccharomyces cerevisiae	None	[48]

lower from 65 to 100°C than recombinant *A. niger* phytase overexpressed in *A. niger.*

Nicotiana tabacum (tobacco) has been utilized to examine the storage stability of transgenic *A. niger* phytase expressed in different plant tissues. Although the yield in leaves was superior, seeds may be preferable for long-term storage. Phytase accounted for 1% of the soluble protein in the seeds and was stable for 1 year at 20°C [71]. In contrast, phytase totaled 14 % of the soluble protein in the leaves, where it remained stable during 2 weeks of senescence [72].

6. FUTURE RESEARCH

Currently, there is much interest in defining the molecular criteria responsible for thermostability, such as glycosylation sites, disulfide bonds, and other stabilizing factors that are dictated by the amino acid sequence. The most thermostable phytase with the best storage stability and *in vivo* activity ultimately may be constructed from parts of two or more different proteins along with some targeted amino acid changes. More work will be needed to determine the effects on thermostability of the components naturally present in plant materials along with a more extensive evaluation of inorganic ions such as calcium and magnesium. The question of how stable phytases are in leaves postharvest both in wet tissue and after drying also needs to be answered. It is not known for certain what molecular characteristics of *A. niger* phytase make it relatively stable in the presence of gastrointestinal enzymes and low pH compared to wheat phytase, or what features of the bacterial phytases determine their resistance to proteases. Perhaps most importantly, studies to directly compare the stability and effectiveness of different phytase products when used in their intended applications in foods or feeds should be performed. These studies may also reveal whether or not differences in the inositol polyphosphate breakdown products from phytases with different specificities have any significant nutritional consequences. The process of finding solutions to these problems will almost certainly generate additional puzzles and ideas to pursue for the next several years.

7. REFERENCES

1. Maenz, D.D. and Classen, H.L. 1998. Phytase Activity in the Small Intestinal Brush Border Membrane of the Chicken. *Poultry Sci.* 77:557–563.
2. Craxton, A., Caffrey, J.J., Burkhart, W., Safrany, S.T., and Shears, S.B. 1997. Molecular Cloning and Expression of a Rat Hepatic Multiple Inositol Polyphosphate Phosphatase. *Biochem. J.* 328:75–81.
3. Mullaney, E.J., Daly, C.B., and Ullah, A.H.J. 2000. Advances in Phytase Research. *Adv. Appl. Microbiol.* 47:157–199.

4. Tomlinson, R.V. and Ballou, C.E. 1962. *Myo*-inositol Polyphosphate Intermediates in the Dephosphorylation of Phytic Acid by Phytase. *Biochem.* 1:166–171.

5. Irving, G.C.J. and Cosgrove, D.J. 1972. Inositol Phosphate Phosphatases of Microbiological Origin: the Inositol Pentaphosphate Products of *Aspergillus ficuum* Phytases. *J. Bacteriol.* 112:434–438.

6. Wyss, M., Brugger, R., Kronenberger, A., Rémy, R., Fimbel, R., Oesterhelt, G., Lehmann, M., and van Loon, A.P.G.M. 1999. Biochemical Characterization of Fungal Phytases (*myo*-Inositol Hexakisphosphate Phosphohydrolases): Catalytic Properties. *Appl. Environ. Microbiol.* 65:367–373.

7. Nelson, T.S., Sheih, T.R., Wodzinski, R.J., and Ware, J.H. 1971. Effect of Supplemental Phytase on the Utilization of Phytate Phosphorus by Chicks. *J. Nutr.* 101:1289–1294.

8. Simons, P.C.M., Versteegh, H.A.J., Jongbloed, A.W., Kemme, P.A., Slump, P., Bos, K.D., Wolters, M.G.E., Beudeker, R.F., and Verschoor, G.J. 1990. Improvement of Phosphorus Availability by Microbial Phytase in Broilers and Pigs. *Brit. J. Nutr.* 64:525–540.

9. Ullah, A.H.J. and Gibson, D.M. 1987. Extracellular Phytase (E.C. 3.1.3.8) from *Aspergillus ficuum* NRRL 3135: Purification and Characterization. *Prep. Biochem.* 17:63–91.

10. Peers, F.G. 1953. The Phytase of Wheat. Biochem. J. 53:102–110.

11. Nakano, T., Joh, T., Tokumoto, E., and Hayakawa, T. 1999. Purification and Characterization of Phytase from Bran of *Triticum aestivum* L. cv. Nourin #61. *Food Sci. Technol. Res.* 5:18–23.

12. Konietzny, U., Greiner, R., and Jany, K.-D. 1995. Purification and Characterization of a Phytase from Spelt. *J. Food Biochem.* 18:165–183.

13. Hayakawa, T., Toma, Y., and Igaue, I. 1989. Purification and Characterization of Acid Phosphatases with or without Phytase Activity from Rice Bran. *Agric. Biol. Chem.* 53:1475–1483.

14. Greiner, R., Konietzny, U., and Jany, K.-D. 1998. Purification and Properties of a Phytase from Rye. *J. Food Biochem.* 22:143–161.

15. Laboure, A.-M., Gagnon, J., and Lescure, A.-M. 1993. Purification and Characterization of a Phytase (*myo*-Inositol-hexakisphosphate Phosphohydrolase) Accumulated in Maize (*Zea mays*) Seedlings during Germination. *Biochem. J.* 295:413–419.

16. Greiner, R. and Alminger, M.L. 1999. Purification and Characterization of a Phytate-Degrading Enzyme from Germinated Oat (*Avena sativa*). *J. Sci. Food Agric.* 79:1453–1460.

17. Greiner, R., Jany, K.-D., and Larsson Alminger, M. 2000. Identification and Properties of *myo*-Inositol Hexakisphosphate Phosphohydrolases (Phytases) from Barley (*Hordeum vulgare*). *J. Cereal Sci.* 31:127–139.

18. Sutardi and Buckle, K.A. 1986. The Characteristics of Soybean Phytase. *J. Food Biochem.,* 10:197–216.

19. Gibson, D.M. and Ullah, A.H.J. 1988. Purification and Characterization of Phytase from Cotyledons of Germinating Soybean Seeds. *Arch. Biochem. Biophys.* 260:503–513.

20. Phillippy, B.Q. 1998. Purification and Catalytic Properties of a Phytase from Scallion (*Allium fistulosum* L.) Leaves. *J. Agric. Food Chem.* 46:3491–3496.

21. Shieh, T.R. and Ware, J.H. 1968. Survey of Microorganisms for the Production of Extracellular Phytase. *Appl. Microbiol.* 16:1348–1351.

22. Shieh, T.R., Wodzinski, R.J., and Ware, J.H. 1969. Regulation of the Formation of Acid Phosphatases by Inorganic Phosphate in *Aspergillus ficuum. J. Bacteriol.* 100:1161–1165.

23. Wyss, M., Pasamontes, L., Rémy, R., Kohler, J., Kusznir, E., Gadient, M., Müller, F., and van Loon, A.P.G.M. 1998. Comparison of the Thermostability Properties of Three Acid Phosphatases from Molds: *Aspergillus fumigatus* Phytase, *A. niger* Phytase, and *A. niger* pH 2.5 Acid Phosphatase. *Appl. Environ. Microbiol.* 64:4446–4451.

24. Ullah, A.H.J. and Cummins, B.J. 1987. Purification, N-Terminal Amino Acid Sequence and Characterization of pH 2.5 Optimum Acid Phosphatase (E.C. 3.1.3.2) from *Aspergillus ficuum. Prep. Biochem.* 17:397–422.

25. Pasamontes, L., Haiker, M., Wyss, M., Tessier, M., and van Loon, A.P.G.M. 1997. Gene Cloning, Purification, and Characterization of a Heat-Stable Phytase from the Fungus *Aspergillus fumigatus. Appl. Environ. Microbiol.* 63:1696–1700.

26. Pasamontes, L., Haiker, M., Henriquez-Huecas, M., Mitchell, D.B., and van Loon, A.P.G.M. 1997. Cloning of the Phytases from *Emericella nidulans* and the Thermophilic Fungus *Talaromyces thermophilus. Biochim. Biophys. Acta.* 1353:217–223.

27. Berka, R.M., Rey, M.W., Brown, K.M., Byun, T., and Klotz, A.V. 1998. Molecular Characterization and Expression of a Phytase Gene from the Thermophilic Fungus *Thermomyces lanuginosus. Appl. Environ. Microbiol.* 64:4423–4427.

28. NOVO NORDISK A/S [DK/DK]; Novo allé, DK-2880 Bagsvaerd (DK). 1998. Peniophora Phytase. International Patent, WO 98/28408.

29. Segueilha, L., Lambrechts, C., Boze, H., Moulin, G., and Galzy, P. 1992. Purification and Properties of the Phytase from *Schwanniomyces castellii. J. Ferment. Bioeng.* 74:7–11.

30. Sano, K., Fukuhara, H., and Nakamura, Y. 1999. Phytase of the Yeast *Arxula adeninivorans. Biotechnol. Lett.* 21:33–38.

31. Nakamura, Y., Fukuhara, H., and Sano, K. 2000. Secreted Phytase Activities of Yeasts. *Biosci. Biotechnol. Biochem.* 64:841–844.

32. Kim, Y.-O., Kim, H.-K., Bae, K.-S., Yu, J.-H., and Oh, T.-K. 1998. Purification and Properties of a Thermostable Phytase from *Bacillus* sp. DS11. *Enzyme Microbiol. Technol.* 22:2–7.

33. Yoon, S.J., Choi, Y.J., Min, H.K., Cho, K.K., Kim, J.W., Lee, S.C., and Jung, Y.H. 1996. Isolation and Identification of Phytase-Producing Bacterium, *Enterobacter* sp.4, and Enzymatic Properties of Phytase Enzyme. *Enzyme Microbiol. Technol.* 18:449–454.

34. Greiner, R., Konietzny, U., and Jany, K.-D. 1993. Purification and Characterization of Two Phytases from *Escherichia coli. Arch. Biochem. Biophys.* 303:107–113.

35. Greiner, R., Haller, E., Konietzny, U., and Jany, K.-D. 1997. Purification and Characterization of a Phytase from *Klebsiella terrigena. Arch. Biochem. Biophys.* 341:201–206.

36. Golovan, S., Wang, G., Zhang, J., and Forsberg, C.W. 2000. Characterization and Overproduction of the *Escherichia coli* appA Encoded Bifunctional Enzyme That Exhibits Both Phytase and Acid Phosphatase Activities. *Can. J. Microbiol.* 46:59–71.

37. Skoglund, E., Larsen, T., and Sandberg, A.-S. 1997. Comparison between Steeping and Pelleting a Mixed Diet at Different Calcium Levels on Phytate Degradation in Pigs. *Can. J. Anim. Sci.* 77:471–477.

38. Igbasan, F.A., Männer, K., Miksch, G., Borriss, R., Farouk, A., and Simon, O. 2000. Comparative Studies of the *in vitro* Properties of Phytases from Various Microbial Origins. *Arch. Anim. Nutr.* 53:353–373.

39. Chen, C.-C., Hunag, C.-T., and Cheng, K.-J. 2001. Improvement of Phytase Thermostability by Using Sorghum Liquor Wastes Supplemented with Starch. *Biotechnol. Lett.* 23:331–333.

40. Kerovuo, J., Lauraeus, M., Nurminen, P., Kalkkinen, N., and Apajalahti, J. 1998. Isolation, Characterization, Molecular Gene Cloning, and Sequencing of a Novel Phytase from *Bacillus subtilis. Appl. Environ. Microbiol.* 64:2079–2085.

41. Han, Y. and Lei, X.G. 1999. Role of Glycosylation in the Functional Expression of an *Aspergillus niger* Phytase (*phy*A) in *Pichia pastoris. Arch. Biochem. Biophys.* 364:83–90.

42. Han, Y., Wilson, D.B., and Lei, X.G. 1999. Expression of an *Aspergillus niger* Phytase Gene (*phy*A) in *Saccharomyces cerevisiae. Appl. Environ. Microbiol.* 65:1915–1918.

43. GIST-BROCADES B.V. NL-2600 MA Delft(NL). 1997. Salt-Stabilized Enzyme Preparations. European Patent, EP 0 758 018 A1.

44. Phillippy, B.Q. 1999. Susceptibility of Wheat and *Aspergillus niger* Phytases to Inactivation by Gastrointestinal Enzymes. *J. Agric. Food Chem.* 47:1385–1388.

45. Kim, D.-H., Oh, B.-C., Choi, W.-C., Lee, J.-K., and Oh, T.-K. 1999. Enzymatic Evaluation of *Bacillus amyloliquefaciens* Phytase as a Feed Additive. *Biotechnol. Lett.* 21:925–927.

46. Phillippy, B.Q. and Mullaney, E.J. 1997. Expression of an *Aspergillus niger* Phytase (*phy*A) in *Escherichia coli*. *J. Agric Food Chem.* 45:3337–3342.

47. Rodriguez, E., Porres, J.M., Han, Y., and Lei, X.G. 1999. Different Sensitivity of Recombinant *Aspergillus niger* Phytase (rPhyA) and *Escherichia coli* pH 2.5 Acid Phosphatase (rAppA) to Trypsin and Pepsin *in Vitro*. *Arch. Biochem. Biophys.* 365:262–267.

48. Wyss, M., Pasamontes, L., Friedlein, A., Rémy, R., Tessier, M., Kronenberger, A., Middendorf, A., Lehmann, M., Schnoebelen, L., Röthlisberger, U., Kusznir, E., Wahl, G., Müller, F., Lahm, H.-W, Vogel, K., and van Loon, A.P.G.M. 1999. Biophysical Characterization of Fungal Phytases (*myo*-Inositol Hexakisphosphate Phosphohydrolases):Molecular Size, Glycosylation Pattern, and Engineering of Proteolytic Resistance. *Appl. Environ. Microbiol.* 65:359–366.

49. Um, J.S. and Paik, I.K. 1999. Effects of Microbial Phytase Supplementation on Egg Production, Eggshell Quality, and Mineral Retention of Laying Hens Fed Different Levels of Phosphorus. *Poultry Sci.* 78:75–79.

50. Sohail, S.S. and Roland, D.A. 1999. Influence of Supplemental Phytase on Performance of Broilers Four to Six Weeks of Age. *Poultry Sci.* 78:550–555.

51. Skoglund, E., Näsi, M., and Sandberg, A.-S. 1998. Phytate Hydrolysis in Pigs Fed a Barley-Rapeseed Meal Diet Treated with *Aspergillus niger* Phytase or Steeped with Whey. *Can. J. Anim. Sci.* 78:175–180.

52. Nasi, M., Partanen, K., and Piironen, J. 1999. Comparison of *Aspergillus niger* Phytase and *Trichoderma reesei* Phytase and Acid Phosphatase on Phytate Phosphorus Availability in Pigs Fed on Maize-Soybean Meal or Barley-Soybean Meal Diets. *Arch. Anim. Nutr.* 52:15–27.

53. Ramseyer, L. Garling, D., Hill, G., and Link, J. 1999. Effect of Dietary Zinc Supplementation and Phytase Pre-Treatment of Soybean Meal or Corn Gluten Meal on Growth, Zinc Status and Zinc-Related Metabolism in Rainbow Trout, *Oncorhynchus mykiss*. *Fish Physiol. Biochem.* 20:251–261.

54. Papatryphon, E., Howell, R.A., and Soares, J.H. 1999. Growth and Mineral Absorption by Striped Bass *Morone saxatilis* Fed a Plant Feedstuff Based Diet Supplemented with Phytase. *J. World Aquacult. Soc.* 30:161–173.

55. Han, Y.M., Yang, F., Zhou, A.G., Miller, E.R., Ku, P.K., Hogberg, M.G., and Lei, X.G. 1997. Supplemental Phytases of Microbial and Cereal Sources Improve Dietary Phytate Phosphorus Utilization by Pigs from Weaning through Finishing. *J. Anim. Sci.* 75:1017–1025.

56. Leeson, S., Namkung, H., Cottrill, M., and Forsberg, C.W. 2000. Efficacy of New Bacterial Phytase in Poultry Diets. *Can. J. Anim. Sci.* 80:527–528.

57. Zyla, K., Ledoux, D.R., Garcia, A., and Veum, T.L. 1995. An *in vitro* Procedure for Studying Enzymic Dephosphorylation of Phytate in Maize-Soybean Feeds for Turkey Poults. *Br. J. Nutr.* 74:3–17.

58. Liu, J., Ledoux, D.R., and Veum, T.L. 1997. *In vitro* Procedure for Predicting the Enzymatic Dephosphorylation of Phytate in Corn-Soybean Meal Diets for Growing Swine. *J. Agric. Food Chem.* 45:2612–2617.

59. Kostrewa, D., Gruninger-Leitch, F., D'Arcy, A., Broger, C., Mitchell, D., and van Loon, A.P.G.M. 1997. Crystal Structure of Phytase from *Aspergillus ficuum* at 2.5 Å Resolution. *Nat. Struct. Biol.* 4:185–190.

60. Kostrewa, D., Wyss, M., D'Arcy, A., and van Loon, A.P.G.M. 1999. Crystal Structure of *Aspergillus niger* pH 2.5 Acid Phosphatase at 2.4 Å Resolution. *J. Mol. Biol.* 288:965–974.

61. Tomschy, A., Tessier, M., Wyss, M., Brugger, R., Broger, C., Schnoebelen, L., van Loon, A.P.G.M., and Pasamontes, L. 2000. Optimization of the Catalytic Properties of *Aspergillus fumigatus* Phytase Based on the Three-Dimensional Structure. *Protein Sci.* 9:1304–1311.

62. Jermutus, L., Tessier, M., Pasamontes, L., van Loon, A.P.G.M., and Lehmann, M. 2001. Structure-Based Chimeric Enzymes as an Alternative to Directed Enzyme Evolution: Phytase as a Test Case. *J. Biotechnol.* 85:15–24.

63. Lehmann, M., Kostrewa, D., Wyss, M., Brugger, R., D'Arcy, A., Pasamontes, L., and van Loon, A.P.G.M. 2000. From DNA Sequence to Improved Functionality: Using Protein Sequence Comparisons to Rapidly Design a Thermostable Consensus Phytase. *Protein Engng.* 13:49–57.

64. Lehmann, M., Lopez-Ulibarri, R., Loch, C., Viarouge, C., Wyss, M., and van Loon, A.P.G.M. 2000. Exchanging the Active Site between Phytases for Altering the Functional Properties of the Enzyme. *Protein Sci.* 9:1866–1872.

65. Lehmann, M., Pasamontes, L., Lassen, S.F., and Wyss, M. 2000. The Consensus Concept for Thermostability Engineering of Proteins. *Biochim. Biophys. Acta.* 1543:408–415.

66. Lim., D., Golovan, S., Forsberg, C.W., and Jia, Z.C. 2000. Crystal Structures of *Escherichia coli* Phytase and Its Complex with Phytate. *Nat. Struct. Biol.* 7:108–113.

67. Ha, N.C., Oh, B.C., Shin, S., Kim, H.J., Oh, T.K., Kim, Y.O., Choi, K.Y., and Oh, B.H. 2000. Crystal Structures of a Novel, Thermostable Phytase in Partially and Fully Calcium-Loaded States. *Nat. Struct. Biol.* 7:147–153.

68. Rodriguez, E., Wood, Z.A., Karplus, P.A., and Lei, X.G. 2000. Site-Directed Mutagenesis Improves Catalytic Efficiency and Thermostability of *Escherichia coli* pH 2.5 Acid Phosphatase/Phytase Expressed in *Pichia pastoris. Arch. Biochem. Biophys.* 382:105–112.

69. Li, J., Hegeman, C.E., Hanlon, R.W., Lacy, G.H., Denbow, D.M., and Grabau, E.A. 1997. Secretion of Active Recombinant Phytase from Soybean Cell-Suspension Cultures. *Plant Physiol.* 114:1103–1111.

70. Rodriguez, E., Han, Y., and Lei, X.G. 1999. Cloning, Sequencing, and Expression of an *Escherichia coli* Acid Phosphatase/Phytase Gene (*app*A2) Isolated from Pig Colon. *Biochem. Biophys. Res. Commun.* 257:117–123.

71. Pen, J., Verwoerd, T.C., van Paridon, P.A., Beudeker, R.F., van den Elzen, P.J.M., Geerse, K., van der Klis, J.D., Versteegh, H.A.J., van Ooyen, A.J.J., and Hoekema, A. 1993. Phytase-Containing Transgenic Seeds as a Novel Feed Additive for Improved Phosphorus Utilization. *Bio/Technology* 11:811–814.

72. Verwoerd, T.C., van Paridon, P.A., van Ooyen, A.J.J., van Lent, J.W.M., Hoekema, A., and Pen, J. 1995. Stable Accumulation of *Aspergillus niger* Phytase in Transgenic Tobacco Leaves. *Plant Physiol.* 109:1199–1205.

CHAPTER 8

Methods for Analysis of Phytate

ERIKA SKOGLUND
ANN-SOFIE SANDBERG

1. INTRODUCTION

IN the early and mid 1900s, the quantitative analysis of phytate ($InsP_6$) was based on precipitation with ferric chloride, first described by Heubner and Stadler in 1914 [1], or purification using anion-exchange chromatography [2–4]. A disadvantage of these methods is the lack of specificity in distinguishing between $InsP_6$ and its degradation products. Because inositol phosphates with three to five phosphate groups ($InsP_3$–$InsP_5$) as well as $InsP_6$ have been shown to be nutritionally significant [5–10], it is of great importance to have a reliable method for the determination of the individual inositol phosphates. There are also difficulties in determining low $InsP_6$ levels using the precipitation and anion-exchange methods. The amount of detectable inositol phosphates should be at least in the nanomolar range, due to the low inositol phosphate concentrations in biological samples. Thus, a sensitive analytical method for the determination of $InsP_6$ is required. With the development of ion-pair HPLC procedures [11–12] and capillary electromigration methods [13–14], it became possible to study $InsP_6$ and some of its hydrolysis products during food processing and digestion. These methods are relatively easy to handle with a short and simple procedure, but they do not differentiate isomeric forms of inositol phosphates. The various isomers have been shown to have different biochemical functions, so the precise stereochemistry of the inositol phosphates is of considerable significance. It is, therefore, desirable to have an analysis method that not only has the capability to separate inositol phosphates with different numbers of phosphate groups, but also to separate the different isomeric forms of the degradation products. During the last few years, a number of isomer-specific ion exchange chromatography (HPIC) methods with gradient elution for determination of inositol phosphates in biological samples have been developed [15–20].

2. SAMPLE PRETREATMENT

The techniques for preparing samples used in most methods for determination of inositol phosphates include liquid- and solid-phase extraction, centrifugation, freezing and thawing (to precipitate gelatinous agents and soluble proteins), and evaporation. For liquid extraction of inositol phosphates from foods or intestinal contents, an acid such as hydrochloric acid (HCl) can be used. Trichloroacetic acid (TCA) might be used in extraction of biological tissues, because it is good for removal of proteins and lipids. The medium used in the ion exchange columns for solid-phase extraction can be either silica-based anion exchange SAX or resin-based anion exchange AG 1-X8. The cleaning on ion exchangers is very useful for separating the ions from most impurities, removing inorganic phosphate and giving essential concentrations of inositol phosphates.

There is a desire for faster preparation of samples. The ion chromatographic methods of Cilliers and van Niekerk [21] and Rounds and Nielsen [17] use extraction, centrifugation, and filtration to prepare the plant, food, or soil samples. In these methods, the ion exchange step is omitted. The ion chromatography method of Skoglund and coworkers [19] includes a cleaning step with anion exchange. Results in our laboratory [22] show that omitting this step and instead filtering 0.5 mL extract with a centrifuge filter (Amicon Microcon 30) gives comparable results in the food samples studied so far.

3. METHODS FOR PHYTATE DETERMINATION

3.1. PRECIPITATION METHODS

Precipitation methods are based on the principle that $InsP_6$ forms an insoluble stable complex with ferric ion in dilute acid and, presumably, is the only phosphate compound with that property [1]. The phosphorus content in the precipitate can be determined after wet ashing or hydrolysis, giving a direct measure of the $InsP_6$ content. A certain molar ratio between iron and $InsP_6$ is required for quantitative precipitation. In indirect methods, a stoichiometric relationship between $InsP_6$ and the unprecipitated ferric ions is determined. Latta and Eskin described in 1980 [23] a colorimetric method for determination of $InsP_6$ that is founded on the reaction between ferric chloride and sulfosalicylic acid, giving a pink-colored reagent with absorbance maximum at 500 nm. When $InsP_6$ is present, the iron becomes bound to the phosphate ester, therefore, unavailable to react with sulfosalicylic acid, resulting in a decrease in pink color intensity. The decrease in absorbance is proportional to the concentration of $InsP_6$ present. The precision in this method may be impaired by the presence of lower inositol phosphates that also complex with iron. Other compounds able to bind to iron, such as polyphenols, are also coprecipitated. It was later suggested by Vaintraub and Lapteva [24] that the purification stage in the Latta and Eskin method [23] be omitted. This reduced the determination time while the precision and sensitivity were increased. Several different methods based on the analysis of phosphorus and iron in ferric $InsP_6$ have been suggested [25–27]. The method by McCane and Widdowson [26] was the one first determining $InsP_6$ directly. A disadvantage of the precipitation methods is the lack of specificity to distinguish between $InsP_6$ and partially dephosphorylated analogues. De Boland and coworkers [28] found that all of the inositol phosphates from $IndP_2$ to $InsP_6$ form insoluble complexes. However, $InsP$, $InsP_2$, and $InsP_3$ are appreciably soluble and not quantitatively precipitated. The methods, furthermore, have a low sensitivity and can thereby not detect low amounts of $InsP_6$.

3.2. ION EXCHANGE METHODS

Because of its simplicity and low cost, ion exchange chromatography is often used for processing a large number of samples, though it is not a rapid procedure. Smith and Clarke [2] introduced the technique for separation of inositol phosphates using a stepwise elution with HCl. About 10 years later, an HCl gradient for qualitative analysis of phosphate esters was used by Cosgrove [3]. Since Harland and Oberleas in 1977 [29] introduced an anion exchange chromatography method with step gradient elution for quantification of $InsP_6$, several modifications to this method have been published [23,30–32]. The eluate is digested, inorganic phosphorus is measured, and $InsP_6$-equivalent is calculated. The $InsP_6$ value from the iron-precipitation method was indicated to be higher than values from the ion exchange method [33]. Ellis and Morris [30] suggested that substances in the acid extract were interfering with $InsP_6$, resulting in the lower values. If EDTA was added to the sample extract and the pH was adjusted to 6, the level of $InsP_6$ determined by ion exchange methods was in accordance with the iron-precipitation method [34]. The effects of particle size and cross-linkage of ion exchange resins on $InsP_6$ recovery were determined by Ellis and Morris [35]. They reported AG1 (X4 resin, 100–200 mesh) to be the most appropriate resin. With the modifications suggested by Ellis and Morris to the method developed by Harland and Oberleas, it was later adopted as an AOAC method [4]. However, contamination of $InsP_6$ with the lower inositol phosphates and nucleotides seemed to interfere with this method [36,37]. In fact, measurements of $InsP_6$, according to AOAC, were found to correlate to the sum of $InsP_3$, $InsP_4$, $InsP_5$, and $InsP_6$ determined by HPLC ion-pair chromatography, suggesting that these inositol phosphates were included in $InsP_6$ determination by the AOAC method (Sandberg, COST 916). Another disadvantage of the method is that it does not detect low levels of $InsP_6$.

3.3. HPLC METHODS

A number of high-performance liquid chromatography (HPLC) techniques have been developed for the analysis of $InsP_6$ in foods [31,38–39]. These methods separate $InsP_6$ from inositol by using reversed phase octadecyl (C-18) stationary phases and aqueous potassium dihydrogen phosphate or sodium acetate mobile phases. A disadvantage of these and other contemporary procedures [40] is elution of $InsP_6$ on the solvent front. $InsP_6$ is only weakly retained on the column resulting in extremely poor resolution.

To purify samples prior to analyses by HPLC, a strong anion exchange resin was used [31,39] after extraction with hydrochloric acid (HCl). The main advantages of this rather time-consuming step are effective concentration of inositol phosphates, removal of inorganic phosphate, and elution of most impurities. HCl is removed by evaporating the extract to dryness.

4. METHODS FOR DETERMINATION OF PHYTATE AND ITS DEGRADATION PRODUCTS

4.1. HPLC ION-PAIR METHODS

Ion pairing is based on the principle that a reagent, added to the mobile phase, is retained by the stationary phase, thus providing the otherwise neutral octadecyl stationary phase with charge to increase retention time on the column. An HPLC procedure was developed for analysis of $InsP_6$ in which error due to coelution with the solvent front was eliminated [41]. In this method, the tetrabutyl ammonium hydroxide (TBA-OH) is used for ion pairing, as it is both hydrophobic and ionized and, therefore, able to bring the octadecyl phase to the $InsP_6$. In 1986, Sandberg and Ahderinne [11] presented a method, later modified to increase sensitivity [8], in which $InsP_6$ and some of its degradation products ($InsP_3$–$InsP_5$) were able to be separated and quantified. The inositol phosphates were purified on strong anion exchange columns, separated using ion-pair HPLC and detected with refractive index detection. In a modification of the ion-pair HPLC procedure for $InsP_3$–$InsP_6$, analysis ultrasonication for extraction- and silica-based SAX (strong anion exchange) columns for purification and separation were used [12,42]. The ion-pair HPLC methods have the advantages of being both sensitive and easy to adopt. However, these methods do not differentiate isomeric forms of inositol phosphates, as gradient elution cannot be performed when using refractive index detection. In addition, the nucleotide adenosine triphosphate has been observed to coelute with $InsP_3$ [43].

4.2. HPIC METHODS

Phillippy and Johnston [36] developed the first high-performance ion chromatography (HPIC) method for determination of InsP$_6$. Since then, several methods have been described using ion chromatography with gradient elution for the separation of inositol phosphates and their different isomers [15,16,18–20,44]. HPIC is an effective and sensitive technique that is able to determine numerous ions simultaneously. The principle underlying this form of chromatography is the attraction between oppositely charged particles. In the separation of analytes, a low-capacity, low-affinity ion exchange column is used, linked to a detector. The eluent, a counterion, is pumped through the column to replace the exchangeable ions in the column and to establish a constant signal from the detector. When the sample ions are injected, they are taken up by the resin column by exchange with an equivalent number of eluent ions. The sample ions are separated as a result of different affinity for the resin exchange sites.

In 1986, Cilliers and van Niekerk [21] presented a method for InsP$_6$ analysis using an ion exchange column with a short and simple preparation procedure that was well suited to the handling of a variety of food samples. Before being injected into the column, the samples were only extracted with 3% TCA, shaken, and centrifuged. Rounds and Nielsen [17] published a method for the quantitation of InsP$_6$-InsP$_2$ with minimal sample preparation and equipment, using only extraction, centrifugation, and filtration. The method was developed to facilitate the analysis of inositol phosphates in plants, foods, and soil. However, some samples, such as intestinal collections, which have a greater concentration of compounds, are difficult to filter. Furthermore, their gradient elution procedure separated the different inositol phosphates but did not resolve their isomers. A number of isomer-specific ion chromatography methods using gradient elution have lately been presented [15,16,19,20,36,44]. The two HPIC methods developed by Skoglund and coworkers [19,20] permit the detection of InsP-InsP$_6$ in food, intestinal contents, and enzymatic hydrolysates with high sensitivity and show a commendable separation of isomers. Figure 8.1 shows chromatographic profiles analyzed according to Skoglund et al. [20] of a sample containing various isomers of inositol phosphates produced by reflux boiling of InsP$_6$ with HCl [Figure 8.1(a)] and a fermented wheat roll sample [Figure 8.1(b)].

4.3. CE METHODS

Since the early 1990s, several methods for the analysis of inositol phosphates have dealt with capillary zone electrophoresis (CZE) and capillary isotachophoresis (CITP) in combination with conductivity detection [13] or indirect UV absorbance detection [14,45,46]. Electrophoresis is defined as the migration of charged compounds through a solution under an electric field. Capillary

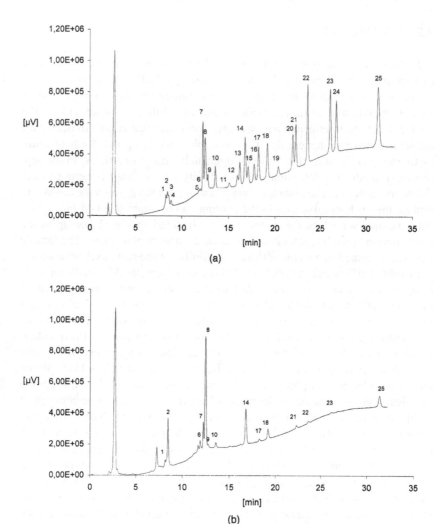

Figure 8.1 Chromatographic profile of (a) InsP$_6$ hydrolyzed by HCl and of (b) a fermented wheat roll sample analyzed according to Skoglund et al. [20]. Unidentified peaks are assigned a star (*). Peaks: (1) DL-Ins(1,6)P$_2$; (2) DL-Ins(1,2)P$_2$; (3) DL-Ins(1,4)P$_2$, DL-Ins(2,4)P$_2$; (4) DL-Ins(4,5)P$_2$; (5)*; (6)*; (7) DL-Ins(1,2,4)P$_3$, Ins(1,3,4)P$_3$; (8) DL-Ins(1,2,6)P$_3$, Ins(1,2,3)P$_3$; (9) DL-Ins(1,4,5)P$_3$; (10) DL-Ins(1,5,6)P$_3$; (11) Ins(4,5,6)P$_3$; (12) Ins(1,2,3,5)P$_4$; (13) DL-Ins(1,2,4,6)P$_4$; (14) DL-Ins(1,2,3,4)P$_4$; (15) Ins(1,3,4,6)P$_4$; (16) DL-Ins(1,2,4,5)P$_4$; (17) DL-Ins(1,3,4,5)P$_4$; (18) DL-Ins(1,2,5,6)P$_4$; (19) Ins(2,4,5,6)P$_4$; (20) DL-Ins(1,4,5,6)P$_4$; (21) Ins(1,2,3,4,6)P$_5$; (22) DL-Ins(1,2,3,4,5)P$_5$; (23) DL-Ins(1,2,4,5,6)P$_5$; (24) Ins(1,3,4,5,6)P$_5$; and (25) InsP$_6$.

electrophoresis (CE) represents a group of electrophoretic techniques, all performed in a narrow-bore capillary. Isotachophoresis (ITP) is a separation method of high performance for ionic components. The separation is achieved through stacking the ions into discrete zones in a capillary tube, in order of their mobilities. The samples are injected between a leading ion with high mobility and a trailing ion with low mobility. When the tube is subjected to an electric field, the ions move at different speeds depending on their mobilities. Blatny and coworkers [13] used capillary ITP with conductivity detection to investigate phytase hydrolysis of InsP$_6$ as a function of time. They concluded ITP to be a rapid and simple method for quantization of InsP to InsP$_6$ and orthophosphate when using two different buffer systems. However, a drawback of capillary electromigration methods is the relatively low sensitivity. Development of concentration techniques and more sensitive detection techniques is required to make the methods useful for bioanalysis of inositol phosphates.

4.4. NMR METHODS

^{31}P-Nuclear magnetic resonance (NMR) can be used to detect InsP$_6$ and its degradation products (InsP$_5$-InsP) and also determine the position of the phosphate groups [47–49]. Because the InsP$_6$ molecule has a plane of symmetry through C-2 and C-5, the P signals owing to C-1 and C-3 are identical, as are those from C-4 and C-6. Therefore, InsP$_6$ produces four resonance peaks in the relation 1:2:2:1. The ^{31}P-NMR technique is useful when investigating the InsP$_6$ amount in plant tissues, its form, and binding to other components. For InsP$_6$ hydrolysis products, more signals appear with increasing degradation, owing to the generation of different isomers. This method for determination of inositol phosphates has many advantages for accuracy and specificity. However, the technique requires expensive equipment. Another drawback is that it has a low sensitivity and, therefore, cannot be used for detection of low amounts of inositol phosphates. Furthermore, the resolution of the signals may be impaired by many P compounds in tissues.

5. DETECTION METHODS

One of the main problems associated with the analysis of inositol phosphates is the detection, in the case that the radiolabel technique is not used, as the ultraviolet light transparency rules out conventional UV absorbance methods. A commonly used detection method for inositol phosphates in HPLC is refractive index detection. Because gradient elution cannot be performed when using refractive index detection, this method does not differentiate isomeric forms of inositol phosphates. Further, the popularity of refractive index measurements has been limited because of the lack of sensitivity and vulnerability to temperature fluctuations. Useful detection techniques in ion chromatography are electrochemical and optical detection. Electrochemical detectors measure the current resulting from the application of a potential across electrodes in a flow cell. The conductivity of the solution or the current caused by the reduction or oxidation of analytes (amperometry) can be measured, depending on how the potential is applied and how the current is measured. A detection limit of picomoles for $InsP-InsP_3$ was demonstrated in the method of Skoglund et al. [20] when using a suppressed conductivity detector. For microcolumn-based separation systems, electrochemical detection is an ideal method; in particular, amperometric methods are very sensitive. This is because the cell volumes can be made very small with no loss in sensitivity, because the detection is based on an electrochemical reaction at the surface of the working electrode. On the contrary, the response is dependent on the path length in optical detectors.

Many species that are not chromophoric or fluorophoric can be post-column derivatized to form complexes that are detectable by UV, visible absorbance, or fluorescence monitoring. Phosphate groups can be derivatized by Fe^{3+} to form complexes for UV detection [50]. This principle has been used for $InsP_6$ and other inositol phosphates [16,20,36,44]. Irth and coworkers [51] used a post-column reaction detection system based on ligand exchange between phosphate ions and an Fe^{3+}-methylcalcein blue (MCB) reagent for fluorescence detection of $Ins(1,2,6)P_3$. Small and Miller Jr. [52] first introduced indirect photometric chromatography (IPC), a technique that may be used to overcome the problem of UV transparency. The principle of IPC is the use of eluent containing light absorbing (usually UV) ions that displace the sample ions from the column, resulting in negative peaks in the baseline absorbance. The technique was first used for $InsP_6$ determination in 1988 [53], although the sensitivity was quite low. Alternatively, an IPC ligand-exchange principle can be applied, based on decreased color intensity of a dye-metal complex when strong chelating agents, such as inositol phosphates, are added. The chelating agents compete with the dye by forming metal complexes. This fundamental principle was applied to $InsP_6$ detection [21] using a red Fe^{3+}-sulfosalicylic acid complex, and to $InsP_2-InsP_6$ detection [15] using the complex between transition metals and 4-(2-pyridylazo)resorcinol (PAR).

6. CONCLUSIONS

All precipitation and some ion exchange methods lack the specificity to distinguish between $InsP_6$ and its hydrolysis products. These techniques, therefore, overestimate the $InsP_6$ content in foods and digesta. Consistent and reproducible values are obtained with HPLC ion-pair chromatography for $InsP_6$, $InsP_5$, $InsP_4$, and $InsP_3$ at levels >0.1 μmol/g food. This method is suitable for studies of mineral bioavailability, as lower inositol phosphates ($InsP_2$ and $InsP_1$) do not negatively affect absorption [9,10,54,55]. Different isomeric forms of inositol phosphates also do not affect the mineral absorption [56]. In the field of physiology, however, the isomeric form of the inositol phosphates is decisive, and accordingly, HPIC methods are promising.

7. REFERENCES

1. Heubner, W. and Stadler, H. 1914. Uber eine titrationmethode zur bestimmung des phytins, *Biochem. Z.* 64:422–437.
2. Smith, D.H. and Clarke, F.E. 1952. Chromatographic separations of inositol phosphourus compounds, *Proc. Soil Sci. Soc. Am.* 16:170–173.
3. Cosgrove, D.J. 1963. The isolation of myo-inositol pentaphosphates from hydrolysates of phytic acid, *Biochem. J.* 89:172–175.
4. Harland, B.F. and Oberleas, D.A. 1986. Anion-exchange method for determination of phytate in foods: Collaborative study, *J. Asso. Off. Anal. Chem,* 69:667–670.
5. Brune, M., Rossander-Hulthén, L., Hallberg, L., Gleerup, A. and Sandberg, A.-S. 1992. Iron absorption from bread in humans: inhibiting effects of cereal fiber, phytate and inositol phosphates with different numbers of phosphate groups, *J. Nutr.* 122:442–449.
6. Han, O., Failla, M.L., Hill, A.D., Morris, E.R. and Smith Jr., J.C. 1994. Inositol phosphates inhibit uptake and transport of iron and zinc by a human intestinal cell line, *J. Nutr.* 124:580–587.
7. Rossander, L., Sandberg, A.-S. and Sandström, B. 1992. The influence of dietary fibre on mineral absorption and utilization, in *Dietary fibre—a Component of Food. Nutritional Function in Health and Disease.* Eds. Schweizer, T. F. and Edwards, C. A. Springer Verlag: London.
8. Sandberg, A.-S., Carlsson, N.-G. and Svanberg, U. 1989. Effects of inositol tri-, tetra-, penta-, and hexaphosphates on *in vitro* estimation of iron availability, *J. Food Sci.* 54:159–161, 186.
9. Sandberg, A.-S., Brune, M., Carlsson, N.-G., Hallberg, L., Skoglund, E. and Rossander-Hulthén, L. 1999. Inositol phosphates with different number of phosphate groups influence iron absorption in humans, *Am. J. Clin. Nutr.* 70:240–246.
10. Sandström, B. and Sandberg, A.-S. 1992. Inhibitory effects of isolated inositol phosphates on zinc absorption in humans, *J. Trace Elem. Electrolytes Health Dis.* 6:(2), 99–103.
11. Sandberg, A.-S. and Ahderinne, R. 1986. HPLC method for determination of inositol tri-, tetra-, penta-, and hexaphosphates in foods and intestinal contents, *J. Food Sci.* 51:547–550.
12. Lehrfeld, J. 1994. HPLC separation and quantitation of phytic acid and some inositol phosphates in foods: problems and solutions, *J. Agric. Food Chem.* 42:2726–2731.
13. Blatny, P., Kvasnicka, F. and Kenndler, E., 1994. Time course of formation of inositol phosphates during enzymatic hydrolysis of phytic acid (myo inositol hexaphosphoric acid) by phytase determined by capillary isotachophoresis, *J. Chromatogr.* A, 679:345–348.
14. Buscher, B.A.P., van der Hoeven, R.A.M., Tjaden, U.R., Andersson, E. and van der Greef, J. 1995. Analysis of inositol phosphates and derivatives using capillary zone electrophoresis-mass spectrometry, *J. Chromatogr.* 712:235–243.
15. Mayr, G.W. 1988. A novel metal-dye detection system permits picomolar-range h.p.l.c. analysis of inositol phosphates from non-radioactively labelled cell or tissue specimens, *Biochem. J.* 254:585–591.
16. Phillippy, B.Q. and Bland, J.M. 1988. Gradient ion chromatography of inositol phosphates, *Anal. Biochem.* 175:162–166.
17. Rounds, M.A. and Nielsen, S.S. 1993. Anion-exchange high-performance liquid chromatography with post-column detection for the analysis of phytic acid and other inositol phosphates, *J. Chromatogr.* A 653:148–152.

18. Guse, A.H., Goldwich, A., Weber, K. and Mayr, G.W., 1995. Nonradioactive, isomer-specific inositol phosphate mass determinations: high-performance liquid chromatography-micrometal-dye detection strongly improves speed and sensitivity of analyses from cells and microenzyme assays, *J. Chromatogr.* 672:189–198.

19. Skoglund, E., Carlsson, N.-G. and Sandberg, A.-S., 1997a. Determination of isomers of inositol mono- to hexaphosphates in selected foods and intestinal contents using high-performance ion chromatography, *J. Agric. Food Chem.* 45:431–436.

20. Skoglund, E., Carlsson, N.-G. and Sandberg, A.-S. 1997b. Analysis of inositol mono- and diphosphate isomers using high-performance ion chromatography and pulsed amperometric detection, *J. Agric. Food Chem.* 45:4668–4673.

21. Cilliers, J.J.L. and van Niekerk, P.J. 1986. LC Determination of phytic acid in food by postcolumn colorimetric detection, *J. Agric. Food Chem.* 34:680–683.

22. Carlsson, N.-G., Bergman, E.L., Skoglund, E., Hasselblad, K. and Sandberg, A.-S. 2001. Rapid analyzes of inositol phosphates, *J. Agric. Food Chem.* 49:1695–1701.

23. Latta, M. and Eskin, M. 1980. A simple and rapid colorimetric method for phytate determination, *J. Agric. Food Chem.* 28:1313–1315.

24. Vaintraub, I.A. and Lapteva, N.A. 1988. Colorimetric determination of phytate in unpurified extracts of seeds and the products of their processing, *Anal. Biochem.* 175:227–230.

25. Crean, D.E.C. and Haisman, D.R. 1963. The interaction between phytic acid and divalent cations during the cooking of dried peas, *J. Sci. Food Agric.* 14:824–833.

26. McCane, R.A. and Widdowson, E.M. 1935. Phytin in human nutrition, *Biochem. J.* 29B: 2694–2699.

27. Schormuller, J., Hohne, R. and Wurdig, G. 1956. Undercuchungen zur bestimmung des phytins, *Dtsch. Lebensm.-Rundsch.* 9:213–224.

28. DeBoland, A.R., Garner, G.B. and O'Dell, B.L. 1975. Identification and properties of 'phytate' in cereal grains and oilseed products, *J. Agric. Food Chem.* 23:1186–1189.

29. Harland, B.F. and Oberleas, D.A. 1977. A modified method for phytate analysis using an ion exchange procedure. Application to textured vegetable proteins, *Cereal Chem.* 54:827–832.

30. Ellis, R. and Morris, E.R. 1982. Comparison of ion-exchange and iron precipitation methods for analysis of phytate, *Cereal Chem.* 59:232–233.

31. Graf, E. and Dintzis, F. 1982a. High-performance liquid chromatographic method for the determination of phytate, *Anal. Biochem.* 119:413–417.

32. O'Neill, I.K., Sargent, M. and Trimble, M.L. 1980. Determination of phytate in foods by phosphorus-31 Fourier transform nuclear magnetic resonance spectrometry, *Anal. Chem.* 52:1288–1291.

33. Cosgrove, D.J. *Inositol Phosphates.* Elsevier Publ. Co., New York, 1980.

34. Ellis, R. and Morris, E.R. 1983. Improved ion-exchange phytate method, *Cereal Chem.* 60: 121–124.

35. Ellis, R. and Morris, E.R. 1985. Appropriate resin selection for rapid phytate analysis by ion exchange chromatography, *Cereal Chem.* 63:58–59.

36. Phillippy, B.Q. and Johnston, M.R. 1985. Determination of phytic acid in foods by ion chromatography with post-column derivatization, *J. Food Sci.* 50:541–542.

37. Phillippy, B.Q., Johnston, M.R., Tao, S.-H. and Fox, M.R.S. 1988. Inositol phosphates in processed foods, *J. Food. Sci.* 53:496–499.

38. Tangendjaja, B., Buckle, K.A. and Wootton, M. 1980. Analysis of phytic acid by high-performance liquid chromatography, *J. Chromatogr.* 197:274–277.

39. Graf, E. and Dintzis, F.R. 1982b. Determination of phytic acid in foods by high-performance liquid chromatography, *J. Agric. Food Chem.* 30:1094–1097.

40. Knuckles, B.E., Kuzmicky, D.D. and Betschart, A.A. 1982. HPLC analysis of phytic acid in selected foods and biological samples, *J. Food Sci.* 47:1257–1258, 1262.

41. Lee, K. and Abendroth, J.A. 1983. High performance liquid determination of phytic acid in foods, *J. Food Sci.* 48:1344–1345, 1351.
42. Lehrfeld, J. 1989. High-performance liquid chromatography analysis of phytic acid on a pH-stable, macroporous polymer column, *Cereal Chem.* 66:510–515.
43. Morris, E.R. and Hill, A.D. 1996. Inositol phosphate content of selected dry beans, peas, and lentils, raw and cooked, *J. Food Comp. Anal.* 9:2–12.
44. Skoglund, E., Carlsson, N.-G. and Sandberg, A.-S. 1998. High-performance chromatographic separation of inositol phosphate isomers on strong anion exchange columns, *J. Agric. Food Chem.* 46:1877–1882.
45. Henshall, A., Harrold, M.P. and Tso, J.M.Y. 1992. Separation of inositol phosphates by capillary electrophoresis, *J. Chromatogr.* 60 S:413–419.
46. Buscher, B.A.P., Irth, H., Andersson, E., Tjaden, U.R. and van der Greef, J. 1994. Determination of inositol phosphates in fermentation broth using capillary zone electrophoresis with indirect UV detection, *J. Chromatogr.* A, 678:145–150.
47. Ersöz, A., Akgün, H. and Aras, N.K. 1990. Determination of phytate in Turkish diet by phosphorus-31 Fourier transform nuclear magnetic resonance spectroscopy, *J. Agric. Food Chem.* 38:733–735.
48. Frølich, W., Drakenberg, T. and Asp, N.-G. 1986. Enzymatic degradation of phytate (myo-inositol hexaphosphate) in whole grain flour suspension and dough. A comparison between ^{31}P NMR spectroscopy and ferric ion method, *J. Cereal Sci.* 4:325–334.
49. Mazzola, E.P., Phillippy, B.Q., Harland, B.F., Miller, T.H., Potemra, J.M. and Katsimpiris, E.W. 1986. Phosphorus-31 nuclear magnetic resonance spectroscopic determination of phytate in foods, *J. Agric. Food Chem.* 34:60–62.
50. Imanari, T., Tanabe, S., Toida, T. and Kawanishi, T. 1982. High-performance liquid chromatography of inorganic anions using Fe^{3+} as a detection reagent, *J. Chromatogr.* 250.55–61.
51. Irth, H., Lamoree, M., De Jong, G.J., Brinkman, U.A.T., Frei, R.W., Kornfeldt, R.A. and Persson, L. 1990. Determination of D-myo-1,2,6-inositol trisphosphate by ion-pair reversed-phase liquid chromatography with post-column ligand exchange and fluorescence detection, *J. Chromatogr.* 499:617–625.
52. Small, H. and Miller Jr. T.E., 1982. Indirect photometric chromatography, *Anal. Chem.* 54:462–469.
53. Matsunaga, A., Yamamoto, A. and Mizukami, E., 1988. Determination of phytic acid in various foods by indirect photometric ion chromatography, *J. Food Hyg. Soc. Jpn.* 29:408–412.
54. Lönnerdal, B., Sandberg, A.-S., Sandström, B. and Kunz, C. 1989. Inhibitory effects of phytic acid and other inositol phosphates on zinc and calcium absorption in suckling rats, *J. Nutr.* 119:211–214.
55. Sandberg, A.-S., Brune, M., Carlsson, N.-G., Hallberg, L., Rossander-Hulthén, L. and Sandström, B. 1993. The effect of various inositol phosphates on iron and zinc absorption in humans. Bioavailability '93. Nutritional, Chemical and Food Processing Implications of Nutrient Availability, Fttlingen, U. Schlemmer, Bundesforschunganstalt für Ernährung, 53–57.
56. Skoglund, E., Lönnerdal, B. and Sandberg, A.-S. 1999. Inositol phosphates influence iron uptake in Caco-2 cells, *J. Agric. Food Chem.* 47:1109–1113.

CHAPTER 9

In vitro and *in vivo* Degradation of Phytate

ANN-SOFIE SANDBERG

1. INTRODUCTION

DEGRADATION of phytate (*myo*-inositol hexakisphosphate, InsP$_6$) occurs during food processing and in the gastrointestinal tract. This degradation is of nutritional importance because the mineral binding strength decreases and the solubility increases when phosphate groups are removed from the inositol ring, resulting in an increased bioavailability of essential dietary minerals [1–4]. In pig breeding, an additional problem is excessive P excretion and pollution, because P in the form of phytate is unavailable for absorption [5]. Major efforts have, therefore, been made to reduce the amount of phytate in foods and feeds by different processes or the addition of enzymes. In contrast to the antinutritional properties, dietary phytate has also been reported to have beneficial effects, such as protection against colon cancer [6,7], arteriosclerosis and coronary heart diseases [8].

Hydrolysis of phytate during biological food/feed processes such as steeping, malting, and fermentation [9–12] is as a result of activity of the enzyme phytase, present naturally in plants and certain microorganisms. Biotechnologically produced fungal phytase preparations are now commercially available and used for feed preparations. In the future, the use of microbial phytases in food processing could be feasible. Moreover, overexpression of endogenous phytase-encoding genes, or heterologous expression of inserted selected genes, can be performed in plants, thereby increasing phytate degradation. Specific isomers of inositol phosphates have shown several important physiological functions in man and animals [13]. Such compounds occur in biologically processed foods/feeds of plant origin.

Hydrolysis of $InsP_6$ may occur in the gastrointestinal tract prior to the intestinal site of absorption. Because most of the essential minerals and trace elements are absorbed in the duodenal or jejunal part of the small intestine, the site and degree of phytate degradation can affect the nutritional value of a high phytate diet. Furthermore, enzymatic hydrolysis of phytate in the gastrointestinal tract leads to formation of specific isomers of inositol phosphates. If biochemically active lower inositol phosphates, formed during food processing, or degradation in the gastrointestinal tract or precursors to these are absorbed in the alimentary tract of humans, it would have important health implications [13].

2. DEGRADATION OF INOSITOL PHOSPHATES

$InsP_6$ can be degraded by enzymatic and nonenzymatic hydrolysis. Enzymatic hydrolysis generally occurs during biological processing and preparation of plant food/feed such as steeping, malting, hydrothermal processing, fermentation, and addition of phytase as well as during degradation in the gastrointestinal tract. Nonenzymatic hydrolysis usually takes place when food/feed is treated with strong acid or high temperature and pressure. The enzymatic degradation is more selective and isomer specific (Figure 9.1).

3. ENZYMATIC HYDROLYSIS OF PHYTATE

Phytases ($InsP_6$ 6-phosphohydrolases) hydrolyze $InsP_6$ to *myo*-inositol and inorganic phosphate via intermediate *myo*-inositol phosphates (penta- to monophosphates) [14]. These enzymes are special kinds of nonspecific acid phosphatases. Those that exhibit the ability to hydrolyze $InsP_6$ can be considered phytases [15].

Phytases are present in plants, microorganisms, and certain animal tissues. Two types of phytases, 3- and 6-phytases, have been recognized as starting the dephosphorylation of $InsP_6$ at different positions (D-3 or L-6) of the inositol ring. The 3-phytase (EC 3.1.3.8) has been considered to be characteristic of microorganisms, while 6-phytase (EC 3.1.3.26) (4-phytase assuming D-configuration) has been considered to be characteristic of the seeds of higher plants [16], but this may not be a general rule. Phytase enzymes have been isolated and characterized from a number of plant sources (the initial preparation of phytase made from rice bran) [17]. It is, however, only recently that some phytases have been isolated and purified to homogenity [18–22].

Considering the *in vitro* degradation of $InsP_6$ by plant phytases, wheat phytase [23–24] and the phytase from rye [18], spelt [22], and barley (Figure 9.1) are reported to be 6-phytases [4]. The major $InsP_5$ in raw soybean was

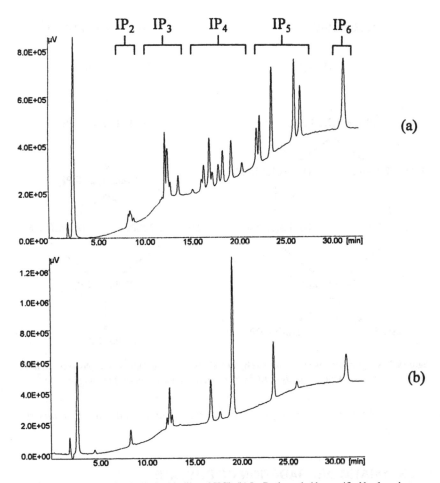

Figure 9.1 (a) InsP$_6$ reflux boiled with diluted HCl. (b) InsP$_6$ degraded by purified barley phytase.

DL-Ins(1,2,4,5,6)P$_5$, and thus, soybean phytase seems to be a 3-phytase [25]. Another leguminous plant, pea, was also found to have a degradation pathway of InsP$_6$ dissimilar to that of cereals [26].

Studies of *in vitro* degradation of InsP$_6$ by microbial phytase show that *A. niger* [27], *K. terrigena* [19], and *S. cerevisae* [28] belong to the 3-phytases, whereas phytases of *E.coli* [20] and *Paramecium* [29] belong to the 6-phytases.

The isomeric patterns during degradation of InsP$_6$ in oats, rye, and barley were found to be similar to those formed by hydrolysis by wheat phytase [26]. The same InsP$_6$ degradation pathway may be assumed in these cereals. From the isomers generated during hydrolysis and in view of earlier data [30], Skoglund et al. [26] suggested the following pathway for hydrolysis by cereal phytase and

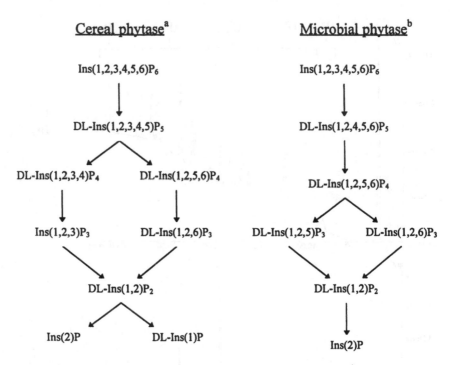

<u>Cereal phytase</u>[a]   <u>Microbial phytase</u>[b]

Figure 9.2 Suggested pathways for phytate hydrolysis by cereal[a] and microbial[b] phytases. [a]Wheat, barley, rye, and oat. [b]*Aspergillus niger, Klebsiella terrigena*, and *Pseudomonas* sp. [26].

microbial phytase (from *A. niger, Klebsiella terrigena*, and *Pseudomonas sp.* (Figure 9.2)).

4. ENZYMATIC DEGRADATION OF PHYTATE DURING FOOD PROCESSING

Phytate hydrolysis can occur during food preparation and production, either by phytase from plants, yeasts, or other microorganisms. Biological process-ing techniques, which increase the activity of native enzymes of cereals and legumes, are soaking, malting, hydrothermal processing, and lactic fermenta-tion. During the germination step of the malting process, enzymes are synthe-sized or activated. Lactic fermentation leads to lowering of pH as a consequence of bacterial production of organic acids, mainly lactic acid, which is favorable for cereal phytase activity. The microorganisms of the starter culture used in fermentation, in some cases, exert phytase activity. Phytase preparations can also be added in the food process. To optimize the food process to increase mineral bioavailability by phytate degradation, it is essential to know optimal conditions for the phytases, which are responsible for phytate degradation in the process. There are differences in optimal conditions for phytase activity between different plant species. Moreover, except for phytate, $InsP_5$ has, in human and animal studies, been identified as an inhibitor of iron and zinc absorption [2–4]. No difference between various isomers of inositol phosphates on iron uptake *in vitro* in Caco-2 cells was shown [31]. Recently, indications were found that $InsP_4$ and $InsP_3$ contribute to the negative effect on iron absorption by interacting with the higher phosphorylated inositol phosphates. These findings suggest that phytate degradation of not only $InsP_5$ and $InsP_6$, but also of $InsP_3$ and $InsP_4$ must occur to improve iron absorption from cereal and soy products [4]. This is probably also true for zinc absorption, as a strong negative correlation was found between zinc absorption and the sum of native $InsP_3$ through $InsP_6$ from cereal and legume meals [32].

4.1. SOAKING, MALTING, AND HYDROTHERMAL PROCESSING

Soaking of wheat bran, whole wheat flour, and rye flour at optimal conditions for wheat phytase activity (pH 4.5–5, 55°C) resulted in complete phytate hydrolysis [11,33] and to a marked increase in *in vitro* iron availability. Sandberg and Svanberg [11] found that the $InsP_6$ and $InsP_5$ contents must be reduced to levels beyond 0.5 μmol/g to give the strong increase in iron solubility at simulated physiological conditions, if no promoting factors were present.

Malting is a process during which the whole grain is soaked and then germinated. The amount of phytate in malted grains of wheat, rye, and oats intended for the production of flour (germinated 3 days at 15°C sprout length 10 mm) was only reduced slightly or not at all. However, when the malted cereals were ground and soaked at optimal conditions for wheat phytase, there was a complete degradation of phytate [9], except for oats, which under the conditions studied, had a low phytase activity. By germination of oats for 5 days at 11°C followed by incubation at 37–40°C, it was possible to reduce phytate content of oats by 98% (to < 0.5 μmol/g) [9]. It is of special importance to find methods of phytate reduction in oats. The high content of phytate in oat combined with a low phytase activity, which is further decreased due to heat treatment of all commercial oat products, suggest that the negative effect on mineral absorption of oat products would be greater than that of other cereals. This assumption was confirmed in recent human absorption studies using radionuclide techniques, which demonstrated a low iron and zinc absorption from breakfast meals containing oat porridge and oat bran bread [34–35]. The iron and zinc absorption from oat porridge made of untreated flour was compared to that from oat porridge made from malted flour, with a phytate reduction of 77%. Iron and particularly zinc absorption was significantly improved from the porridge made from malted flour [36].

Bergman et al. [37] developed optimal conditions for *hydrothermal processing* of whole barley kernels to degrade phytate and to increase the content of free *myo*-inositol using a multivariate design. The hydrothermal treatment was comprised of two wet steeps (1 and 2), where different concentrations of lactic acid were used, and two dry steeps followed by successive drying. At optimal conditions (temperature 1 = 48°C, temperature 2 = 50°C, 0.8% lactic acid), the phytate content of the whole kernels was reduced by 95–96%.

4.2. FERMENTATION

Lactic fermentation is an old method for food processing and preservation. Today, defined starter cultures and controlled conditions are used. Due to the production of lactic acid and other organic acids, the pH is lowered, and the phytase activity is increased. Lactic fermentation of maize, soybeans, and sorghum reduces the phytate [38,39]. Svanberg et al. [12] have shown that combined germination and lactic fermentation of white sorghum and maize gruels can yield an almost complete degradation of phytate. At optimal conditions, the effect on *in vitro* estimation of iron availability was an almost 10-fold increase.

The presence of tannins in sorghum not only decreases the availability of iron per se, but also inactivates the phytase enzyme. The use of lactic acid fermentation for high-tannin sorghum was, therefore, reported to be less effective in reducing phytate content [12].

The acidity of the dough is of great importance for phytate degradation during scalding and sourdough fermentation of bread [40,41]. Larsson and Sandberg [42] investigated the phytate reduction in scalded bread and bread with varying amounts of sourdough baked with rye bran or oat bran addition. The most marked phytate reduction of 96–97% occurred in bread made with 10% sourdough (pH 4.6) or in the breads in which the pH had been adjusted in the mild scalding with lactic acid, resulting in a pH between 4.4 and 5.1 in both dough and bread. The percentage iron absorption from breadmeals containing sourdough-fermented bread was (with 10% sourdough) similar to that of white bread not containing inhibitory factors [1]. Consequently, the amounts of iron absorbed from whole-meal bread with its high content of iron, would be greater than from white bread, provided that the fermentation was optimized.

In another study of bread making, Türk and Sandberg [43] investigated the effects of different dough additives. Bread was made using whole wheat flour and flour of 60% extraction rate, yeast was added, and phytate degradation was studied after different fermentation times. The addition of milk to the dough inhibited enzymatic phytate hydrolysis resulting in depressed human iron absorption from the bread [44]. Fermented milk did not significantly affect enzymatic hydrolysis during fermentation, probably depending on the presence of lactic acid lowering the pH [43]. The addition of acetic acid or lingonberries to the dough increased the phytate reduction to 96% and 83%, respectively, compared to 55% in control bread without additives [24]. The pH of the doughs with 96% reduction was between 4.5 and 5. This degradation was mainly a result of the activity of phytase in the flour, and the contribution of phytase activity from Baker's yeast during bread fermentation is very small. It was concluded that conditions during bread fermentation disfavor yeast phytase expression [28].

4.3. ADDITION OF ENZYMES

An alternative to activation of the intrinsic enzymes of foods is the addition of phytase during food processing. Microbial phytase enzyme preparations are now available commercially, making their use in food processing technically feasible. Moreover, the phytase-encoding gene (*phy*A) *A. niger* has been cloned and overexpressed, resulting in a more than 10-fold increase in phytase activity compared to the wild-type strain [45].

Phytase from *A. niger* 1500 PU/g flour (one phytase unit, PU, is the amount of enzyme which liberates under standard conditions, pH 5.0, 37°C, 1 nmol of inorganic phosphate from sodium phytate in 1 min) was added to the doughs prepared using whole wheat flour and flour of 60% extraction rate. The phytate hydrolysis in the dough was, however, not complete (maximum 88% hydrolysis), unless the pH was lowered to 3.5, which is close to the lower pH optimum [24]. *A. niger* phytase, therefore, may be useful in the production of lactic acid fermented cereals and legumes (which normally have a low pH value) aimed at high mineral availability. A very effective phytate degradation was achieved by adding *A. niger* phytase to an oat-based nutrient solution fermented by *Lactobacillus plantarum*, but the added enzyme had a negative influence on the viable counts of *Lactobacilli* as well as the aroma [46].

The effect of reducing the phytate in soy-protein isolates on iron absorption was investigated in humans [47]. The addition of *A. niger* phytase in amounts that almost completely removed phytate in soy isolates increased iron absorption four fold to five fold.

5. HYDROLYSIS OF PHYTATE IN THE GUT

A number of factors have to be considered when discussing phytate degradation in the gut, e.g., if there is an effect of dietary phytase, intestinal mucosa phytase, or phytase produced by the intestinal microflora. It has also been suggested that an adaptation to increased phytate degradation may occur in addition. Balance studies in humans indicate that an increasing dietary calcium level reduces the degree of phytate degradation in humans [48–50].

5.1. DEGRADATION OF PHYTATE IN THE STOMACH AND SMALL INTESTINE

Few studies have been performed to determine hydrolysis of phytate in the stomach. It is likely that consumption of raw plant foods may lead to some hydrolysis in the stomach, as the enzyme phytase is present, and in particular, cereal phytase activity increases when the pH is lowered. Investigating hydrolysis of inositol phosphates in the stomach of slaughtered pigs showed that approximately 50% of the feed $InsP_6$ was hydrolyzed [51]. This hydrolysis was probably a result of activity of the native feed phytase, as raw barley was included in the feed. Lantzsch et al. [52] found little hydrolysis in the stomach of pigs in the absence of feed phytase. By removing some of the phosphate groups in the stomach, the inositol phosphates with a lower degree of phosphorylation formed are probably more easily hydrolyzed further by enzymes located in the intestine.

Soluble sodium phytate is hydrolyzed by homogenized mucosal tissue from a wide range of species. Crushing the cell of the mucosa may, however, cause the release of intracellular phytase, which normally would have no contact with the intestinal contents. The situation in the intact intestine may, therefore, differ. Using this type of methodology, Bitar and Reinhold [53] claimed that humans also had a mucosal phytase, an observation that was later confirmed by Iqbal et al. [54]. Davies and Flett [55] found mucosal phytase in the rat and showed that it was concentrated in the brush border. Factors influencing the hydrolysis of $InsP_3$-$InsP_6$ by mucosal homogenates from the pig were recently reported by Hu et al. [56]. Activities toward substrates increased with a decreasing number of phosphate groups. This suggests that hydrolysis in the intestine may be enhanced by earlier phytase activity that had led to production of some lower inositol phosphates in the stomach. Phytate disappearance from digesta has been demonstrated when soluble phytate was incubated in loops of rat duodenum, jejunum, and ileum [55]. In all natural diets, however, calcium and magnesium are present, and reaction with phytate is to be expected, the strength of which depends on the number of phosphate groups on the inositol molecule [57]. The main part of the phytate in the diet and ileal digesta is, therefore, present as an insoluble complex.

For determination of degradation of feed or food components in the stomach and small intestine, studies in ileostomized pigs [51,58] and humans have been performed [59,60]. The ileostomy model was used for investigation of the extent of phytate hydrolysis [61,62] and factors affecting this hydrolysis. Specific analyses of $InsP_6$ and its hydrolysis products ($InsP_3$-$InsP_5$) were performed using HPLC ion-pair chromatography [63,64].

The effect of dietary phytase and intestinal phytase on $InsP_6$ hydrolysis was investigated [61]. Hydrolysis of $InsP_6$ from raw wheat bran and extruded bran was compared. Seven ileostomy subjects were studied during two four-day periods with a constant low-fiber diet and the addition of 54 g of raw mixture

TABLE 9.1. Recovery of Dietary Phytate in Ileostomy Contents.

Diet	$InsP_6$ Recovery %
A) Raw wheat bran	$42 \pm 12\ (n = 7)$
B) Extruded wheat bran	$95 \pm 9.8\ (n = 7)$
C) Phytase deactivated wheat bran	$95 \pm 8.8\ (n = 8)$

Mean $\pm$ SD, $n =$ number of subjects.

of bran, gluten, and starch during one period and 54 g of the corresponding extruded product during the other period. Sandberg et al. [61] found that 60% of the $InsP_6$ from raw bran were hydrolyzed during the passage through the stomach and small intestine and that hydrolysis products were formed, but this did not occur when the subjects were fed extruded bran (Table 9.1). During extrusion cooking, the intrinsic bran phytase was deactivated. Sandberg and Andersson [62] extended the study using phytase-deactivated wheat bran and found no hydrolysis of $InsP_6$ and no formation of hydrolysis products. Further, Sandberg et al. [61,62] concluded that dietary phytase is of significance for phytate hydrolysis in the stomach and small intestine, while intestinal phytase, if present in humans, does not play a significant role. It was recently demonstrated that very low phytase activity occurs in the human small intestine, thus confirming that the human small intestine has very limited ability to digest $InsP_6$ [54]. These conclusions were further supported by analysis of the specific isomers formed in the ileal content during hydrolysis. Wheat phytase starts the hydrolysis at the L-6 position forming specific isomer $DL\text{-}Ins(1,2,3,4,5)P_5$, which was found in the intestinal content following consumption of raw wheat bran. Hydrolysis of $InsP_6$ by microbial or intestinal phytase would result in formation of $DL\text{-}Ins(1,2,4,5,6)P_5$, which was not found [65].

The ileal content from an ileostomy subject consuming raw wheat bran is shown in Figure 9.3. As shown, the $DL\text{-}Ins(1,2,3,4,5)P_5$ peak rises above the peaks of the other $InsP_5$. This indicates hydrolysis of $InsP_4$ by cereal phytase. The ileal content was composed of high amounts of $InsP_6$, with the peak of $DL\text{-}Ins(1,2,5,6)P_4$ and $DL\text{-}Ins(1,2,4,5)P_4$ predominant. The latter isomer has been proposed to be a second messenger [66]. However, this hypothesis is controversial [67]. Further analysis of the lower inositol phosphates in the intestinal contents showed that only small amounts of $InsP_1$ isomers were present and that the dominating $InsP_2$ isomer was $DL\text{-}Ins(1,2)P_2$ [26].

Indirect evidence for phytate hydrolysis in the human stomach and small intestine by *A. niger* phytase was reported in a iron absorption study by Sandberg [68]. Adding *A. niger* phytase to the phytate-containing meal markedly increased iron absorption. Thus, both phytases from wheat and *A. niger* are active in the human gut. As the contents of the stomach acidify during the digestive

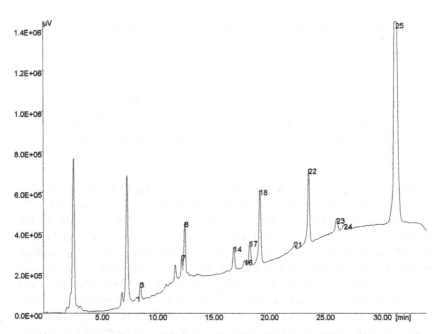

Figure 9.3 Inositol phosphates in ileostomy contents of a human subject consuming raw wheat bran [65]. The predominating isomer of inositol pentaphosphate [DL-Ins(1,2,3,4,5)IP$_5$], formed by cereal phytase hydrolysis of inositol hexaphosphate, was identified in the ileostomy contents. Peaks: (1) DL-Ins(1,6)P$_2$; (3) DL-Ins(1,4)P$_2$, DL-Ins(2,4)P$_2$; (7) DL-Ins(1,2,4)P$_3$, Ins(1,3,4)P$_3$; (8) DLIns(1,2,6)P$_3$, Ins(1,2,3)P$_3$; (14) DL-Ins(1,2,3,4)P$_4$; (16) DL-Ins(1,2,4,5)P$_4$; (17) DL-Ins(1,3,4,5)P$_4$; (18) DL-Ins(1,2,5,6)P$_4$; (21) Ins(1,2,3,4,6)P$_5$; (22)DL-Ins(1,2,3,4,5)P$_5$; (23)DL-Ins(1,2,4,5,6)P$_5$; (24) Ins(1,3,4,5,6)P$_5$; (25) IP$_6$.

process, the pH would appear to favor *A. niger* phytase considerably over wheat phytase. In addition, it was recently reported that wheat phytase is more susceptible to inactivation by gastrointestinal enzymes [69].

A possible adaptation to increased phytate degradation in the small intestine after a longer period of high phytate intake was investigated (Sandberg et al., unpublished results). Nine ileostomy subjects consumed oat bran added to a low-fiber diet during a three-week period. Ileostomy contents from days 3 and 17 were analyzed (Sandberg et al., unpublished results). A complete recovery of inositol InsP$_6$ was found in ileostomy subjects during days 3 and 17. No increased phytate degradation was found after 17 days of consumption, showing that no adaptation occurred during this period (Figure 9.4). Furthermore, a complete recovery of dietary inositol penta-, tetra-, and triphosphates was obtained.

In these ileostomy studies, the dietary level of calcium was relatively high (around 30 mmol/day). However, in another ileostomy study [70], where degradation of phytate from soy flour, soy concentrates, and soy isolate was investigated, different dietary calcium levels were used. The calcium:phytate

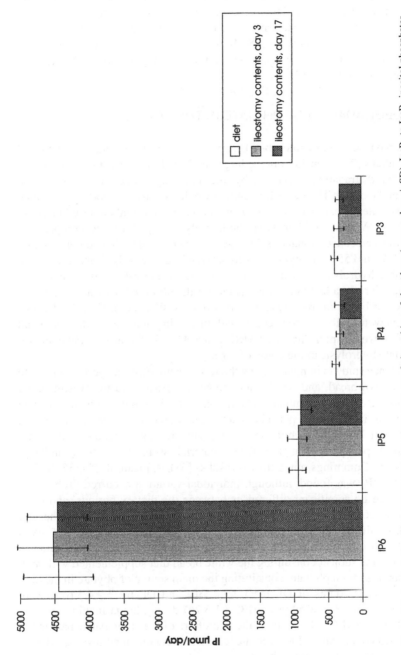

Figure 9.4 Inositol phosphates in diets and ileostomy contents of nine subjects fed oat bran (mean values ± SD). $InsP_3$ to $InsP_6$ inositol phosphates with three to six phosphate groups per inositol residue [68].

molar ratios in this study were 20, 22, 28, and 4.5. Independent of these molar ratios, an almost complete $InsP_6$ recovery was obtained, suggesting that these differences in the dietary calcium level did not affect the degradation of $InsP_6$ in the stomach and small intestine [68]. Similar results were obtained in pigs, indicating no effect of dietary calcium level on phytate degradation in the stomach and small intestine [51,58].

5.2. DEGRADATION OF PHYTATE IN THE COLON

The phytase level of microbial origin is very low in the small intestine of mammalians [71]. Considering the large number of microorganisms present in the colon, degradation of phytate by microbial phytase could be expected to occur in the colon. The germ-free animal would be a useful model with which to investigate the extent of phytate degradation by microbial phytase *in vivo*. In a study by Wise and Gilburt [72], the hydrolysis of phytate was compared in germ-free and conventional rats. In conventional rats fed high- and low-calcium diets, 22% and 55% of phytate was hydrolyzed, respectively. Considering that negligible hydrolysis was observed in the germ-free rats, it was concluded that the microflora had been responsible for the observed hydrolysis. Indirect support for bacterial hydrolysis was reported by Wise et al. [73]. Hydrolysis products were found in the caecum, with increasing amounts in the colon and feces followed by phosphorous NMR. Lantzsch et al. [52] also found evidence of phytate hydrolysis in the colon of pigs.

A balance study in humans during three consecutive 24-day periods on white bread, brown bread, and whole-meal bread was performed to investigate the effect of dietary fiber on mineral absorption [74]. The phytate content was held constant during the three periods by addition of sodium phytate to the diet in periods with brown and white bread to the same level as in whole-meal bread.

Inositol phosphates in the feces material were analyzed (Sandberg, Andersson, Cummings, unpublished results) [68]. A mean of 25–35% hydrolysis of $InsP_6$ was found, although individual variation occurred (Table 9.2). There were no significant differences between the dietary periods, and there was no increased degradation after the third period compared with the first period. As the diet did not contain phytase activity, the results suggest that a degradation of phytate occurs in the colon of humans. In the feces samples from a human subject consuming the white bread diet supplemented with high amounts of sodium phytate, constituting the main source of phytate in the diet, almost equal amounts of DL-Ins(1,2,3,4,5)P_5 and DL-Ins(1,2,4,5,6)P_5 were found [65]. Both cereal phytase (EC 3.1.3.26) during bread making and microbial phytase (EC 3.1.3.8) from the microflora of the colon participate in the degradation of phytate. Moreover, the feces samples contained several different isomers of inositol tetraphosphates.

TABLE 9.2. Recovery of Phytate in Feces.

| Subjects | Recovery InsP$_6$ % | | |
	White bread + phytate	Brown bread + phytate	Whole-meal bread
1	102	87	81
2	60	73	79
3	82	71	64
4	82	48	52
5	93	73	63
6	42	53	46
Mean ± SD	76.8 ± 20	67.5 ± 13	64.2 ± 13

The calcium level of this study was rather high (30–40 mmol/day). In a study with pigs, Sandberg et al. [58] found that the dietary calcium level markedly affected the phytate degradation in the colon. Total gastrointestinal degradation of InsP$_6$ in pigs fed a rapeseed diet (not containing phytase activity) was 97, 77, and 42% ($p < 0.001$) when calcium intakes were 4.5, 9.9, and 15 g/day, while no differences in phytate hydrolysis of ileal digesta were found between periods. Similar to these results, Skoglund et al. [51] found an impaired InsP$_6$ degradation in the colon when the pig feed was supplemented with calcium.

6. CONCLUSIONS

Phytate can be degraded using biological food/feed processing techniques increasing the activity of the naturally occurring enzyme phytase by soaking, malting, hydrothermal treatment, fermentation, or by addition of phytases. Degradation of phytate improves the bioavailability of iron, zinc, and phosphorus in man and animals. However, to substantially increase iron absorption, degradation of phytate must be virtually complete.

Phytate degradation in the stomach and small intestine of humans occurs as a result of activity of dietary phytase of plant or microbial origin. The plant phytase is less stable than microbial phytase at the physiological conditions of the gastrointestinal tract. The activity of intestinal mucosa phytase is low in humans and does not lead to any significant phytate degradation in the intestine when consuming a normal diet. No adaptation to increased phytate degradation in the stomach and small intestine occurs after a period of consumption of a high-phytate diet. The phytate degradation in the colon could be considered an effect of microbial phytase produced by the microflora present. This degradation is reduced by increasing the calcium intake, probably due to formation of insoluble calcium phytate complexes. It is not known, however, which and to what extent inositol phosphates, formed by degradation of phytate, may be absorbed by the intestinal epithelia and potentially exert a physiological response. Further work is needed on the positive and/or negative health effects of different inositol phosphates and their stereoisomers. Biotechnologically produced phytase, today used in animal feeding, may tomorrow be used in food processing as well as gene modified plants and starter cultures with improved phytase activity. Some of the degradation products of InsP$_6$ have been identified as mineral absorption inhibitors, and some may have health benefits. A controlled degradation of InsP$_6$ forming desired specific isomers of inositol phosphates with health benefits is a challenge for the future.

7. REFERENCES

1. Brune, M., Rossander-Hulthén, L., Hallberg, L., Gleerup, A. and Sandberg, A.-S. 1992. Iron absorption from bread in humans: inhibiting effects of cereal fiber, phytate and inositol phosphates with different numbers of phosphate groups. *J. Nutr.* 122:442–449.
2. Lönnerdal, B., Sandberg, A.-S., Sandström, B. and Kunz, C. 1989. Inhibitory effects of phytic acid and other inositol phosphates on zinc and calcium absorption in suckling rats. *J. Nutr.* 119:211–214.
3. Sandström, B. and Sandberg, A.-S. 1992. Inhibitory effects of isolated inositol phosphates on zinc absorption in humans. *J. Trace Elements and Electrolytes in Health and Disease* 6:99–103.
4. Sandberg, A.-S., Brune, M., Carlsson, N.-G., Hallberg, L., Skoglund, E. and Rossander-Hulthén, L. 1999. Inositol phosphates with different number of phosphate groups influence iron absorption in humans. *Amer. J. Clin. Nutr.* 70:240–246.
5. Jongbloed, A.W. and Lenis, N.P. 1992. Alteration of nutrition as a means to reduce environmental pollution by pigs. *Livestock Prod. Sci.* 31:75–94.
6. Graf, E. and Eaton, J.W. 1985. Dietary suppression of colonic cancer. Fiber or phytate? *Cancer* 56:717–718.
7. Shamsuddin, A.M. 1995. Inositol phosphates have novel anticancer function. *J. Nutr.* 125:725–732.
8. Jariwalla, R.J., Sabin, R., Lawson, S. and Heman, Z.S. 1990. Lowering of serum cholesterol and triglycerides and modulation of divalent cations by dietary phytase. *J. Appl. Nutr.* 42:18–28.
9. Larsson, M. and Sandberg, A.-S. 1992. Phytate reduction in oats during malting. *J. Food Sci.* 57:994–997.
10. Nayini, N.R. and Markakis, P. 1983. Effect of fermentation on the inositol phosphates of bread. *J. Food Sci.* 48:262–263.
11. Sandberg, A.-S. and Svanberg, U. 1991. Phytate hydrolysis by phytase in cereals. Effects on *in vitro* estimation of iron availability. *J. Food Sci.* 56:1330–1333.
12. Svanberg, U., Lorri, W. and Sandberg, A.-S. 1993. Lactic fermentation of non-tannin and high-tannin cereals: effect on *in vitro* estimation of iron availability and phytate hydrolysis. *J. Food Sci.* 58:408–412.
13. Holub, B.J. 1987. The cellular forms and functions of the inositol phospholipids and their metabolic derivatives. *Nutr. Rev.* 45:65–71.
14. Cosgrove, D.J. 1966. The chemistry and biochemistry of inositol polyphosphates. *Rev. Pure Appl. Chem.* 16:209–224.

15. Gibson, D.M. and Ullah, A.B. 1990. Phytases and their action on phytic acid. In *Inositol Metabolism in Plants*. D.J. Morre, W.F. Boss and F.A. Loewus. Wiley-Liss Inc., New York.

16. Cosgrove, D.J. 1980. Inositol phosphates. Their chemistry, biochemistry and physiology. Elsevier Scientific Publishing Company, Amsterdam.

17. Suzuki, U., Yoshimura, K. and Takaishi, M. 1907. Ueber ein enzym phytase das anhydro-oxy-methilen diphosphorusäure spaltet. *Bull. Coll. Agric. Tokyo Imp. Univ.* 7:503–512.

18. Greiner, R., Koneitzny, U. and Jang, K.-D. 1998. Purification and characterisation of a phytase from rye. *Food Biochem.* 22:143–161.

19. Greiner, R., Haller, E., Konietzny, U. and Jang, K.D. 1997. Purification and characterisation of a phytase from *Klebsiella terrigena*. *Arch. Biochem. Biophys.* 341:201–206.

20. Greiner, R., Konietzny, U. and Jang, K.D. 1993. Purification and characterization of two phytases from *Escherichia coli*. *Arch. Biochem. Biophys.* 303:107–113.

21. Greiner, R. and Larsson-Alminger, M. 1999. Purification and characterisation of a phytate degrading enzyme from oat (*Avena Sativa*). *J. Sci. Food Agric.* 79:1453–1460.

22. Konietzny, U., Greiner, R. and Jany, K.J. 1994. Characterization of a phytase from spelt. Presented at The International Euro food Tox IV Conference, Olsztyn, Poland.

23. Phillippy, B.Q. 1989. Identification by two-dimensional NMR of *myo*-inositol tris- and tetrakis(phosphates) formed from phytic acid by wheat phytase. *J. Agric. Food Chem.* 37:1261–1265.

24. Türk, M., Carlsson, N.-G. and Sandberg, A.-S. 1996. Reduction in the levels of phytate during wholemeal bread making; effect of yeast and wheat phytases. *J. Cereal. Sci.* 23:257–264.

25. Phillippy, B.Q. and Bland, J.M. 1988. Gradient ion chromatography of inositol phosphates. *Anal. Biochem.* 175:162–166.

26. Skoglund, E., Carlsson, N.-G. and Sandberg, A.-S. 1997. Analysis of inositol mono- and diphosphate isomers using high-performance ion chromatography and pulsed amperometric detection. *J. Agric. Food Chem.* 45:4668–4673.

27. Irving, G.C.J. and Cosgrove, D.J. 1972. Inositol Phosphate Phosphatases of Microbiological Origin: the Inositol Phentaphosphatate Products of *Aspergillus ficuum* Phytases. *J. Bacteriol.* 112:434–438.

28. Türk, M., Sandberg, A.-S., Carlsson, N.-G. and Andlid, T. 2000. IP$_6$ hydrolysis by Baker's yeast. Capacity, kinetics and degradation. *J Agric. Food Chem.* 48:100–104.

29. Kaay, J.V.D. and Haasert, J.M.V. 1995. Stereospecificity of inositol IP$_6$ dephosphorylation by *Paramecium phytase*. *Biochem. J.* 312:907–910.

30. Cosgrove, D.J. 1970. Inositol phosphatases of microbiological origin. Inositol phosphate intermediates of the dephosphorylation of the IP$_6$ of *myo*-inositol and D-chiro-inositol by a bacterial (*Pseudomonas sp.*) phytase. *Aust. J. Biol. Sci.* 23:1207–1220.

31. Skoglund, E., Lönnerdal, B. and Sandberg, A.-S. 1999. Inositol phosphates influence iron uptake in caco-2 cells. *J. Agric. Food. Chem.* 47:1109–1113.

32. Sandberg, A.-S. 1991. The effect of food processing on phytate hydrolysis and availability of iron and zinc. Proceedings of Nutritional and Toxicological Consequences of Food Processing AIN Symposium, Washington, 1990. *Adv. Exp. Med. Biol.* 289:499–508.

33. Mellanby, E. 1950. Some points in the chemistry and biochemistry of phytic acid and phytase. In *A Story of Nutritional Research*. 248–282. Baltimore: Williams and Wilkins.

34. Sandström, B., Almgren, A., Kivistö, B. and Cederblad, A. 1987. Zinc absorption from meals based on rye, barley, oatmeal, triticale and whole-wheat. *J. Nutr.* 117:1898–1902.

35. Rossander-Hulthén, L., Gleerup, A. and Hallberg, L. 1990. Inhibitory effect of oat products on non-haem iron absorption in man. *Eur. J. Clin. Nutr.* 44:783–791.

36. Larsson, M., Rossander-Hulthén, L., Sandström, B. and Sandberg, A.-S. 1996. Improved zinc and iron absorption from malted oats with reduced phytate content. *Br. J. Nutr.* 76:677–688.

37. Bergman, E.-L., Fredlund, K., Reinikainen, P. and Sandberg, A.-S. 1999. Hydrothermal processing of barley (*Cv Blenheim*): Optimisation of phytate degradation and increase of free *myo*-inositol. *J. Cereal Sci.* 29:261–272.

38. Sudarmadji, S. and Markakis, P. 1977. The phytate and phytase of soybean tempeh. *J. Sci. Food Agric.* 28:381–383.

39. Lopez, Y., Gordon, D.T. and Fields, L. 1983. Release of phosphorus from phytate by natural lactic acid fermentation. *J. Food Sci.* 48:952–954.

40. Saalovaara, H. and Göransson, M. 1983. Nedbrytning av fytinsyra vid framställning av surt och osyrat rågbröd. *Näringsforskning* 27:97–101.

41. Bartnik, M. and Florysiak, J. 1988. Phytate hydrolysis during bread-making in several sorts of Polish bread. *Die Nahrung* 32:37–42.

42. Larsson, M. and Sandberg, A.-S. 1991. Phytate reduction in bread containing oat flour, oat bran or rye bran. *J. Cereal Sci.* 14:141–149.

43. Türk, M. and Sandberg, A.-S. 1992. Phytate hydrolysis during bread-making: effect of addition of phytase from *Aspergillus niger. J. Cereal Sci.* 15:281–294.

44. Hallberg, L., Brune, M., Erlandsson, M., Sandberg, A.-S. and Rossander-Hultén, L. 1991. Calcium effect of different amounts on nonheme and heme iron absorption in humans. *Amer. J. Clin. Nutr.* 53:112–119.

45. Hartingsveldt, W.V., Zeijl, C.M.V., Harteveld, G.M., Gouka, R.J., Suykerbuyk, M.G.E., Luiten, R.G.M., Parington, P.A.V., Selten, G.C.M., Veenstra, A.E., Gorcom, R.F.M.V. and Hondel, C.A.M.V.D. 1993. Cloning, characterisation and overexpression of the phytase-encoding gene (*phyA*) of *Aspergillus niger. Gene* 127:87–94.

46. Marklinder, I.M., Larsson, M., Fredlund, K. and Sandberg, A.-S. 1995. Degradation of phytate by using varied sources of phytases in an oat-based nutrient solution fermented by *Lactobacillus plantarum* 2991. *Food Microbiol.* 12:487–495.

47. Hurrell, R.F., Juillerat, M.-A., Reddy, M.B., Lynch, S.R., Dassenko, S.A. and Cook, J.D. 1992. Soy protein, phytate and iron absorption in humans. *Amer. J. Clin. Nutr.* 56:573–578.

48. Cruickshank, E.W.H., Duckworth, J., Kosterlitz, H.W. and Warnock, G.M. 1945. The digestibility of the phytate-P of oatmeal in adult man. *J. Physiol.* 104:41–46.

49. Walker, A.R.P., Fox, F.W. and Irving, K.T. 1948. Studies in human mineral metabolism. 1. The effect of bread rich in phytate-phosphorus on the metabolism of certain mineral salts with special reference to calcium. *Biochem. J.* 42:452–462.

50. Ellis, R., Morris, E.R., Hill, A.D., Andersson, H.L.L. and McCarron, P.B. 1986. Effect of level of calcium intake on *in vivo* hydrolysis of dietary phytate. *Fed. Proc.* (Abstract) 374.

51. Skoglund, E., Larsen, T. and Sandberg, A.-S. 1997. Comparison between steeping and pelleting a mixed diet at different calcium levels on phytate degradation in pigs. *Can. J. Anim. Sci.* 77:471–477.

52. Lantzsch, H.J., Scheuermann, S.E. and Menke, K.H. 1988. Gastrointestinal hydrolysis of phytate from wheat, barley and corn in young-pigs. *Z. Tierphysiol. Tierernahr. Futtermittelkd.* 59:273–284.

53. Bitar, K. and Reinhold, J.G. 1972. Phytase and alkaline phosphatase activities in intestinal mucosae of rat, chicken, calf and man. *Biochim. Biophys. Acta* 268:442–452.

54. Iqbal, T.H., Lewis, K.O. and Cooper, B.T. 1994. Phytase activity in the human and rat small intestine. *Gut.* 35:1233–1236.

55. Davies, N.T. and Flett, A.A. 1978. The similarity between alkaline phosphatase (EC 3.1.3.1) and phytase (EC 3.1.3.8) activities in rat intestine and their importance in phytate-induced zinc deficiency. *Br. J. Nutr.* 39:307–316.

56. Hu, H.L., Wise, A. and Henderson, C. 1996. Hydrolysis of phytate and inositol tri-, tetra-, and penta-phosphates by the intestinal mucosa of the pig. *Nutr. Res.* 16:781–787.

57. Xu, P., Price, J., Wise, A. and Aggett, P.J. 1992. Interaction of inositol phosphates with calcium, zinc and histidine. *J. Inorg. Biochem.* 47:119–130.

58. Sandberg, A.-S., Larsen, T. and Sandström, B. 1993. High dietary calcium level decreases colonic phytate degradation in pigs fed a rapeseed diet. *J. Nutr.* 123:559–566.

59. Sandberg, A.-S., Andersson, H., Hallgren, B., Hasselblad, K., Isaksson, B. and Hultén, L. 1981. Experimental model for *in vivo* determination of dietary fibre and its effect on the absorption of nutrients in the small intestine. *Br. J. Nutr.* 45:283–294.

60. Sandberg, A.-S., Ahderinne, R., Andersson, H., Hallgren B. and Hultén, L. 1983. The effect of citrus pectin on the absorption of nutrients in the small intestine. *Human Nutrition: Clinical Nutrition* 37C:171–183.

61. Sandberg, A.-S., Andersson, H., Carlsson, N.-G. and Sandström, B. 1987. Degradation products of bran phytate formed during digestion in the human small intestine. Effect of extrusion cooking on digestibility. *J. Nutr.* 117:2061–2065.

62. Sandberg, A.-S. and Andersson, H. 1988. The effect of dietary phytase on the digestion of phytate in the stomach and small intestine of humans. *J. Nutr.* 118:469–473.

63. Sandberg, A.-S. and Ahderinne, R. 1986. HPLC method for determination of inositol, tri-tetra-, penta- and IP₆ in foods and intestinal contents. *J. Food Sci.* 51:547–550.

64. Sandberg, A.-S., Carlsson, N.-G. and Svanberg, U. 1989. Effects of inositol tri, tetra-, penta- and IP₆ on *in vitro* estimation of iron availability. *J. Food Sci.* 54:159–161.

65. Skoglund, E., Carlsson, N.-G. and Sandberg, A.-S. 1997. Determination of isomers of inositol mono- to IP₆ in selected foods and intestinal contents using high-performance ion chromatography. *J. Agric. Food Chem.* 45:431–436.

66. Batty, I.R., Nahorsky, S.R. and Irvine, R.F. 1985. Rapid formation of inositol, 1,3,4,5-tetrakisphosphate following muscarine receptor stimulation of rat cerebral cortical slices, *Biochem. J.* 232:211–215.

67. Cullen, R.J., Chung, S.-K., Chang, Y.-T., Dayson, A.P. and Irvine, R.F. 1995. Specificity of the purified inositol (1,3,4,5) tetrakisphosphate-binding protein from porcine platelets. *FEBS Lett.* 358:240–242.

68. Sandberg, A.-S. 1996. Degradation of inositol phosphates at different sites. In: *Dietary Fibre and Fermentation in the Colon.* Proceedings of COST 92, Espoo, Finland. June 15–17, 1995. Luxembourg: Office for Official Publications of the European Communities, 1996. 191–197.

69. Phillippy, B.Q. 1999. Susceptibility of wheat and *Aspergillus niger* phytases to inactivation by gastrointestinal enzymes. *J. Agric. Food Chem.* 47:1385–1388.

70. Sandström, B., Andersson, H., Kivistö, B. and Sandberg, A.-S. 1986. Apparent small intestinal absorption of nitrogen and minerals from soy and meat protein based diets. *J. Nutr.* 116: 2209–2218.

71. Cooper, J.R. and Gowing, H.S. 1983. Mammalian small intestinal phytase (*EC* 3.1.3.8). *Br. J. Nutr.* 50:673–678.

72. Wise, A. and Gilburt, D.J. 1982. Phytate hydrolysis by germ-free and conventional rats. *Appl. Environ. Microbiol.* 43:753.

73. Wise, A., Richards, C.P. and Trimble, M.L. 1983. Phytate hydrolysis in the gastrointestinal tract of the rat followed by phosphorus-31 Fourier transform nuclear magnetic resonance spectroscopy. *Appl. Environ. Microbiol.* 45:313–314.

74. Andersson, H., Nävert, B., Bingham, S.R., Englyst, H.N. and Cummings, J.H. 1983. The effects of breads containing similar amounts of phytate but different amounts of wheat bran on calcium, zinc and iron balance in man. *Br. J. Nutr.* 50:503–510.

CHAPTER 10

Influence of Processing Technologies on Phytate and Its Removal

SHRIDHAR K. SATHE
MAHESH VENKATACHALAM

1. INTRODUCTION

G LOBALLY, cereals and legumes are important sources of total calories, carbohydrates, proteins, and several other nutrients in the diets of many populations. Cereals and legumes are also the major source of dietary phytate intake. In this chapter, the term phytate is used to collectively describe phytic acid and its salts (unless otherwise noted). Because phytates are chemically reactive under a variety of conditions, understanding how phytates interact with other food components is critical in optimizing the nutritional quality of cereals and legumes, especially with respect to bioavailability of minerals and proteins. Phytate removal may be desirable because it potentially forms complexes with minerals and dietary proteins and decreases their bioavailability. Because phytates are heat stable, they are not easily removed by cooking, autoclaving, roasting, or any of the conventional heat processing methods. Solubility of phytates in aqueous solvents can, however, be used to reduce/eliminate food phytates when desired. Acid hydrolysis of phytates as well as ability of endogenous and/or added enzymes to effect phytate hydrolysis are additional means to reduce/eliminate food phytates. The choice of a method for phytate reduction is largely dependent on the type of food and the final product form in which that food is consumed.

Phytates can chelate minerals such as calcium, zinc, and iron, resulting in insoluble complexes. Such reduction in mineral solubility may lead to a decrease in mineral absorption and bioavailability. Certain minerals such as iron and copper catalyze oxidative enzymes that generate free radicals, resulting in undesirable oxidative damage such as cell membrane damage (produces leaky cells). Because phytates have the ability to chelate minerals that participate in undesirable oxidative reactions, phytates have been suggested to have protective effects. The ability of phytates to chelate the divalent minerals makes them a natural antioxidant. For these and several other reasons, phytate reduction or elimination from a food may not always be desirable. The role of phytic acid in health and disease has been recently reviewed [1], and several chapters in this book cover the role of phytate in human health. Influences of different processing methods on phytates are reviewed in this chapter. An attempt is made to emphasize data published in the last ten years (1990–2000), because the majority of the data published prior to 1990 was reviewed earlier [2].

2. SOAKING

Dry beans and cereals have tough seed coats that are not highly permeable to water. Hence, soaking is often used as a pretreatment to facilitate grain processing. Soaking treatment may last for a short period (such as 15–20 min) or for very long period (usually 12–16 h). The amount of soaking medium may also change substantially depending on the type of grain and type of processing needed after the seeds are soaked. In household situations, cereals and legumes are typically soaked in water overnight (12–14 h). If salts, acids, or alkalies are added to the soaking medium, the rate of water imbibition can change significantly. For example, using alkali salts, such as sodium bicarbonate/carbonate, typically increases the water imbibition rate, because alkalies not only improve permeability of the seed coat but also loosen it. Phytate is water soluble, and therefore, significant phytate reduction can be realized when soak water is discarded (Table 10.1). The soaking temperature may have a significant effect on not only the rate of water imbibition but also on whether endogenous phytases are active or not. If a soaking step is carried out at temperatures above 45°C but below 60°C, a significant percent of phytate hydrolysis can take place due to activation of endogenous phytases and acid phosphatases. For example, Bergman et al. [3] have recently shown that barley (cv. Blenheim) flour phytase was optimally active at pH 4.8 and temperature 57°C, while the extracted phytase from the same flour was optimally active at pH 5.2 and temperature of 47°C. Because it is often difficult to separate phytases from phosphatases, it is important to recognize and account for the activities of phosphatases in understanding phytate hydrolysis. A recent extensive evaluation of plant feedstuffs [4] suggests that there is a significant, positive correlation between phytase and acid phosphatase activities for cereals, cereal by-products, and oilseeds. Viveros et al. [4] also noted that the ratio of acid

TABLE 10.1. Effect of Various Soaking Treatments on Phytate Contents of Legumes and Cereals.

Common Name/Scientific Name	Phytate Reduction (%)	Reference(s)
Legumes		
Indian tribal pulse (*Mucuna monosperma*)		
Distilled water (18 h, RT[a])	28.0–35.0	[17]
0.02% NaHCO$_3$ (18 h, RT)	25.0–28.0	[17]
Indian tribal pulse (*Mucuna pruriens*)		
Distilled water (9 h)	27.0[b]	[18]
0.02% NaHCO$_3$ (9 h)	17.0[b]	[18]
Prosopis chilensis (Molina) Stunz		
Distilled water (12 h)	14.0[b]	[19]
0.02% NaHCO$_3$ (12 h)	8.0[b]	[19]
Vigna aconitifolia		
Distilled water (6 h)	44.0[b]	[20]
0.02% NaHCO$_3$ (6 h)	42.0[b]	[20]
Vigna sinensis		
Distilled water (6 h, RT)	38.0	[20]
0.02% NaHCO$_3$ (6 h, RT)	32.0	[20]
Lentils		
Distilled water	27.0	[21]
0.07% NaHCO$_3$ solution	23.0	[21]
0.1% Citric acid solution	37.0	[21]
Faba beans		
Distilled water (9–12 h, 30–37°C)	0.0–32.7	[22[b],23[b],24,25]
0.07% NaHCO$_3$ (9 h)	0.0[b]	[22,23]
0.1% Citric acid (9 h)	0.0[b]	[22,23]
Lathyrus beans		
Drinking water (12 h, 37°C)	6.0–10.0	[26]
Freshly boiled water (2 h, 37°C)	18.0–28.0	[26]
Tamarind solution (6 h, 37°C)	30.0	[26]
0.5% CaOH$_2$ (6 h, 37°C)	28.0–39.0	[26]
Great Northern beans		
Distilled water (18 h, 22°C)	69.6	[27]
Mixed salt solution[c] (18 h, 22°C)	8.7	[27]
Kidney beans		
Tap water (12 h, 24°C)	8.2	[28]
Distilled water (12–18 h, 22–30°C)	5.7–51.7	[25,27]
Mixed salt solution[c] (18 h, 22°C)	22.4	[27]
Pinto beans		
Distilled water (18 h, 22°C)	52.7	[27]
Mixed salt solution[c] (18 h, 22°C)	18.2	[27]
Brown beans[e]		
Distilled water (4 h, 21–55°C, pH 6.4)	0.0–19.0	[29]
Distilled water (2 h, 37°C, pH 6.4 + 2 h, 37°C, pH 8.0[d])	7.0	[29]

(continued)

TABLE 10.1.   (continued).

Common Name/Scientific Name	Phytate Reduction (%)	Reference(s)
Distilled water (4–17 h, 21–55°C, pH 6.4) [presoaked 17 h, 37°C]	33.0–85.0	[29]
Citrate buffer (4–17 h, 21–55°C, pH 4.5)	10.0–34.0	[29]
Citrate buffer (4–17 h, 37°C, pH 6.0)	2.0	[29]
Citrate buffer (2 h, 37°C, pH 6.0 + 2 h, 37°C, pH 8.0[d])	53.0	[29]
Citrate buffer (4 h, 37°C, pH 6.0) [presoaked 17 h, 37°C]	17.0	[29]
Tris buffer (4–17 h, 21–55°C, pH 7.0–8.0)	0.0–98.0	[29]
Tris buffer (4–17 h, 55°C, pH 7.0) [presoaked 17 h, 37°C]	51.0–98.0	[29]
Distilled water + tannase (4 h, 37°C, pH 6.4)	55.0	[29]
Citrate buffer + tannase (4 h, 37°C, pH 6.0)	4.0	[29]
Bambara groundnut		
Water (24 h)	20.0[b]	[30]
Pigeon pea		
Water (24 h)	18.0[b]	[30]
Chinese legume (*Phaseolus angularis*)		
Distilled water (18 h, 30°C)	26.9	[31]
Chinese legume (*Phaseolus calcaratus*)		
Distilled water (18 h, 30°C)	44.6	[31]
Dry beans		
Distilled water (12–18 h, 20–22°C)	3.9–7.0	[32,33]
Pinto beans		
Distilled water (18 h, 21°C)	1.9	[34]
2% NaHCO$_3$ (12 h, 21°C)	4.4	[34]
Mixed salt solution[c] (12 h, 21°C)	6.9	[34]
Sanilac beans		
Distilled water (18 h, 21°C)	1.0	[34]
2% NaHCO$_3$ (12 h, 21°C)	3.9	[34]
Mixed salt solution[c] (12 h, 21°C)	5.3	[34]
Cranberry beans		
Distilled water (18 h, 21°C)	1.4	[34]
2% NaHCO$_3$ (12 h, 21°C)	4.1	[34]
Mixed salt solution[c] (12 h, 21°C)	7.2	[34]
Viva pink beans		
Distilled water (18 h, 21°C)	0.9	[34]
2% NaHCO$_3$ (12 h, 21°C)	0.3	[34]
Mixed salt solution[c] (12 h, 21°C)	8.3	[34]
Black-eyed beans		
Tap water (12 h, 24°C)	7.6	[28]
Pink beans		
Tap water (12 h, 24°C)	20.1	[28]

TABLE 10.1. (continued).

Common Name/Scientific Name	Phytate Reduction (%)	Reference(s)
Chickpeas		
Distilled water (12 h, 37°C)	11.0–22.7	[35,36]
2% NaHCO$_3$ (4 h)	7.5[b]	[36]
Cowpeas		
DI water (72 h, 27°C)	20.9–28.3	[37]
Kidney beans		
DI water (12 h, 30°C)	5.7	[25]
Peas (Spanish)		
DI water (12 h, 30°C)	4.6–11.3	[38]
Moth bean		
Tap water (12 h, 37°C)	46.0–50.0	[39]
Mixed salt solution[c] (12 h, 37°C)	47.0–50.0	[39]
Soybeans		
Distilled water (5–72 h, RT-60°C)	0.0–65.2	[37,40–42]
Tap water (12 h, 25–50°C)	48.3–58.7	[42]
Distilled water (12 h, 25–50°C)	73.3–82.3	[42]
(γ irradiated 1.0 kGy)		
Tap water (12 h, 25–50°C)	67.0–71.7	[42]
(γ irradiated 1.0 kGy)		
California small white beans		
Distilled water (3 h, 23–60°C)	0.0–49.3	[40]
Distilled water (3 h, 70–90°C)	11.5 19.0	[40]
Acetate buffer (24 h, 50°C,	32.6–61.5	[40]
pH 4.0–5.8)		
Mung beans		
Water (5–18 h, RT-60°C).	2.3–30.0	[28,40,43]
Black gram		
Water (12 h, 37°C)	25.0–30.0	[35,44]
Amphidiploids (black gram ×		
mung bean)		
Water (18 h, 37°C)	31.0–37.0	[45]
Lima beans		
Water (5–72 h, 27–60°C)	0.0–44.7	[37,40]
Rape seed		
Water (24 h, 22°C)	22.0	[46]
Cereals		
Wheat		
Water (5–24 h, RT-60°C)	4.2–46.0	[40,47]
Acetate buffer (24 h, 55°C, pH 4.8)	91.0	[47]
Wheat Bran		
Water (8 h, 21°C)	19.5	[48]
Rye (whole grain)		
Water (24 h, 55°C)	57.0	[47]
Acetate buffer (24 h, 55°C, pH 4.8)	92.0	[47]
Barley		
Hulled grains		
Water (24 h, 55°C)	56.0	[47]
Acetate buffer (24 h, 55°C, pH 4.8)	89.0	[47]

(*continued*)

TABLE 10.1.   (continued).

Common Name/Scientific Name	Phytate Reduction (%)	Reference(s)
Dehulled grains		
Water (24 h, 55°C)	77.0	[47]
Acetate buffer (24 h, 55°C, pH 4.8)	99.0	[47]
Naked grains		
Water (12 h, 53–55°C)	41.0	[47]
Acetate buffer (12 h, 53–55°C, pH 4.8)	61.0	[47]
Oats		
Whole hulled grains		
Water (24 h, 37°C)	13.0	[47]
Acetate buffer (24 h, 37°C, pH 4.8)	26.0	[47]
Whole dehulled grains		
Water (37–40°C, 8 h) (initial wet-steeping 53–57°C, 20 min)	1.0	[47]
Ground dehulled grains		
Water (17 h, 20–37°C) (with or without preincubation of whole grain at 8 h, 37–40°C)	72.0–77.0	[47]
Naked whole grains		
Water (24 h, 37°C)	7.0	[47]
Water (8 h, 20–55°C) (initial wet-steeping 53–57°C, 20 min)	9.0–19.0	[47]
Ground naked grains		
Water (17 h, 20–37°C) (with or without preincubation of whole grain at 8 h, 37–40°C)	88.0–94.0	[47]
Sorghum		
Tap water (24 h)	16.3–21.3[b]	[49]
Pearl millet		
DI water (15 h, 28°C)	14.8	[50]
0.2 N HCl (0.3–.75 h, 28°C) [dehulled 8.1–15.84%]	60.0–74.0	[50]

[a]RT—Room temperature (actual temperature not indicated).
[b]Soaking temperature not indicated.
[c]Mixed salt solution—2.5% sodium chloride, 1.5% sodium bicarbonate, 0.5% sodium carbonate and 1% sodium tripolyphosphate in distilled water.
[d]pH adjusted with NaOH.
[e]Ground beans.

phytases to phytase activity ranged from 3.1 for wheat bran to 633 for beet pulp.

3. COOKING

During cooking, food is exposed to elevated temperatures for a certain period of time. Factors that influence cooking temperature and time vary considerably and include the following:

(1) Final desired texture of a cooked product, which varies somewhat with the consumer
(2) Type of food being cooked
(3) Physical state and chemical composition of the food being cooked
(4) Cooking medium, especially presence/absence of salts, acids, and alkalies in the cooking medium
(5) Cooking method (e.g., atmospheric versus pressure cooking)
(6) Cooking equipment and heat and mass transfer rates during cooking
(7) Amount of food that is cooked

Regardless of cooking method, tissue softening is one of the desirable changes that takes place during cooking. Softening of the texture is partly due to loss of cell membrane integrity. Cell membranes and middle lamellae contain pectic substances often insolubilized by cross-linkages afforded by divalent minerals such as calcium and magnesium. Phytates, being metal chelators, bind with calcium and magnesium and may, therefore, inhibit such cross-linkages by minerals that lead to weakened cell membranes and middle lamellae, resulting in shorter cooking times. High phytate peas were observed to have lower cooking times than their low-phytate counterparts [5]. Similarly, Moscoso et al. [6] observed higher cooking times for foods with low phytate content, indicating that low phytate content may promote more cross-linking between middle lamellae. On the other hand, several studies have failed to establish a clear relationship between phytate content and cooking time of food [7–11]. Perhaps part of the difficulty in explaining such mixed results is due to a third possibility that the middle lamellar layers may coagulate more easily (in the absence of calcium and magnesium cross-linkages) during cooking. Such collapse of the cementing material may lead to formation of larger and thicker layers surrounding the cells impeding the effective heat transfer, thereby increasing cooking time in low-phytate grains and legumes.

Because phytate is heat stable, significant phytate reduction during cooking is not expected unless either the cooking water was discarded or the food received additional processing treatment such as soaking (soaking water discarded), germination, fermentation, etc. Another possible mechanism by which phytate reduction can occur during cooking is phytate hydrolysis due to activation of the endogenous phytases or phosphatases during the early part of the cooking phase. The presence of acids and alkalies during cooking may also contribute to phytate hydrolysis. Typically, cooking allows a significant reduction (up to 50%, in most cases) in phytate. Higher phytate reduction is possible when cooking is combined with certain other treatments (Table 10.2). Pressure cooking (autoclaving) typically causes higher phytate reduction than cooking under atmospheric pressure. A potentially important variation of the standard

TABLE 10.2. **Effect of Heat Processing on Phytate Contents of Legumes, Cereals, and Their Food Products.**

Common Name/Scientific Name	Phytate Reduction (%)	Reference(s)
Legumes		
Black-eyed beans		
Cooking[a] (3 h, 100°C, 3% NaCl)	13.3	[28]
Pressure cooking (3 h, 115.5°C, 3% NaCl)	91.5	[28]
Red kidney beans		
Cooking (90 min, 100°C) [soaked[c] 18 h, 22°C]	57.8	[27]
Cooking (15 min, 100°C) [soaked in mixed salt solution[g] 18 h, 22°C]	31.3	[27]
Cooking (3 h, 100°C, 3% NaCl)	7.7	[28]
Pressure cooking (3 h, 115.5°C, 3% NaCl)	69.7	[28]
Great Northern beans		
Cooking (90 min, 100°C) [soaked[c] 18 h, 22°C]	76.1	[27]
Cooking (15 min, 100°C) [soaked in mixed salt solution[g] 18 h, 22°C]	13.0	[27]
Pinto beans		
Cooking (90 min, 100°C) [soaked[c] 18 h, 22°C]	61.5	[27]
Cooking (15 min, 100°C) [soaked in mixed salt solution[g] 18 h, 22°C]	25.6	[27]
Pink beans		
Cooking (3 h, 100°C, 3% NaCl)	26.4	[28]
Pressure cooking (3 h, 115.5°C, 3% NaCl)	74.9	[28]
Cowpeas		
Cooking (30–45 min, 100°C)	17.1–65.7	[51,52[b],53]
Cooking (30–45 min, 100°C) [soaked 6–12 h, 25°C]	38.5–58.6	[53,54]
Cooking (35 min, 100°C, 0.1% NaHCO₃)	39.6	[53]
Cooking (35 min, 100°C, 0.1% Kanwa rock salt)	27.5	[53]
Pressure cooking (15–20 min, 105–121°C)	5.4–27.9	[37,53,55]
Pressure cooking (15 min, 121°C) [soaked 10 h and germinated 24 h, 25–30°C]	14.40	[55]
Pressure cooking (15 min, 121°C) [roasted 5 min, 160°C]	21.1	[55]
Pressure cooking (15 min, 121°C, 0.1% Kanwa rock salt)	10.7	[53]

TABLE 10.2. (continued).

Common Name/Scientific Name	Phytate Reduction (%)	Reference(s)
Pressure cooking (15 min, 121°C, 0.1% NaHCO$_3$)	26.8	[53]
Roasting (5 min, 160°C)	19.8	[53]
Chickpeas		
Cooking (100°C)	7.0–35.7[b]	[35,51,56]
Cooking (30 min, 100°C) [soaked 4–12 h, 37°C]	20.0–46.9	[35[b],36[b,d],54]
Cooking (100°C) [soaked 4 h, 2% NaHCO$_3$]	12.5[d]	[36]
Cooking (100°C) [soaked 12 h, 37°C and germinated 68 h, 25°C]	35.0–40.0[b]	[35]
Cooking (15–20 min, 100°C) [fresh immature seeds]	60.0	[36]
Pressure cooking (15 min–1 h, 120–121°C) [whole seed]	16.9–64.8	[55,57,58]
Pressure cooking (1 h, 120°C) [cotyledon]	3.7–54.9	[58]
Pressure cooking (15 min, 121°C) [soaked 12 h, 37°C]	22.0–29.0	[35]
Pressure cooking (15 min, 121°C) [soaked 10 h, 25–30°C and germinated 24 h, 25–30°C]	51.7	[55]
Pressure cooking (15 min, 121°C) [roasted 5 min, 160°C]	56.3	[55]
Roasting (2–5 min, 160–200°C) [whole seed]	16.1–60.5	[36[b,d],55,57,58[b,d]]
Roasting (6 min) [soaked 12 h and rested 2–6 h]	27.3–31.8[d,e]	[36]
Roasting [cotyledon]	17.0–68.8[b,e]	[58]
Roasting (2–3 min) [immature seeds]	46.6[e]	[36]
Roasting (2–3 min) [immature seeds in pods]	40.0[e]	[36]
Frying[e] (2 min) and cooking (15–20 min, 100°C) [immature seeds]	53.3	[36]
Frying (2 min)[e] and cooking (5 min, 100°C) [presoaked[d] 4 h and cooked[b] 100°C]	37.5	[36]
Balila, Falafel	32.0–57.0	[59]
Lima beans		
Cooking (1–2 h, 100°C)	15.4–44.0	[60]
Pressure cooking (15–20 min, 105–121°C)	5.6–8.5	[37]

(*continued*)

TABLE 10.2. (continued).

Common Name/Scientific Name	Phytate Reduction (%)	Reference(s)
Bambara groundnut		
Cooking (100°C)	14.0[b]	[30]
Normal cooking (100°C) [soaked 24 h]	22.0[b,d]	[30]
Normal cooking (100°C) [soaked 23 h and dehulled]	23.0[b,d]	[30]
Cooking (100°C) [soaked 24 h and germinated 72 h]	56.0[b,d]	[30]
Pressure cooking (15 min, 121°C)	24.0	[30]
Pressure cooking (15 min, 121°C) [soaked 24 h]	32.0[d]	[30]
Pressure cooking (15 min, 121°C) [soaked 23 h and dehulled]	33.0[d]	[30]
Pigeon pea		
Cooking (100°C)	16.0[b]	[30]
Cooking (100°C) [soaked 24 h]	25.0[b,d]	[30]
Cooking (100°C) [soaked 24 h and dehulled]	28.0[b,d]	[30]
Normal cooking (100°C) [soaked 24 h and germinated 72 h]	53.0[b,d]	[30]
Pressure cooking (15 min, 121°C)	27.0–38.5	[30,57]
Pressure cooking (15 min, 121°C) [soaked 24 h]	30.0[d]	[30]
Pressure cooking (15 min, 121°C) [soaked 23 h and dehulled]	36.0[d]	[30]
Roasting (2 min, 200°C)	33.3	[57]
Lentils		
Cooking (30–35 min, 100°C) [soaked 9–12 h, RT]	39.0–58.6	[21,54]
Cooking (35 min, 100°C) [soaked in 0.07% $NaHCO_3$ 9 h, RT]	29.0	[21]
Cooking (35 min, 100°C) [soaked in 0.1% citric acid 9 h, RT]	32.0	[21]
Pressure cooking (15 min, 121°C) [roasted 5 min, 160°C]	45.8	[55]
Pressure cooking (15 min, 121°C) [soaked 10 h, 25–30°C and germinated 24 h, 25–30°C]	38.0	[55]
Roasting (5 min, 160°C)	44.6	[55]
Black beans		
Cooking (30 min, 100°C) [fresh harvested beans]	11.0	[61]
Cooking (15 min, 100°C, mineral salt solutions[f]) [fresh harvested beans]	7.0–9.0	[61]

TABLE 10.2. (continued).

Common Name/Scientific Name	Phytate Reduction (%)	Reference(s)
Cooking (30 min, 100°C) [stored 6 months, 5–40°C, 50–80% RH]	17.0–27.0	[61]
Cooking (15 min, 100°C, mineral salt solutions[f]) [stored 6 months, 5–40°C, 50–80% RH]	11.0–22.0	[61]
Cooking (20 min, 100°C) [germinated 48 h, 28°C]	14.3	[62]
Pressure cooking (15 min, 121°C)	12.1	[55]
Pressure cooking (15 min, 121°C) [roasted 5 min, 160°C]	30.9	[55]
Pressure cooking (15 min, 121°C) [soaked 10 h, 25–30°C and germinated 24 h, 25–30°C]	34.7	[55]
Roasting (5 min, 160°C)	46.2	[55]
Navy bean		
Cooking (30 min, 100°C) [soaked 12 h]	40.7[d]	[54]
Indian tribal pulse (*Mucuna monosperma*)		
Cooking (3 h, 100°C)	31.0–39.0	[17]
Pressure cooking (90 min, 121°C)	25.0–32.0	[17]
Dry beans		
Cooking (1.5 h, 100°C)	8.7	[32]
Cooking (1.5–3 h, 100°C) [soaked 12–18 h, 20°C]	31.6–51.0	[32,33]
Soybean		
Cooking (30 min, 100°C)	16.9[b]	[52]
Cooking (30 min, 100°C) [germinated 48 h, 28°C]	12.5	[62]
Pressure cooking (15–20 min, 105–121°C)	6.4–30.1	[37,57]
Roasting (2 min, 200°C)	17.0	[57]
Frying (5 min, 190°C) [soaked 24 h and dehulled]	89.8[d]	[63]
Chinese indigenous legumes (*Phaseolus angularis*)		
Cooking (30 min, 100°C)	35.5	[31]
Pressure cooking (20 min, 121°C)	54.4	[31]
Chinese indigenous legumes (*Phaseolus calcaratus*)		
Cooking (30 min, 100°C)	41.6	[31]
Pressure cooking (20 min, 121°C)	58.0	[31]
Indian tribal pulse (*Mucuna pruriens*)		
Cooking (90 min, 100°C)	18.0	[18]
Pressure cooking (45 min, 121°C)	44.0	[18]

(*continued*)

TABLE 10.2. (continued).

Common Name/Scientific Name	Phytate Reduction (%)	Reference(s)
Prosopis chilensis (Molina) Stunz		
Cooking (120 min, 100°C)	32.0	[19]
Pressure cooking (45 min, 121°C)	35.0	[19]
Vigna aconitifolia		
Cooking (30 min, 100°C)	44.0	[20]
Pressure cooking (20 min, 121°C)	25.0	[20]
Vigna sinensis		
Cooking (60 min, 100°C)	31.0	[20]
Pressure cooking (30 min, 121°C)	13.0	[20]
Polish pea		
Frying (5 min, 190°C)	82.9	[63]
Moth bean		
Cooking (100°C) [soaked 12 h, 37°C]	47.0–52.0[b]	[39]
Cooking (100°C) [soaked 12 h, 37°C and germinated 60 h, 25°C)	50.0–58.0[b]	[39]
Pressure cooking (15 min, 121°C)	24.2	[55]
Pressure cooking (20 min, 121°C) [soaked 12 h, 37°C]	48.0–53.0	[39]
Pressure cooking (20 min, 121°C) [soaked in mixed salt solution[g] 12 h, 37°C]	52.0–57.0	[39]
Pressure cooking (15 min, 121°C) [soaked 10 h, 25–30°C and germinated 24 h, 25–30°C]	45.3	[55]
Pressure cooking (15 min, 121°C) [roasted 5 min, 160°C]	45.3	[55]
Roasting (5 min, 160°C)	47.29	[55]
Lathyrus beans		
Cooking (70–80°C) [soaked 12 h, 37°C]	19.0–22.0[b]	[26]
Pressure cooking (15 min, 121°C) [soaked 12 h, 37°C]	23.0–37.0	[26]
Mung bean		
Cooking (100°C)	15.0–44.0[b]	[43,51]
Cooking (3 h, 100°C, 3% NaCl)	36.3	[28]
Cooking (100°C) [soaked 12 h, 37°C]	20.0[b]	[43]
Pressure cooking (3 h, 115.5°C, 3% NaCl)	67.6	[28]
Pressure cooking (10–15 min, 121°C)	17.6–18.0	[43,57]
Pressure cooking (15 min, 121°C) [soaked 12 h, 37°C]	25.0	[43]
Roasting (2 min, 200°C)	21.6	[57]

TABLE 10.2. **(continued).**

Common Name/Scientific Name	Phytate Reduction (%)	Reference(s)
Black gram		
Cooking (100°C)	5.0–9.0[b]	[35,44]
Cooking (100°C) [soaked 12 h, 37°C]	29.0–40.0[b]	[35,44]
Cooking (100°C) [soaked 12 h, 37°C and germinated 48 h, 25°C)	48.0–52.0[b]	[35]
Pressure cooking (10–15 min, 121°C)	8.0–27.0	[44,55,57]
Pressure cooking (15 min, 121°C) [soaked 12 h, 37°C]	33.0–43.0	[35,44]
Pressure cooking (15 min, 121°C) [soaked 10 h, 25–30°C and germinated 24 h, 25–30°C]	28.7[d]	[55]
Pressure cooking (15 min, 121°C) [roasted 5 min, 160°C]	37.8	[55]
Roasting (2–5 min, 160–200°C)	30.4–40.9	[55,57]
Amphidiploids (black gram x mung bean)		
Cooking (100°C)	3.0–15.0[b]	[45]
Cooking (100°C) [soaked 12 h, 37°C]	15.0–22.0[b]	[45]
Pressure cooking (15 min, 121°C)	11.0–22.0	[45]
Pressure cooking (15 min, 121°C) [soaked 12 h, 37°C]	18.0–27.0	[45]
African oil bean seed		
Cooking (8–10 h, 100°C)	42.9	[108]
Faba/broad bean		
Cooking (35–45 min, 100°C) [soaked 9–12 h, 25°C]	0.0–30.76	[22[d],24,64]
Cooking (35–45 min, 100°C) [soaked 9–12 h, 25°C]	32.0–35.0	[24]
Cooking (35 min, 100°C) [soaked in 0.07% NaHCO$_3$ 9 h]	0.0[d]	[22,23]
Cooking (35 min, 100°C) [soaked in 0.1% citric acid 9 h]	28.9–35.0[d]	[22,23]
Pressure cooking (25–30 min, 121°C) [soaked 12 h, 25–37°C]	20.0–41.0	[24,64]
Pressure cooking (25–30 min, 121°C) [soaked 12 h, 25°C and dehulled]	53.0–55.0	[24]
Dry heating (15 min, 120°C)	36.0–40.0	[22,23]
Frying (15 min, 190°C)	67.7	[63]
Foul jerra, Foul mudames [Saudi Arabic food]	7.0–70.0	[59]
Lupin seeds		
Cooking (20 min, 100°C) [germinated 48 h, 28°C]	0.0	[62]

(*continued*)

TABLE 10.2.   (continued).

Common Name/Scientific Name	Phytate Reduction (%)	Reference(s)
Rice bean		
Cooking (50 min, 100°C)	13.0–15.0	[65]
Pressure cooking (15–20 min, 121°C)	2.0–18.0	[65,66]
Cereals		
Rye		
Extrusion cooking (100–170°C)	0.0–23.0	[67]
Fresh corn		
Cooking (100°C)	18.1[b]	[68]
Roasting (sand bath)	46.7[b,e]	[68]
Roasting (charcoal)	41.9[b,e]	[68]
Cooking (hot powdered charcoal)	29.2[b,e]	[68]
Lime cooking (55–75 min) [Nixtamalized]	35.0	[69]
Dried corn		
Roasting (sand bath)	23.7[b,e]	[68]
Cooking (30 min, 100°C)	16.3–18.9[b]	[52,68]
Popping	11.5[b,e]	[68]
Rice		
Cooking (10 min, 100°C) [distilled deionized water]	1.5–3.6	[70]
Cooking (10 min, 100°C) [tap water]	67.9–71.9	[70]
Kabsa, Saleeg, Rus Abiedh, Baryani [Saudi Arabic foods]	11.0–65.0	[59]
Kuwaiti rice preparations	0.0–82.0	[71]
Quinoa seeds		
Cooking (25 min, 100°C)	15.3–19.7	[72]
Cooking (25 min, 100°C) [soaked 12–14 h, 20°C]	68.5–76.7	[72]
Cooking (25 min, 100°C) [soaked 12–14 h, 20°C and germinated 30 h, 30°C]	48.6–49.5	[72]
Cooking (10 min, 100°C) [*Lactobacillus* fermented 16–18 h, 30°C]	82.5–88.3	[72]
Cooking (10 min, 100°C) [soaked 12–14 h, 20°C and germinated 30 h, 30°C and *Lactobacillus* fermented 16–18 h, 30°C]	96.5–98.2	[72]
Sorghum		
Cooking (100°C)	17.9–37.5[b]	[49,52]
Pearl millet		
Pressure cooking (15 min, 121°C)	36.3	[73,74]

TABLE 10.2. **(continued).**

Common Name/Scientific Name	Phytate Reduction (%)	Reference(s)
Barley		
Pressure cooking (15 min, 121°C)	8.0	[75]
Wheat		
Roasting (traditional method)	25.0[b,e]	[76]
Dalya (porridge)	87.9[b,e]	[76]
Margoog, Gerish, Harees, Kibbah	41.0–80.0	[59]
[Saudi Arabic foods]		

[a]Cooking is carried out at atmospheric pressure and unless otherwise specified is done in water.
[b]Time of cooking is not specified, and in most cases, cooking is done until the legume or cereal attains the desired softness.
[c]Soaking is carried out in water unless otherwise specified.
[d]Soaking temperature is not specified.
[e]Temperature of cooking/roasting not specified.
[f]Salt solutions: (1) 2.0% sodium chloride, 0.75% sodium bicarbonate, 0.25% sodium carbonate and 1% sodium tripolyphosphate in DI water; and (2) 1.0% sodium chloride, 0.8% sodium citrate in DI water.
[g]Mixed salt solution—1.0% sodium chloride, 0.8% sodium bicarbonate, 0.5% sodium carbonate and 1% sodium tripolyphosphate in DI water.

cooking method, as far as one can determine, that has not yet been systematically investigated, is the extent of phytate removal if one discarded and changed the cooking water several times. Such treatment in addition to increased phytate removal may, of course, result in additional nutrient losses.

4. GERMINATION

Sprouting and germination are commonly used processing methods for improving the eating quality of cereals and legumes. At home, cereals and legumes are often germinated (or sprouted) prior to their use in salads and other cooked dishes. In preparing alcoholic beverages from certain cereals such as barley, wheat, and sorghum, germination precedes the malting and fermentation steps. During germination, reserve nutrients are mobilized to provide for the growth of roots and shoots. Phytate, being a major source of phosphorus, is hydrolyzed by endogenous enzymes to release inorganic phosphorus. As can be seen from Table 10.3, phytate removal may be quantitative or may be marginal (<3%) depending on the seed, germination conditions, and the duration of germination.

The variable degree of phytate hydrolysis during germination is not only due to germination conditions but also to variable amounts of endogenous enzymes (phytases and phosphatases) capable of hydrolyzing phytate. Although not reported in many studies, the temperature during germination may significantly influence the extent of phytate hydrolysis (especially by phytases), because most phytases have optimum temperatures >40°C.

TABLE 10.3. **Effects of Germination and Malting on Phytate Content of Legumes and Cereals.**

Common Name/Scientific Name	Phytate Reduction (%)	Reference(s)
LEGUMES		
Pigeon pea		
Germination (120 h, 25°C)	20.2	[5]
Germination (72 h, 25°C)	30.0–65.8	[57,30[a,c]]
[soaked[b] 12–24 h, 25°C]		
Germination (72 h)	33.0[a,c]	[30]
[soaked 24 h and dehulled]		
Chickpea		
Germination (24– 68 h, 25–30°C)	18.9–64.1	[35,55,57,
[soaked 5–12 h, 25– 37°C]		77,78[c]]
Lentils		
Germination (24–144 h, 20–30°C)	31.0–81.8	[21,55,77,79]
[soaked 6–10 h, 25–30°C]		
Cowpea		
Germination (72 h, 27°C)	37.7–40.7	[37]
Germination (24–90 h, 25–32°C)	0.0–67.6	[55,77,80]
[soaked 6–10 h, 25–32°C]		
Garden pea		
Germination (120 h, 25°C)	69.4	[77]
Dwarf Grey pea		
Germination (120 h, 22°C)	21.7	[81]
Early Alaska pea		
Germination (120 h, 22°C)	47.9	[81]
Soybeans		
Germination (72 h, 27°C)	30.9–33.3	[37]
Germination (48–120 h, 20–35°C)	5.2–76.3	[57,78[c],81,
[soaked 3–12 h, 25–30°C]		82,83]
Germination (120 h, 20–35°C)	59.2–90.3[c]	[83]
(γ irradiated 0.05–0.2 kGy and		
soaked 3 h)		
Black gram		
Germination (24–240 h, 25–37°C)	34.5–54.0	[35,44,55,
[soaked 10–12 h, 25–37°C]		57,84]
Mung beans		
Germination (48–120 h, 24–32°C)	30.2–69.5	[28,43,57,80]
[soaked 6–12 h, 24–37°C]		
Amphidiploids (black gram—		
mung bean)		
Germination (60 h, 25°C)	39.0–49.0	[45]
[soaked 12h, 37°C]		
Black-eyed bean		
Germination (120 h, 24°C)	77.5	[28]
[soaked 12 h, 24°C]		
Black bean		
Germination (24–48 h, 25–30°C)	30.6–64.1	[55,78[c]]
[soaking 5–10 h, 25–30°C]		

TABLE 10.3. (continued).

Common Name/Scientific Name	Phytate Reduction (%)	Reference(s)
Lima beans		
Germination (72 h, 27°C)	26.8–32.9	[37]
Red kidney beans		
Germination (120 h, 24°C)	35.9	[28]
[soaked 12 h, 24°C]		
Pink beans		
Germination (120 h, 24°C)	46.5	[28]
[soaked 12 h, 24°C]		
Great Northern beans		
Germination (120 h, 25°C)	57.8	[85]
[soaked 16 h, 25°C]		
Pea		
Germination (72–192 h, 25°C)	19.8–80.0	[38,77,86]
Faba bean		
Germination (48–144 h, 20–37°C)	45.0–69.0	[22,24c,64]
(soaked 6–12 h, 25–37°C)		
Germination (72–192 h, 25°C)	15.0–78.0	[25,86]
Kidney beans		
Germination (72 h, 25°C)	30.2	[25]
Chinese indigenous legume		
(*Phaseolus angularis*)		
Germination (48 h, 30°C)	42.6	[31]
[soaked 12 h, 30°C]		
Chinese indigenous legume		
(*Phaseolus angularis*)		
Germination (48 h, 30°C)	55.4	[31]
[soaked 12 h, 30°C]		
Moth bean		
Germination (24–60 h, 25–30°C)	48.0–66.4	[39,55]
[soaked 10–12 h, 25–37°C]		
Bambara groundnut		
Germination (72–120 h, 25°C)	23.0–58.5c	[30a,82]
[soaked 6–24 h]		
Germination (72 h)	25.0a,c	[30]
[soaked 23 h and dehulled]		
Rapeseed		
Germination (48–240 h, 20–22°C)	13.0–68.9	[46,87]
[soaked 24 h, 20–24°C]		
Lupin seeds		
Germination (96–192 h, 20–25°C)	13.0–42.0	[86,88]
Lathyrus beans		
Germination (36 h, 30°C)	33.0	[26]
[soaked 12 h, 37°C]		

(*continued*)

TABLE 10.3.   (continued).

Common Name/Scientific Name	Phytate Reduction (%)	Reference(s)
Cereals		
Rice		
Germination (96 h, 32°C)	66.3	[80]
[soaked 6 h, 32°C]		
Germination (18 days, 14°C)	94.3	[89]
Pearl millet		
Germination (24 h, 30°C)	62.1	[90]
[soaked 12 h, 30°C]		
Germination (48 h, 25–35°C)	24.5–91.3	[91]
Finger millet		
Germination (96 h, 30°C)	94.4	[92]
Germination (24 h, 30°C)	5.7	[93]
[soaked 12 h, 30°C]		
Wheat		
Malting (2 h, 55°C, pH 4.5)	95.0	[94]
[soaked 9 h, 23°C and germinated		
30–40 h, 15°C]		
Sorghum		
Germination (96 h)	68.3–86.9[a,c]	[49]
(soaked 12 h)		
Corn		
Germination (96 h, 32°C)	59.79	[80]
[soaked 6 h, 32°C]		
Barley		
Germination (24–48 h, 30°C)	4.1–16.0[c]	[78]
[soaked 5 h]		
Malting (2 h, 55°C, pH 4.5)	99.0	[94]
[soaked 9 h, 23°C and germinated		
30–40 h, 15°C]		
Oats		
Malting (20 h, 55°C, pH 4.5)	64.0	[94]
[soaked 9 h, 23°C and germinated		
30–40 h, 15°C]		
Malting (17 h, 37–40°C)	98.0	[94]
[soaked 7 h, 23°C and germinated		
120 h, 11°C]		
Rye		
Malting (2 h, 55°C, pH 4.5)	95.0	[94]
[soaked 9 h, 23°C and germinated		
30–40 h, 15°C]		

[a]Incubation temperature for germination is not specified.
[b]Soaking is done in water unless otherwise specified.
[c]Soaking temperature is not specified.

TABLE 10.4. Effect of Extrusion Processing on Phytate Content of Legumes and Cereals.

Common Name/Scientific Name	Phytate Reduction (%)	Reference(s)
Legumes		
Pea, Spanish	5.9–13.5	[38]
Faba beans	26.7	[25]
Kidney beans	21.4	[25]
Navy beans	34.5–48.1	[54]
Chickpeas	21.4–70.5	[54]
Cowpeas	63.5–76.8	[54]
Lentils	46.8–59.3	[54]
Cereals		
Oat bran	0.0	[95]
Rice bran	0.0	[95]
Wheat bran	0.0	[95]

5. EXTRUSION

Extrusion cooking is a widely used technique in food manufacture, especially for cereal processing (such as snack foods and breakfast cereals), and is extremely versatile with respect to automation, production capacities, ingredient selection, and shapes and textures of extruded products [12]. Food material is typically exposed to very high temperatures and shear forces during extrusion. One may, therefore, expect a significant hydrolysis of phytate during extrusion processing. However, because the duration of exposure of food to high temperature and shear is short (usually only a few minutes), phytate reduction (<30%) during extrusion is usually small (Table 10.4).

6. DEHULLING

The effectiveness of dehulling in removing phytate is dependent on the type of seed that is being processed as well as the morphological distribution of seed phytate. For example, wheat and rice endosperms are almost devoid of phytate, as phytate is concentrated in the germ and aleurone layers (pericarp) of the cells of the kernel. On the other hand, in corn, the majority of phytate (88%) is present in the germ. In pearl millets, phytate is distributed in germ and bran fractions. In legumes, the majority of phytate is distributed mainly in the cotyledons (mainly within the protein bodies). Consequently, simple dehulling may be effective in removing significant amounts of phytate only in seeds, where most of the phytate is present in bran or seed coat (Table 10.5). Removing the germ portion is an effective way to remove a significant amount of phytate from corn. Such methods are obviously not very effective in removing phytate

TABLE 10.5. Effects of Dehulling on Phytate Content of Legumes and Cereals.

Common Name/Scientific Name	Phytate Reduction (%)	Reference(s)
Legumes		
Pea, Spanish	−6.7−−13.7	[38]
Faba beans	−9.7	[25]
Kidney beans	−1.89	[25]
Sanilac beans	−6.9	[96]
Great Northern beans	−59.8	[96]
Small white beans	−40.5	[96]
Cranberry beans	−28.9	[96]
Viva pink beans	−34.7	[96]
Pinto beans	−7.6	[96]
Light red kidney beans	−31.9	[96]
Dark red kidney beans	−28.3	[96]
Small red beans	−47.3	[96]
Black beauty beans	−23.2	[96]
Chickpea	−21.5−0.0	[56]
Cereals		
Pearl millet	30.2–36.6	[97]
Proso millet	27.0–53.0	[98]
Sorghum (80% extraction)	29.5–40.4	[49]

Note: the negative sign indicates the increase in phytate content.

from legumes. As the seed coat contributes substantial weight to the whole seed weight, the removal of the seed coat can result in an increase in phytate content on a unit weight basis (Table 10.5). In legumes, milling followed by protein bodies separation (air-classification) may remove a substantial amount of phytate. Such treatment may also contribute to significant protein losses.

7. FERMENTATION AND BREAD MAKING

Many cereals and legumes are extensively used in the preparation of a variety of fermented foods [13,14]. Microbes used in these fermentations may be natural microflora commonly found in the cereal/legume that are fermented or specially cultivated cultures designed to bring about specific changes in the cereal/legume that is being fermented. The type of microorganism, the fermentation conditions used, and the starting amount of phytate present in the raw material significantly affect the extent of phytate removal (Table 10.6). The efficiency of the microbial enzymes controls the rate and extent of phytate removal during fermentation. Numerous studies indicate that phytate hydrolysis during fermentation significantly improves bioavailability of minerals (calcium, magnesium, copper, zinc, and iron). Phytate hydrolysis occurs throughout the different stages of bread making and obviously depends on the type of bread

TABLE 10.6. Effects of Fermentation and Bread Making on Phytate Content of Legumes and Cereals.

Common Name/Scientific Name	Phytate Reduction (%)	Reference(s)
Legumes		
Mung beans		
Lactobacillus fermentation (24 h, 30°C) (soaked[b] 16 h, 25°C)	26.4	[57]
Soybeans		
Lactobacillus fermentation (24 h, 30°C) (soaked 16 h, 25°C)	32.3	[57]
Natural fermentation (72 h, AT[a])	61.5	[52]
Tempeh	30.0–38.5	[41,99,100,119]
Akla/Koose (fermented doughnut)	82.1	[52]
Pigeon pea		
Lactobacillus fermentation (24 h, 30°C) (soaked 16 h, 25°C)	53.8	[57]
Chickpea		
Lactobacillus fermentation (24 h, 30°C) (soaked 16 h, 25°C)	39.1	[57]
Dhokla[g]	97.18	[101]
Khaman[g]	58.18	[101]
Locust beans		
Dawadawa (African fermented food)	39.2	[102]
Dry beans		
Lactobacillus fermentation (72 h, 37°C)	20.1	[32]
Lactobacillus fermentation (72 h, 37°C) [soaked 12 h, 20°C]	10.7	[32]
Lactobacillus fermentation (72 h, 37°C) [cooked 1.5 h, 100°C]	37.5	[32]
Lactobacillus fermentation (72 h, 37°C) [soaked 12 h, 20°C and cooked 1 h, 100°C]	41.48	[32]
Tempeh	42.9	[103]
Cowpea		
Natural fermentation (72 h, AT)	64.7	[52]
Akla/Koose (fermented doughnut)	74.9	[52]
Black gram		
Lactobacillus fermentation (24 h, 30°C) (soaked 16 h, 25°C)	30.4	[57]
Natural fermentation (28 h, 20–35°C)	0.8–42.6	[104,]
Natural fermentation (45–60 h, 30–35°C) [soaked 2–12 h, 37°C]	13.3–54.0	[41,105[c]]
Black gram and rice (1:1 blend)		
Natural fermentation (45 h, 30°C) [soaked 2 h]	48.8[c]	[105]
Idli[g]	40.0	[101]
Brown beans		
Natural fermentation (24–72 h, 21–37°C) [ground beans soaked 17 h, 37°C]	66.0–74.0	[29]

(*continued*)

TABLE 10.6. (continued).

Common Name/Scientific Name	Phytate Reduction (%)	Reference(s)
Natural fermentation (24–72 h, 21–37°C) [whole beans soaked 17 h in 37°C]	63.0–87.0	[29]
Lentils		
Natural fermentation (96 h, 28–42°C)	15.0–70.2	[106,107]
African oil bean seed		
Natural fermentation (120 h) [cooked 8–10 h, 100°C and dehulled]	75.9[e]	[108]
Lathyrus beans		
Natural fermentation [soaked 12 h, 37°C and pressure cooked 15 min, 121°C]	20.0–30.0[d,e]	[26]
Peanut		
Oncom (peanut press cake)		
R. oligosporus fermentation (72 h, 30°C)	96.3	[109]
Neurospora fermentation (72 h, 30°C).	48.5	[109]
Cereals		
Rice		
Natural fermentation (8–72 h, 20°C)	80.4	[52]
Natural fermentation (8 h, 30°C) [soaked 2 h]	100.0[c]	[105]
Wheat		
Yeast fermentation (2 h, 50°C, pH 5.2) [flour extracted with 2.4% HCl,1 h]	0.0–20.1	[110]
Rabadi[f] fermentation (48 h, 30–40°C)	62.5–75.0	[111]
Finger millet		
Natural fermentation (72 h, 30°C) [soaked 12 h, 30°C and germinated 24 h, 30°C]	60.0	[93]
Pearl millet		
Natural fermentation (14 h, 37°C)	43.0–44.0	[97]
Natural fermentation (72 h, 28–30°C) [Damirga process]	86.0–93.0	[97]
Lactobacillus fermentation (72 h, 30°C) (pressure cooked 15 min, 121°C)	43.9–46.9	[73]
Yeast fermentation (72 h, 30°C) (pressure cooked 15 min, 121°C)	40.9–42.4	[73]
Lactobacillus + Yeast mixed fermentation (72 h, 30°C) [pressure cooked 15 min, 121°C]	41.4–51.5	[73]
Lactobacillus + Yeast mixed fermentation (72 h, 30°C) [soaked 12 h, 30°C, germinated 24 h, 30°C and pressure cooked 20 min, 121°C]	98.5–100.0	[90]
Yeast + Lactobacillus sequential fermentation (72 h, 30°C) [pressure cooked 15 min, 121°C]	50.7–100.0	[74]
Rabadi fermentation (9 h, 35°C)	27.0–30.0	[112]

TABLE 10.6. (continued).

Common Name/Scientific Name	Phytate Reduction (%)	Reference(s)
Barley		
Rabadi fermentation (48 h, 30–40°C) [raw flour]	53.0–74.0	[75]
Rabadi fermentation (48 h, 30–40°C) [pressure cooked 15 min, 121°C]	56.0–78.9	[75]
Sorghum		
Natural fermentation (12–72 h, AT) (starter culture)	52.0–60.0	[49[d],52]
Koko (fermented porridge)	86.3	[52]
Corn		
Natural fermentation (72 h, AT)	55.2–58.5	[52]
Kenkey (fermented dumpling)	17.4–60.0	[52,113]
Koko (fermented porridge)	65.1–85.7	[52]
Agidi/Eko (fermented pudding)	83.7–86.5	[52]
Bread making		
Brown breads[k,h]	22.0–58.0	[114]
White breads[h]	66.0–100.0	[2,114[k,l]]
Bran breads (unleavened)	0.0	[114]
Whole-meal breads (unleavened)	8.9	[114]
Whole-meal breads[l,h]	30.0–48.0	[114]
Whole wheat pup loaves	22.0	[115]
White pup loaves	66.0	[115]
Pakistani flat breads[j] (Chapati, Roti, Nan, puri)	44.2–85.0	[68,76]
Soy-fortified bread (yeast 0-15 g/lb loaf)	45.6–87.9	[116]
Saudi Arabic breads[i]	11.0–46.0	[59]
Arabic high bran breads	23.0–32.4	[117]
Polish wheat bread	63.5	[118]
Polish rye bread	79.5	[118]
Polish high gluten bread	22.2	[118]
Polish high fiber bread	37.8	[118]
Iranian flat breads	13.0–39.0	[2]

[a]AT refers to ambient temperature (actual temperature is not specified).
[b]Soaking is done in water unless otherwise specified.
[c]Soaking temperature is not specified.
[d]Fermentation temperature is not specified.
[e]Fermentation time is not indicated.
[f]Rabadi—traditional fermented Indian food (cereal + butter milk).
[g]Traditional Indian foods (fermented and steamed).
[h]Bread making include mixing, kneading, proving, baking and cooling.
[i]Saudi Arabic breads—Mafrood, Samouli, Toast, Tamees, Burr.
[j]Unleavened and leavened breads.
[k]Homemade yeast breads with proprietary yeast bread mixes.
[l]Homemade yeast breads with conventional ingredients.

being made (Table 10.6). Some of the factors that significantly affect phytate hydrolysis include the following:

(1) *Flour*: type, freshness, extraction rate
(2) *Yeast*: presence/absence (depending on type of bread), amount
(3) *Dough*: pH, water content, fermentation time
(4) *Baking conditions*: leavening time, temperature
(5) *Additives*: calcium and magnesium salts, sodium bicarbonate

The ability of different microorganisms to hydrolyze phytate has been investigated to some extent [2,15], but much remains to be done. For example, to date, no one knows which microorganisms produce the most efficient and economical phytases/phosphatases that can be used in these fermented foods. Molecular understanding of such phytases/phosphatases in microorganisms may help in the development of treatments that can effect phytate removal from a certain food without the need for fermentation. Identification and development of a thermostable enzyme capable of withstanding temperatures of 100°C or higher may be especially useful in this regard, because cereals and legumes are often cooked or otherwise exposed to high temperatures (such as in extrusion processing) prior to consumption. The availability of thermostable phytate hydrolyzing enzymes will enable the consumer/processor to include the enzymes in a uniform manner with the food being processed as it is exposed to heat treatment, obviating the need to incur extra processing steps. With the availability of rapid and inexpensive molecular techniques, it is now possible to screen several enzymes simultaneously to permit identification of thermostable phytases. However, such investigations have not yet been undertaken. When phytate removal is desired, the use of tailor-made phytate hydrolyzing enzymes may prove to be the most robust, efficient, and economical means.

8. STORAGE

A number of studies indicate an appreciable decrease in phytate content during storage (Table 10.7). The extent of such reduction depends on the type of seed, storage conditions (especially relative humidity and temperature), and the age of the seed (fresh pods versus dry seeds). There are three possible mechanisms that may explain phytate reduction during storage.

(1) Phytate may form insoluble complexes with other food components (such as proteins), and therefore, phytate extraction may be incomplete, resulting in lower values.
(2) If the water activity is sufficient and if the storage temperature is sufficiently high, endogenous seed phytases/phosphatases may be activated, which may account for partial loss of phytate.

TABLE 10.7. Effects of Storage Conditions on Phytate Content
of Legumes and Cereals.

Common Name/Scientific Name	Phytate Reduction (%)	Reference(s)
Legumes		
Chickpea		
Storage (3–12 months, 5– 37°C)	2.6–56.4	[120,121]
Pigeon pea		
Storage (3–12 months, 5– 37°C)	7.2–42.3	[120]
Mung bean		
Storage (3–12 months, 5– 37°C)	2.9–37.3	[120]
Black gram		
Storage (3–12 months, 5– 37°C)	2.6–40.3	[120]
Soybean		
Storage (3–12 months, 5– 37°C)	2.5–29.0	[120]
Tempeh		
Fresh stored (2 weeks, 5°C)	72.5–76.0	[100]
Fried (2 weeks, 5°C)	72.2–76.3	[100]
Beans (*Phaseolus vulgaris*)		
Fresh whole/ground beans (14 months, RT[a], AH[b])	0.0	[122]
Fresh whole/ground beans (14 months, –23°C)	0.0	[122]
Whole/ground beans (4 months, 41°C, 75% RH) [prestored 7–11 months, RT]	17.0–28.7	[122]
Whole/ground beans (4 months, 41°C, 75% RH) [prestored 7–11 months, –23°C]	22.4–27.6	[122]
Whole/ground beans (4 months, 41°C, dry state) [prestored 7–11 months, RT]	22.8	[122]
Whole/ground beans (4 months, 41°C, 75% RH) [prestored 7–11 months, –23°C]	21.4	[122]
Kidney beans		
Storage (9 months, 2–32°C, 12.5–17.9% moisture content)	0.0–65.0	[6]
Dry beans		
Storage (2–6 months, 4–5°C)	16.5–20.6	[123]
Storage (1–6 months, 30–41°C, 75% RH)	9.3–63.4	[33,123]
Storage (2–6 months, 4–5°C) [γ irradiated, 2 kGy]	19.1–33.3	[123]
Storage (2–6 months, 30°C, 75% RH) [γ irradiated, 2 kGy]	8.6–49.5	[123]
Storage (2–6 months, 4–5°C) [microwave treated, 2 min]	21.3–31.5	[123]
Storage (2–6 months, 30°C, 75% RH) [microwave treated, 2 min]	13.9–38.9	[123]
Cereals		
Barley		
Kernels (3 months, 41°C, 75% RH)	4.7–9.9	[122]
Kernels (3 months, 41°C, dry)	0.7–2.0	[122]

[a]RT refers to room temperature (actual temperature is not indicated).
[b]AH refers to ambient humidity.

TABLE 10.8. Effects of Radiation on Phytate Content of Soybeans.

Common Name/Scientific name	Phytate Reduction (%)	Reference(s)
Gamma irradiation (0.05–1.0 kGy)	3.3–51.8	[42,83]
Gamma irradiation (20–100 kGy) [seeds with 7.5% moisture content]	1.6–4.0	[16]
Gamma irradiation (56–65 kGy) [seeds with 22.48 and 30.47% moisture content]	1.1–6.0	[16]
Microwave heating (15 min) [seeds with 7.5% moisture content]	45.9	[16]

(3) At sufficiently high temperature and water activity, microbial growth on seeds (especially yeasts and molds) may permit phytate hydrolysis in infected seeds due to microbial phytases/phosphatases.

Seed storage under dry and cool conditions is normally not expected to affect seed phytate content in any manner, perhaps with the exception of partial loss of solubility on extended storage.

9. RADIATION

Few studies have reported radiation to be effective in phytate removal from seeds and grains (Table 10.8). A closer examination of these studies indicates that the removal of phytate probably stems from the other treatments (such as germination, soaking, etc.) given to the test samples. For example, Hafez et al. [16] noted that phytate reduction was <6% when soybeans (moisture content 7.5–30.5%) were exposed to γ irradiation doses ranging from 20–100 kGy. In the same study, microwave treatment of the same soybeans (7.5% moisture content) caused a significant reduction (46%) in phytate. Microwave treatment causes the samples to heat up, while γ irradiation typically does not raise the sample temperature significantly. Consequently, any reduction in phytate as a result of exposure to radiation must be evaluated with caution.

10. REFERENCES

1. Zhou, J.R. and Erdman, J.W. 1995. Phytic acid in health and disease. *Crit. Rev. Food Sci. Nutr.* 35:495–508.
2. Reddy, N.R., Pierson, M.D., Sathe, S.K. and Salunkhe, D.K. 1989. *Phytates in Cereals and Legumes.* CRC Press Inc., Boca Raton, Florida.

3. Bergman, E.L., Autio, K. and Sandberg, A.S. 2000. Optimal conditions for phytate degradation, estimation of phytase activity, and localization of phytate in barley (cv. Blenheim). *J. Agric. Food Chem.* 48:4647–4655.

4. Viveros, A., Centeno, C., Brenes, A., Canales, R. and Lozano, A. 2000. Phytase and acid phosphatase activities in plant feedstuffs. *J. Agric. Food Chem.* 48:4009–4013.

5. Mattson, S. 1946. The cookability of yellow peas: a colliod chemical and biochemical study. *Acta Agric. Sci.* 2:185.

6. Moscoso, W., Bourne, M.C. and Hood, L.F. 1984. Relationships between the hard-to-cook phenomenon in red kidney beans and water-absorption, puncture force, pectin, phytic acid, and minerals. *J. Food Sci.* 49:1577–1583.

7. Crean, D.E.C. and Haisman, D.R. 1963. The interaction between phytic acid and divalent cations during the cooking of dried peas. *J. Sci. Food Agric.* 14:824–833.

8. El-Saied, H.M. and Abdel-Hamid, A.E. 1981. Cooking-quality of broad-bean varieties as influenced by some physicochemical measurements. *Z. Ernahrungswiss.* 20:200–207.

9. Henderson, H.M. and Ankrah, S.A. 1985. The relationship of endogenous phytase, phytic acid and moisture uptake with cooking time in *Vicia faba*-minor cv aladin. *Food Chem.* 17: 1–11.

10. Proctor, J.P. and Watts, B.M. 1987. Effect of cultivar, growing location, moisture and phytate content on the cooking times of freshly harvested navy beans. *Can. J. Plant Sci.* 67:923–926.

11. Rosenbaum, T.M. and Baker, B.E. 1969. Constitution of leguminous seeds VII. Ease of cooking field peas (*Pisum sativum*) in relation to phytic acid content and calcium diffusion. *J. Sci. Food Agric.* 20:709–712.

12. Cheftel, J.C. 1990. Extrusion cooking: operating principles, research trends and food applications. 1. In *Processing and Quality of Foods High Temperature/Short Time (HTST) Processing: Gaurantee for High Quality Food with Long Shelflife*, 1.60–1.73 Elsevier Applied Science, London.

13. Reddy, N.R., Pierson, M.D. and Salunkhe, D.K. 1986. *Legume-Based Fermented Foods*. CRC Press, Boca Raton, Florida.

14. Steinkraus, K.H. 1983. *Handbook of Indigenous Fermented Foods*. Marcel Dekker, New York, New York.

15. Lopez, H.W., Ouvry, A., Bervas, E., Guy, C., Messager, A., Demigne, C. and Remesy, C. 2000. Strains of lactic acid bacteria isolated from sour doughs degrade phytic acid and improve calcium and magnesium solubility from whole wheat flour. *J. Agric. Food Chem.* 48:2281–2285.

16. Hafez, Y.S., Mohammed, A.I., Perera, P.A., Singh, G. and Hussein, A.S. 1989. Effects of microwave-heating and gamma-irradiation on phytate and phospholipid contents of soybean (*Glycine max* L). *J. Food Sci.* 54:958–962.

17. Vijayakumari, K., Siddhuraju, P. and Janardhanan, K. 1996. Effect of soaking, cooking and autoclaving on phytic acid and oligosaccharide contents of the tribal pulse, *Mucuna monosperma* DC ex Wight. *Food Chem.* 55:173–177.

18. Vijayakumari, K., Siddhuraju, P. and Janardhanan, K. 1996. Effect of different post-harvest treatments on antinutritional factors in seeds of the tribal pulse, *Mucuna pruriens* (L) DC. *Int. J. Food Sci. Nutr.* 47:263–272.

19. Vijayakumari, K., Siddhuraju, P. and Janardhanan, K. 1997. Effect of domestic processing on the levels of certain antinutrients in *Prosopis chilensis* (Molina) Stunz seeds. *Food Chem.* 59:367–371.

20. Vijayakumari, K., Siddhuraju, P., Pugalenthi, M. and Janardhanan, K. 1998. Effect of soaking and heat processing on the levels of antinutrients and digestible proteins in seeds of *Vigna aconitifolia* and *Vigna sinensis*. *Food Chem.* 63:259–264.

21. Vidal-Valverde, C., Frias, J., Estrella, I., Gorospe, M.J., Ruiz, R. and Bacon, J. 1994. Effect of processing on some antinutritional factors of lentils. *J. Agric. Food Chem.* 42:2291–2295.

22. Vidal-Valverde, C., Frias, J., Sotomayor, C., Diaz-Pollan, C., Fernandez, M. and Urbano, G. 1998. Nutrients and antinutritional factors in faba beans as affected by processing. *Z. Lebensm. Unters. Forsch.* 207:140–145.

23. Fernandez, M., Aranda, P., LopezJurado, M., GarciaFuentes, M.A. and Urbano, G. 1997. Bioavailability of phytic acid phosphorus in processed *Vicia faba* L. var. Major. *J. Agric. Food Chem.* 45:4367–4371.

24. Sharma, A. and Sehgal, S. 1992. Effect of processing and cooking on the antinutritional factors of faba bean (*Vicia-faba*). *Food Chem.* 43:383–385.

25. Alonso, R., Aguirre, A. and Marzo, F. 2000. Effects of extrusion and traditional processing methods on antinutrients and *in vitro* digestibility of protein and starch in faba and kidney beans. *Food Chem.* 68:159–165.

26. Srivastava, S. and Khokhar, S. 1996. Effects of processing on the reduction of beta-ODAP (beta-N-oxalyl-L-2,3-diaminopropionic acid) and anti-nutrients of khesari dhal, *Lathyrus sativus. J. Sci. Food Agric.* 71:50–58.

27. Iyer, V., Salunkhe, D.K., Sathe, S.K. and Rockland, L.B. 1980. Quick-cooking beans (*Phaseolus vulgaris* L). 2. Phytates, oligosaccharides, and anti-enzymes. *Qualitas Plantarum-Plant Foods Hum. Nutr.* 30:45–52.

28. Tabekhia, M.M. and Luh, B.S. 1980. Effect of germination, cooking, and canning on phosphorus and phytate retention in dry beans. *J. Food Sci.* 45:406–408.

29. Gustafsson, E.L. and Sandberg, A.S. 1995. Phytate reduction in brown beans (*Phaseolus vulgaris* L). *J. Food Sci.* 60:149–152, 156.

30. Igbedioh, S.O., Olugbemi, K.T. and Akpapunam, M.A. 1994. Effects of processing methods on phytic acid level and some constituents in bambara groundnut (*Vigna-subterranea*) and pigeon pea (*Cajanus cajan*). *Food Chem.* 50:147–151.

31. Chau, C.F. and Cheung, P.C.K. 1997. Effect of various processing methods on antinutrients and *in vitro* digestibility of protein and starch of two Chinese indigenous legume seeds. *J. Agric. Food Chem.* 45:4773–4776.

32. Barampama, Z. and Simard, R.E. 1994. Oligosaccharides, antinutritional factors, and protein digestibility of dry beans as affected by processing. *J. Food Sci.* 59:833–838.

33. Bernal-Lugo, I., Castillo, A., Deleon, F.D., Moreno, E. and Ramirez, J. 1991. Does phytic acid influence cooking rate in common beans? *J. Food Biochem.* 15:367–374.

34. Deshpande, S.S. and Cheryan, M. 1983. Changes in phytic acid, tannins, and trypsin inhibitory activity on soaking of dry beans (*Phaseolus vulgaris* L). *Nutr. Rep. Int.* 27:371–377.

35. Duhan, A., Chauhan, B.M., Punia, D. and Kapoor, A.C. 1989. Phytic acid content of chickpea (*Cicer arietinum*) and black gram (*Vigna mungo*)—varietal differences and effect of domestic processing and cooking methods. *J. Sci. Food Agric.* 49:449–455.

36. Khan, N., Zaman, R. and Elahi, M. 1988. Effect of processing on the phytic acid content of bengal grams (*Cicer arietinum*) products. *J. Agric. Food Chem.* 36:1274–1276.

37. Ologhobo, A.D. and Fetuga, B.L. 1984. Distribution of phosphorus and phytate in some Nigerian varieties of legumes and some effects of processing. *J. Food Sci.* 49:199–201.

38. Alonso, R., Orue, E. and Marzo, F. 1998. Effects of extrusion and conventional processing methods on protein and antinutritional factor contents in pea seeds. *Food Chem.* 63:505–512.

39. Khokhar, S. and Chauhan B.M. 1986. Antinutritional factors in moth bean (*Vigna aconitifolia*): Varietal differences and effects of methods of domestic processing and cooking. *J. Food Sci.* 51:591–594.

40. Chang, R., Schwimmer, S. and Burr, H.K. 1977. Phytate: Removal from whole dry beans by enzymatic hydrolysis and diffusion. *J. Food Sci.* 42:1098–1101.

41. Sudarmadji, S. and Markakis, P. 1977. The phytate and phytase of soybean tempeh. *J. Sci. Food Agric.* 28:381–383.

42. Sattar, A., Neelofar and Akhtar, M.A. 1990. Effect of radiation and soaking on phytate content of soybean. *Acta. Aliment. Hung.* 19:331–336.

43. Kataria, A., Chauhan, B.M. and Punia, D. 1989. Antinutrients and protein digestibility (*in vitro*) of mungbean as affected by domestic processing and cooking. *Food Chem.* 32:9–17.
44. Kataria, A., Chauhan, B.M. and Gandhi, S. 1988. Effect of domestic processing and cooking on the antinutrients of black gram. *Food Chem.* 30:149–156.
45. Kataria, A., Chauhan, B.M. and Punia, D. 1989. Antinutrients in amphidiploids (black gram × mung bean)—varietal differences and effect of domestic processing and cooking. *Plant Foods Hum. Nutr.* 39:257–266.
46. Mahajan, A. and Dua, S. 1997. Nonchemical approach for reducing antinutritional factors in rapeseed (*Brassica campestris* var. Toria) and characterization of enzyme phytase. *J. Agric. Food Chem.* 45:2504–2508.
47. Fredlund, K., Asp, N.G., Larsson, M., Marklinder, I. and Sandberg, A.S. 1997. Phytate reduction in whole grains of wheat, rye, barley and oats after hydrothermal treatment. *J. Cereal Sci.* 25:83–91.
48. Tyagi, P.K., Tyagi, P.K. and Verma, S.V.S. 1998. Effect of water soaking wheat-bran on phytate phosphorus autolysis and its feeding value to chicks. *Indian J. Anim. Sci.* 68:669–671.
49. Mahgoub, S.E.O. and Elhag, S.A. 1998. Effect of milling, soaking, malting, heat treatment and fermentation on phytate level of four Sudanese sorghum cultivars. *Food Chem.* 61:77–80.
50. Pawar, V.D. and Parlikar, G.S. 1990. Reducing the polyphenols and phytate and improving the protein quality of pearl-millet by dehulling and soaking. *J. Food Sci. Technol.* (India) 27:140–143.
51. Kumar, K.G., Venkataraman, L.V., Jaya, T.V. and Krishnamurthy, K.S. 1978. Cooking characteristics of some germinated legumes—changes in phytins, ca^{++}, mg^{++} and pectins. *J. Food Sci.* 43:85–88.
52. Marfo, E.K., Simpson, B.K., Idowu, J.S. and Oke, O.L. 1990. Effect of local food processing on phytate levels in cassava, cocoyam, yam, maize, sorghum, rice, cowpea, and soybean. *J. Agric. Food Chem.* 38:1580–1585.
53. Uzogara, S.G., Morton, I.D. and Daniel, J.W. 1990. Changes in some antinutrients of cowpeas (*Vigna unguiculata*) processed with kanwa alkaline salt. *Plant Foods Hum. Nutr.* 40:249–258.
54. Ummadi, P., Chenoweth, W.L. and Uebersax, M.A. 1995. The influence of extrusion processing on iron dialyzability, phytates and tannins in legumes. *J. Food Process. Preserv.* 19:119–131.
55. Agte, V., Joshi, S., Khot, S., Paknikar, K. and Chiplonkar, S. 1998. Effect of processing on phytate degradation and mineral solubility in pulses. *J. Food Sci. Technol.* (India) 35:330–332.
56. Attia, R.S., Shehata, A.M.E., Aman, M.E. and Hamza, M.A. 1994. Effect of cooking and decortication on the physical properties, the chemical composition and the nutritive value of chickpea (*Cicer arietinum* L). *Food Chem.* 50:125–131.
57. Chitra, U., Singh, U. and Rao, P.V. 1996. Phytic acid, *in vitro* protein digestibility, dietary fiber, and minerals of pulses as influenced by processing methods. *Plant Foods Hum. Nutr.* 49:307–316.
58. Hussain, B., Khan, S., Ismail, M. and Sattar, A. 1989. Effect of roasting and autoclaving on phytic acid content of chickpea. *Nahrung* 33:345–348.
59. Almana, H.A. 2000. Extent of phytate degradation in breads and various foods consumed in Saudi Arabia. *Food Chem.* 70:451–456.
60. Egbe, I.A. and Akinyele, I.O. 1990. Effect of cooking on the antinutritional factors of lima beans (*Phaseolus lunatus*). *Food Chem.* 35:81–87.
61. Sievwright, C.A. and Shipe, W.F. 1986. Effect of storage conditions and chemical treatments on firmness, *in vitro* protein digestibility, condensed tannins, phytic acid and divalent cations of cooked black beans (*Phaseolus vulgaris*). *J. Food Sci.* 51:982–987.
62. Trugo, L.C., Donangelo, C.M., Trugo, N.M.F. and Knudsen, K.E.B. 2000. Effect of heat treatment on nutritional quality of germinated legume seeds. *J. Agric. Food Chem.* 48:2082–2086.

63. Abd El-Moniem, G.M., Honke, J. and Bednarska, A. 2000. Effect of frying various legumes under optimum conditions on amino acids, *in vitro* protein digestibility, phytate and oligosaccharides. *J. Sci. Food Agric.* 80:57–62.

64. Khalil, A.H. and Mansour, E.H. 1995. The effect of cooking, autoclaving and germination on the nutritional quality of faba beans. *Food Chem.* 54:177–182.

65. Saikia, P., Sarkar, C.R. and Borua, I. 1999. Chemical composition, antinutritional factors and effect of cooking on nutritional quality of rice bean [*Vigna umbellata* (Thunb; Ohwi and Ohashi)]. *Food Chem.* 67:347–352.

66. Verma, P. and Mehta, U. 1988. Study of physical characteristics, sensory evaluation and the effect of sprouting, cooking and dehulling on the antinutritional factors of rice bean (*Vigna umbellata*). *J. Food Sci. Technol.*(India) 25:197–200.

67. Fretzdorff, B. and Weipert, D. 1986. Phytic acid in cereals. 1. Phytic acid and phytase in rye and rye products. *Z. Lebensm. Unters. Forsch.* 182:287–293.

68. Khan, N., Zaman, R. and Elahi, M. 1991. Effect of heat treatments on the phytic acid content of maize products. *J. Sci. Food Agric.* 54:153–156.

69. Urizar-Hernandez, A.L. and Bressani, R. 1997. The effect of lime cooking of corn on the phytic acid, calcium, total and ionizable iron content. *Arch. Latinoam. Nutr.* 47:217–223.

70. Toma, R.B. and Tabekhia, M.M. 1979. Changes in mineral elements and phytic acid contents during cooking of 3 California rice varieties. *J. Food Sci.* 44:619–621.

71. Mameesh, M.S. and Tomar, M. 1993. Phytate content of some popular kuwaiti foods. *Cereal Chem.* 70:502–503.

72. Valencia, S., Svanberg, U., Sandberg, A.S. and Ruales, J. 1999. Processing of quinoa (*Chenopodium quinoa*, Wild): effects on *in vitro* iron availability and phytate hydrolysis. *Int. J. Food Sci. Nutr.* 50:203–211.

73. Khetarpaul, N. and Chauhan, B.M. 1989. Effect of fermentation by pure cultures of yeasts and lactobacilli on phytic acid and polyphenol content of pearl millet. *J. Food Sci.* 54:780–781.

74. Khetarpaul, N. and Chauhan, B.M. 1991. Sequential fermentation of pearl-millet by yeasts and lactobacilli effect on the antinutrients and *in vitro* digestibility. *Plant Foods Hum. Nutr.* 41:321–327.

75. Gupta, M. and Khetarpaul, N. 1993. Effect of rabadi fermentation on phytic acid and *in vitro* digestibility of barley. *Nahrung* 37:141–146.

76. Khan, N., Zaman, R. and Elahi, M. 1986. Effect of processing on the phytic acid content of wheat products. *J. Agric. Food Chem.* 34:1010–1012.

77. Belavady, B. and Banerjee, S. 1953. Studies on the effect of germination on the phosphorus values of some common Indian pulses. *Food Res.* 18:223.

78. Trugo, L.C., Muzquiz, M., Pedrosa, M.M., Ayet, G., Burbano, C., Cuadrado, C. and Cavieres, E. 1999. Influence of malting on selected components of soya bean, black bean, chickpea and barley. *Food Chem.* 65:85–90.

79. Ayet, G., Burbano, C., Cuadrado, C., Pedrosa, M.M., Robredo, L.M., Muzquiz, M., delaCuadra, C., Castano, A. and Osagie, A. 1997. Effect of germination, under different environmental conditions, on saponins, phytic acid and tannins in lentils (*Lens culinaris*). *J. Sci. Food Agric.* 74:273–279.

80. Marero, L.M., Payumo, E.M., Aguinaldo, A.R., Matsumoto, I. and Homma, S. 1991. Antinutritional factors in weaning foods prepared from germinated cereals and legumes. *Lebensm. Wiss. U. Technol.* 24:177–181.

81. Chen, L.H. and Pan, S.H. 1977. Decrease of phytates during germination of pea seeds (*Pisium sativa*). *Nutr. Rep. Int.* 16:125–131.

82. Akpapunam, M.A., Igbedioh, S.O. and Aremo, I. 1996. Effect of malting time on chemical composition and functional properties of soybean and bambara groundnut flours. *Int. J. Food Sci. Nutr.* 47:27–33.

83. Sattar, A., Neelofar and Akhtar, M.A. 1990. Irradiation and germination effects on phytate, protein and amino-acids of soybean. *Plant Foods Hum. Nutr.* 40:185–194.
84. Reddy, N.R., Balakrishnan, C.V. and Salunkhe, D.K. 1978. Phytate phosphorus and mineral changes during germination and cooking of black gram (*Phaseolus mungo*) seeds. *J. Food Sci.* 43:540–543.
85. Sathe, S.K., Deshpande, S.S., Reddy, N.R., Goll, D.E. and Salunkhe, D.K. 1983. Effects of germination on proteins, raffinose oligosaccharides, and antinutritional factors in the great northern beans (*Phaseolus vulgaris* L). *J. Food Sci.* 48:1796–1800.
86. Honke, J., Kozlowska, H., Vidal-Valverde, C., Frias, J. and Gorecki, R. 1998. Changes in quantities of inositol phosphates during maturation and germination of legume seeds. *Z. Lebensm. Unters. Forsch.* 206:279–283.
87. Thompson, L.U. and Serraino, M.R. 1985. Effect of germination on phytic acid, protein and fat-content of rapeseed. *J. Food Sci.* 50:1200.
88. delaCuadra, C., Muzquiz, M., Burbano, C., Ayet, G., Calvo, R., Osagie, A. and Cuadrado, C. 1994. Alkaloid, alpha-galactoside and phytic acid changes in germinating lupin seeds. *J. Sci. Food Agric.* 66:357–364.
89. Kikunaga, S., Katoh, Y. and Takahashi, M. 1991. Biochemical changes in phosphorus compounds and in the activity of phytase and alpha-amylase in the rice (*Oryza sativa*) grain during germination. *J. Sci. Food Agric.* 56:335–343.
90. Khetarpaul, N. and Chauhan, B.M. 1990. Effects of germination and pure culture fermentation by yeasts and lactobacilli on phytic acid and polyphenol content of pearl-millet. *J. Food Sci.* 55:1180–1182.
91. Kumar, A. and Chauhan, B.M. 1993. Effects of phytic acid on protein digestibility (*in vitro*) and HCl-extractability of minerals in pearl-millet sprouts. *Cereal Chem.* 70:504–506.
92. Mbithi-Mwikya, S., Van Camp, J., Yiru, Y. and Huyghebaert, A. 2000. Nutrient and antinutrient changes in finger millet (*Eleusine coracana*) during sprouting. *Lebensm. Wiss. U. Technol.* 33:9–14.
93. Sripriya, G., Antony, U. and Chandra, T.S. 1997. Changes in carbohydrate, free amino acids, organic acids, phytate and HCl extractability of minerals during germination and fermentation of finger millet (*Eleusine coracana*). *Food Chem.* 58:345–350.
94. Larsson, M. and Sandberg, A.S. 1992. Phytate reduction in oats during malting. *J. Food Sci.* 57:994–997.
95. Gualberto, D. G., Bergman, C.J., Kazemzadeh, M. and Weber, C.W. 1997. Effect of extrusion processing on the soluble and insoluble fiber, and phytic acid contents of cereal brans. *Plant Foods Hum. Nutr.* 51:187–198.
96. Deshpande, S.S., Sathe, S.K., Salunkhe, D.K. and Cornforth, D.P. 1982. Effects of dehulling on phytic acid, polyphenols, and enzyme inhibitors of dry beans (*Phaseolus vulgaris* L). *J. Food Sci.* 47:1846–1850.
97. Abdalla, A.A., El Tinay, A. H., Mohamed, B.E. and Abdalla, A.H. 1998. Effect of traditional processes on phytate and mineral content of pearl millet. *Food Chem.* 63:79–84.
98. Lorenz, K. 1983. Tannins and phytate content in proso millets (*Panicum miliaceum*). *Cereal Chem.* 60:424–426.
99. Sutardi and Buckle, K.A. 1985. Phytic acid changes in soybeans fermented by traditional inoculum and 6 strains of *Rhizopus oligosporus*. *J. Appl. Bacteriol.* 58:539–543.
100. Sutardi and Buckle, K.A. 1985. Reduction in phytic acid levels in soybeans during tempeh production, storage and frying. *J. Food Sci.* 50:260–262, 263.
101. Reddy, N.R. and Pierson, M.D. 1994. Reduction in antinutritional and toxic components in plant foods by fermentation. *Food Res. Int.* 27:281–290.
102. Eka, O.U. 1980. Effect of fermentation on the nutrient status of locust beans. *Food Chem.* 5:303–308.

103. Paredes-Lopez, O. and Harry, G.I. 1989. Changes in selected chemical and antinutritional components during tempeh preparation using fresh and hardened common beans. *J. Food Sci.* 54:968–970.

104. Yadav, S. and Khetarpaul, N. 1994. Indigenous legume fermentation—effect on some antinutrients and *in vitro* digestibility of starch and protein. *Food Chem.* 50:403–406.

105. Reddy, N.R. and Salunkhe, D.K. 1980. Effects of fermentation on phytate phosphorus and mineral content in black gram, rice, and black gram and rice blends. *J. Food Sci.* 45:1708–1712.

106. Kozlowska, H., Honke, J., Sadowska, J., Frias, J. and Vidal-Valverde, C. 1996. Natural fermentation of lentils: Influence of time, concentration and temperature on the kinetics of hydrolysis of inositol phosphates. *J. Sci. Food Agric.* 71:367–375.

107. Cuadrado, C., Ayet, G., Robredo, L.M., Tabera, J., Villa, R., Pedrosa, M.M., Burbano, C. and Muzquiz, M. 1996. Effect of natural fermentation on the content of inositol phosphates in lentils. *Z. Lebensm. Unters. Forsch.* 203:268–271.

108. Kingsley, M.O. 1995. Effect of processing on some antinutritive and toxic components and on the nutritional composition of the African oil bean seed (*Pentaclethra macrophylla benth*). *J. Sci. Food Agric.* 68:153–158.

109. Fardiaz, D. and Markakis, P. 1981. Degradation of phytic acid in oncom (fermented peanut press cake). *J. Food Sci.* 46:523–525.

110. Harland, B.F. and Frolich, W. 1989. Effects of phytase from 3 yeasts on phytate reduction in Norwegian whole wheat-flour. *Cereal Chem.* 66:357–358.

111. Gupta, M. and Khetarpaul, N. 1993. HCl extractability of minerals from rabadi—a wheat—flour fermented food. *J. Agric. Food Chem.* 41:125–127.

112. Dhankher, N. and Chauhan, B.M. 1987. Effect of temperature and fermentation time on phytic acid and polyphenol content of rabadi—a fermented pearl-millet food. *J. Food Sci.* 52:828–829.

113. Amoa, B. and Muller, H.G. 1976. Studies on kenkey with particular reference to calcium and phytic acid. *Cereal Chem.* 53:365–375.

114. Mckenzieparnell, J.M. and Davies, N.T. 1986. Destruction of phytic acid during home bread-making. *Food Chem.* 22:181–192.

115. Tangkongchitr, U., Seib, P.A. and Hoseney, R.C. 1981. Phytic acid. 2. Its fate during bread-making. *Cereal Chem.* 58:229–234.

116. Ranhotra, G.S., Loewe, R.J. and Puyat, L.V. 1974. Phytic acid in soy and its hydrolysis during bread making. *J. Food Sci.* 39:1023–1025.

117. Dagher, S.M., Shadarevian, S. and Birbari, W. 1987. Preparation of high bran arabic bread with low phytic acid content. *J. Food Sci.* 52:1600–1603.

118. Bartnik, M. and Florysiak, J. 1988. Phytate hydrolysis during breadmaking in several sorts of Polish bread. *Nahrung* 32:37–42.

119. Suparmo and Markakis, P. 1987. Tempeh prepared from germinated soybeans. *J. Food Sci.* 52:1736–1737.

120. Chitra, U. and Singh, U. 1998. Effect of storage on cooking quality characteristics of grain legumes. *J. Food Sci. Technol.* (India) 35:51–54.

121. Gulati, N. and Sood, D.R. 1998. Effect of storage on pectin, phytic acid and minerals in chickpea (*Cicer arietinum* L.). *J. Food Sci. Technol.* (India) 35:342–345.

122. Ockenden, I., Falk, D.E. and Lott, J.N.A. 1997. Stability of phytate in barley and beans during storage. *J. Agric. Food Chem.* 45:1673–1677.

123. Cunha, M.F., Sgarbieri, V.C. and Damasio, M.H. 1993. Effects of pretreatment with gamma-rays or microwaves on storage stability of dry beans. *J. Agric. Food Chem.* 41:1710–1715.

CHAPTER 11

The Antioxidant Effects
of Inositol Phosphates

JOHN R. BURGESS
FENG GAO

1. INTRODUCTION

PHYTATE (*myo*-inositol hexakisphosphate, $InsP_6$) is a food component that is considered an antinutrient by virtue of its ability to chelate divalent minerals and prevent their absorption, but it has also been shown to have anticancer activity [1] and antioxidant activity [2]. It forms an iron chelate that suppresses lipid peroxidation by blocking iron-driven hydroxyl radical generation. This chapter will focus on the antioxidant properties that $InsP_6$ exhibits *in vivo*.

2. ANTIOXIDANT PROPERTIES OF INOSITOL PHOSPHATES *IN VITRO*

Phytate is a very stable and potent chelating agent that exhibits the ability to complex a variety of divalent and trivalent ions. At physiological pH, $InsP_6$ forms complexes with Cu^{2+}, Zn^{2+}, Co^{2+}, Mn^{2+}, Fe^{2+}, and Ca^{2+} [3]. Because phytate binds essential minerals and can prevent their absorption, most human nutritionists view the compound negatively. However, its unique chelating action with iron provides phytate with antioxidant characteristics. In the iron-assisted Haber-Weiss reaction, the formation of •OH requires the availability of at least one reactive iron coordination site, as well as iron solubility.

$$O_2^- + Fe^{3+} \Rightarrow Fe^{2+} + O_2$$

$$Fe^{2+} + H_2O_2 \Rightarrow Fe^{3+} + •OH + OH^- \quad \text{(Fenton Reaction)}$$

Graf and coworkers [4] postulated that phytate would suppress iron-catalyzed oxidative reactions by forming a unique chelate with Fe (III) occupying all six of the coordination sites and displacing water. This complete coordination was confirmed by titration with azide. These researchers also showed that the Fe(III)-phytate complex that forms inhibited lipid peroxidation and hydroxyl radical formation [2]. Additionally, they found that the phytate-iron complexes were much slower than unbound iron at oxidizing ascorbate and that phytate accelerates the oxidation of Fe $^{2+}$ to Fe $^{3+}$. Rimbach and Pallauf [5] recently confirmed inhibition of hydroxyl radical formation by phytate using electron spin resonance spectroscopy with 5,5-dimethyl-1-pyrroline-N-oxide as a spin trap. Although the optimal ratio of phytate to iron for complete inhibition differs between the two research groups, >0.25 for Graf et al. [4] and >5 for Rimbach and Pallauf [5], the fact that phytate is a potent inhibitor of iron-catalyzed hydroxyl radical formation is indisputable.

Evidence that all six phosphates on *myo*-inositol are not required for the inhibition of hydroxyl radical formation was presented by Hawkins et al. [6]. They showed that Ins(1,2,3,4,6)P_5 and DL-Ins(1,2,3,4,5)P_5 were effective at inhibiting hydroxyl radical formation to a greater degree than Ins(1,3,4,5,6)P_5 and DL-Ins(1,2,4,5,6)P_5 and concluded that the 1,2,3-trisphosphate grouping contained the essential binding site. Phillippy and Graf [7] measured the free coordination sites and inhibition of arachidonic acid oxidation by inositol tris- and tretrakisphosphates and deduced that the structure of the most potent inhibitory complex of ferric-Ins(1,2,3)P_3 would possess the three phosphates in axial, equatorial, and axial positions, respectively. These researchers [7] reasoned that this configuration orients the three phosphate groups in the right three-dimensional confirmation to allow bonding of all six iron coordination sites. *Myo*-inositols phosphorylated at the 1, 2, and 3 positions can be formed via the degradation of InsP$_6$ degradation by plant phytases. For example, Ins(1,2,3,5,6)P_5, Ins(1,2,3,6)P_4, and Ins(1,2,3)P_3 are produced by phytases from wheat [8] and rice bran [9]. Interestingly, Phillippy and Graf [7] also showed that hydrolysis of 1,2,3-trisphosphate by bacterial alkaline phosphatase was inhibited up to 50% in the presence of iron, whereas other inositol tris- and tetrakisphosphates were not, suggesting that iron binding might spare the antioxidant forms of the inositol phosphates over other forms in the gastrointestinal tract.

3. ANTIOXIDANT PROPERTIES OF INOSITOL PHOSPHATES *IN VIVO*

A question that remains is whether this antioxidant property of phytate might function *in vivo*, and whether it mediates the potential disease-sparing characteristics of phytate that have been observed in cancer studies [10]. To test this hypothesis, researchers have used cells in culture or animal models and measured the ability of phytate to attenuate the oxidative stress induced by various means. To date, the literature on this question is equivocal, with some studies indicating that InsP$_6$ provides some antioxidant protection, whereas other studies have found no effect or a pro-oxidant effect. It is important to recognize that oxidative stress protection in biological systems is a complex and multifaceted process [11,12], thus the ability of phytate to contribute as an antioxidant *in vivo* may only be evident under conditions in which stress is invoked.

Rao and coworkers [13] tested the ability of InsP$_6$ to protect Sprague-Dawley rats from damage caused by experimentally induced ischemia-reperfusion injury. They found that phytate injected intravenously, at concentrations of 7.5 and 15 mg/100 g BW, prior to cardiac excision and testing, protected the myocardium from damage as indicated by markers of muscle damage, heart function, blood flow, and lipid peroxidation. Kamp et al. [14] studied the effect of cigarette smoke extracts in combination with asbestos on oxidant-stress-induced injury in alveolar epithelial cells and found that phytate at 500 μM attenuated DNA damage and ATP depletion in one cell type (A549) but not a second (WI-26). Also, phytate did not prevent oxidant-induced glutathione (GSH) depletion in either cell type. Porres et al. [15] tested the effect of intrinsic dietary phytate to protect the liver and colon mucosa of pigs from oxidative stress induced by moderately high intakes of iron. Microbial phytase was added to diets, containing either 140 or 845 mg/kg iron, to remove the intrinsic dietary phytate (1.1–1.3%) in a 2^2 factorial design. Their results indicated that phytate may be protective for some tissues under the stress induced by high dietary intake of iron but not under conditions in which iron intake is normal.

In contrast to the previously discussed studies in which some evidence of antioxidant function *in vivo* was presented, other studies have not observed a protective effect of InsP$_6$ against oxidative stress. Rimbach and Pallauf [5] fed growing albino rats various diets containing either marginal (30 mg/kg) or enriched (300 mg/kg) iron, deficient or adequate vitamin E (dl-a-tocopheryl acetate, 0 or 50 mg/kg), and supplemental or no supplemental phytate (10 or 0 g/kg) in a 2^3 factorial design and assessed iron status and oxidative stress by measuring thiobarbituric acid reactive substances (TBARS) and protein carbonyls. After 28 days of diet consumption they found no protective effect by phytate on the higher liver TBARS generated or greater content of liver protein carbonyls in rats deficient in vitamin E compared to rats adequate in this essential lipid-soluble antioxidant. They also observed a negative effect of phytate intake, at 1% of the diet on iron absorption and liver iron concentration in rats consuming marginal levels of iron. These authors and others [16] have cautioned that the negative effects of significant phytate intake on mineral bioavailability might compromise its positive effect as an antioxidant. Rimbach and Pallauf [17] tested the effects of increasing dietary phytate intake (0, 7.5, and 15 g/kg diet) on magnesium (Mg) absorption and status, as well as markers of oxidative stress and antioxidant protection in the liver. They found that dietary phytate, at 0.75 and 1.5%, impaired Mg absorption and decreased Mg concentration in the plasma and femur. The livers of the animals fed higher amounts of phytate showed more evidence of oxidative stress, and liver homogenates were more susceptible to iron-induced lipid peroxidation than those from the rats given no phytate. Taken together, the results from these studies suggest that phytate has the potential to contribute to antioxidant defenses *in vivo* by chelating iron and preventing reactive oxygen species (ROS) formation, however, its nonspecific divalent cation binding properties can compromise the absorption of minerals that themselves participate, directly or indirectly, in antioxidant defense.

4. PROTECTION FROM VITAMIN E AND SELENIUM DEFICIENCY BY PHYTATE

We conducted an experiment to test whether the concentration and the form of inositol phosphates were important factors in determining whether phytate would exhibit antioxidant or pro-oxidant effects *in vivo*. Our lab has used a model of chronic oxidative stress induced by deficiency in vitamin E and selenium to study the adaptive responses invoked to protect rats from damage. Deficiency of these two nutrients leads to a gradual increase in oxidative stress, indicated by greater tissue concentrations of lipid oxidation products, as well as adaptations that lead to upregulation of many components of the cell's multifaceted defense system [18–21]. To determine whether phytate and $Ins(1,2,3,6)P_4$ would exhibit antioxidant properties *in vivo*, these inositol phosphates were fed to rats to test whether their presence would prevent the effects of vitamin E and selenium deficiency. Inositol phosphates were prepared enzymatically as described by Phillippy and Graf [7]. Long Evans hooded rats (40–50 g) were randomly assigned to either a diet sufficient in both vitamin E and Se (control) or one deficient in these two nutrients (-ESe) [18,19], and were fed these diets for two weeks. Then, 12 rats that were consuming -ESe diet were switched to the deficient diet supplemented with phytate (0.1%, 1 g/kg), or $Ins(1,2,3,6)P_4$ (0.1%), or $Ins(1,2,5,6)P_4$ (0.1%) (four rats in each group). The rats consumed these diets for an additional four weeks during which time body weights were measured. At the conclusion of this six-week time period, the animals were killed, and tissues were taken for analysis. Malondialdehyde (MDA) concentrations of rat small intestine and colon were determined as previously described [18]. The effects of inositol phosphates in this model were compared to those of the synthetic antioxidants butylated hydroxytoluene (BHT) and tert-butyl hydroquinone (TBHQ). In these experiments, weanling (40–50 g) Long-Evans hooded rats were randomly assigned to torula yeast-based diets adequate in vitamin E (100 mg α-tocopherol equivalents/kg diet) and selenium (0.228 mg Se/kg diet) (control), deficient in vitamin E (<0.1 mg/kg diet) and selenium (<0.025 mg/kg diet) (-ESe), or deficient in vitamin E and selenium with either 0.02% TBHQ or 0.02% BHT added to the tocopherol-stripped corn oil (10%) as the lipid source.

Weight gain as an indicator of the effect of oxidative stress on growth is illustrated in Figure 11.1. Rats consuming deficient diets gained about 30% less weight over the course of the six-week treatment period in comparison to rats consuming diets sufficient in both vitamin E and selenium. Rats consuming deficient diets supplemented with 0.1% phytate gained significantly more weight than rats consuming deficient diets, but supplementation with either inositol tetrakisphosphate did not improve weight gain in animals consuming deficient diets. This result suggests that phytate provides partial protection against the effects of oxidative stress on weight gain induced by the dietary deficiency. In comparison, rats consuming deficient diets supplemented with the potent synthetic antioxidant TBHQ also gained significantly more weight than animals consuming the deficient diets without supplementation, whereas supplementation of deficient diets with BHT provided no benefit in terms of weight gain [Figure 11.1(b)].

To test whether the partial protective effect of phytate on weight gain was associated with its antioxidant properties, the content of MDA that could be extracted from the small intestine and colon was measured. The amount of MDA in the small intestine and colon of rats fed the -ESe diet was about 40–100% more than that measured in the intestines of the control group [Figure 11.2(a),(b)]. Supplementation with phytate (0.1%) led to significantly less MDA in both the small intestine and the colon. The MDA content of small intestine and colon for rats supplemented with Ins(1,2,3,6)P_4 was less than that of the deficient animals but more than that for phytate and the control. These data indicate that both phytate and Ins(1,2,3,6)P_4 provide some antioxidant defense protection in the gastrointestinal tract of rats consuming the deficient diets. This is verified by comparison to the results obtained by feeding synthetic antioxidants [Figure 11.2(b)]. In these experiments, TBHQ exhibited complete protection from MDA formation in the small intestine but only partial protection in the colon, whereas BHT provided only slight and nonsignificant protection in both tissues. As the results from previous studies would suggest [7], Ins(1,2,5,6)P_4 supplementation provided no protection from MDA formation in the small intestine and colon. Moreover, this inositol tetrakisphosphate may have exhibited a slight pro-oxidant effect, as the mean MDA content in the colon of rats fed this inositol tetraphosphate was greater, although not statistically significant, than that of the deficient animals.

5. CONCLUSIONS AND QUESTIONS TO ADDRESS IN FUTURE STUDIES

The published results and those presented here support the concept that the chemical form of inositol polyphosphate present in the gastrointestinal tract influences its antioxidant properties *in vivo*. The question of whether the amount

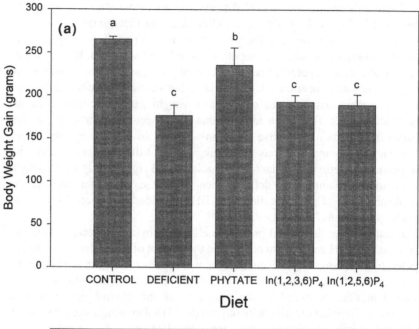

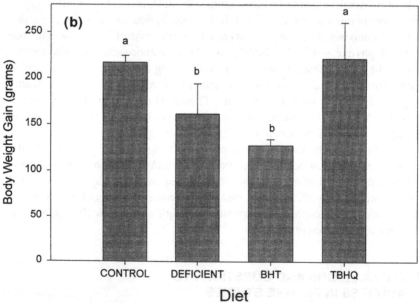

Figure 11.1 Body weight gain in rats consuming diets sufficient in vitamin E and selenium or deficient in these two nutrients and supplemented with (**a**) inositol phosphates or (**b**) synthethic antioxidants. Bar values are means with error bars representing standard deviations. Diet effects were tested using one-way ANOVA with Fisher LSD as the post hoc test. Bar values with different letters are significantly different ($P < 0.05$).

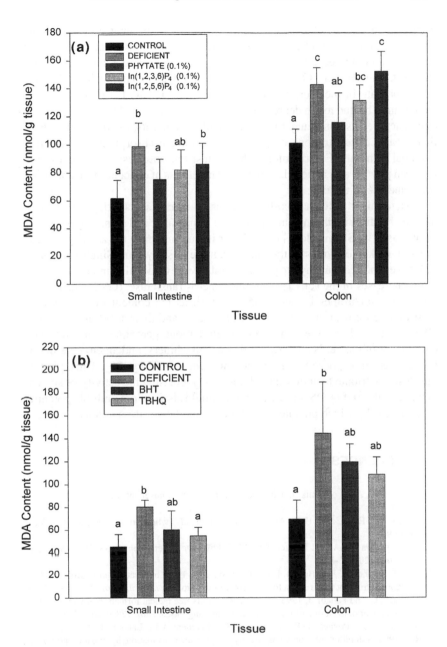

Figure 11.2 Malondialdehyde concentration of small intestine and colon for rats consuming diets sufficient in vitamin E and selenium or deficient in both nutrients and supplemented with (**a**) inositol phosphates or (**b**) synthetic antioxidants. Bar values are means with error bars representing standard deviations. Diet effects were tested using one-way ANOVA with Fisher LSD as the post hoc test. Bar values with different letters are significantly different ($P < 0.05$).

of phytate in the diet alters its antioxidant potential is still unanswered. Phytate supplementation to diets, at 0.1%, provided some antioxidant protection against the combined effects of vitamin E and selenium deficiency, whereas Rimbach and Pallauf [17] found no protection against vitamin E deficiency at 1% phytate in the diet. The differences in the results between these two studies may be due to the degree of oxidative stress imposed by single nutrient deficiency versus deficiency of two antioxidant nutrients, as well as the effect that greater concentrations of phytate have on the absorption of important essential minerals. Future studies should test the dose-response relationship between phytate and antioxidant/pro-oxidant effects at concentrations between 0.1% and 1% of the diet.

Phytate might influence oxidative stress by mechanisms independent of its hydroxyl radical inhibiting characteristics. It may alter cell signaling pathways as proposed by Shamsuddin [1,22,23], or it may influence the activity and expression of key enzymes in the antioxidant defense system. Singh and Singh [24] have reported that phytate given by oral gavage to lactating mice for 21 days led to increases in the activity of glutathione-S-transferase, (GST), the protein levels of cytochromes b_5 and P-450, as well as the content of acid-soluble sulfhydryl groups in the livers of both the dams and their suckling pups. The GST family includes the non-selenium glutathione peroxidase, an important component of the cell's antioxidant defense machinery, and an enzyme that is upregulated in response to antioxidant nutrient deficiency [25]. Thus, phytate may also contribute to antioxidant defense by increasing the activity of key enzymes that detoxify ROS. Future research should also explore multiple potential mechanisms by which phytate contributes to oxidative stress defense *in vivo*.

6. REFERENCES

1. Shamsuddin, A.M., Vucenik, I. and Cole, K.E. 1997. IP6: a novel anti-cancer agent. *Life Sci.* 61:343–354.
2. Graf, E. and Eaton, J.W. 1990. Antioxidant functions of phytic acid. *Free Radic. Biol. Med.* 8:61–69.
3. Vohra, P., Gray, G.A. and Kratzer, F.H. 1965. Phytic acid-metal complexes. *Proc. Soc. Exp. Biol. Med.* 120:447–449.
4. Graf, E., Mahoney, J.R., Bryant, R.G. and Eaton, J.W. 1984. Iron-catalyzed hydroxyl radical formation. Stringent requirement for free iron coordination site. *J. Biol. Chem.* 259:3620–3624.
5. Rimbach, G. and Pallauf, J. 1998. Phytic acid inhibits free radical formation *in vitro* but does not affect liver oxidant or antioxidant status in growing rats. *J. Nutr.* 128:1950–1955.
6. Hawkins, P.T., Poyner, D.R., Jackson, T.R., Letcher, A.J., Lander, D.A. and Irvine, R.F. 1993. Inhibition of iron-catalysed hydroxyl radical formation by inositol polyphosphates: a possible physiological function for myo-inositol hexakisphosphate. *Biochem. J.* 294:929–934.
7. Phillippy, B.Q. and Graf, E. 1997. Antioxidant functions of inositol 1,2,3-trisphosphate and inositol 1,2,3,6-tetrakisphosphate. *Free Radic. Biol. Med.* 22:939–946.

8. Phillippy, B.Q. 1989. Identification by two-dimensional NMR of inositol tris- and tetrakis (phosphates) formed from phytic acid and wheat phytase. *J. Agric. Food Chem.* 37:1261–1265.

9. Hayakawa, T., Suziki, K., Miura, H., Ohno, T. and Igaue, I. 1990. Myo-inositol polyphosphate intermediates in the dephosphorylation of phytic acid by phosphatase with phytase activity from rice bran. *Agric. Biol. Chem.* 54:279–286.

10. Nelson, R.L. 1992. Dietary iron and colorectal cancer risk. *Free Radic. Biol. Med.* 12:161–168.

11. Jadhav, S.J., Nimbalkar, S.S., Kulkarni, A.D. and Madhavi, D.L. 1996. Lipid oxidation in biological and food systems. In *Food Antioxidants: Technological, Toxicological and Health Perspectives*, Madhavi, D.L., Deshpande, S.S. and Salunkhe, D.K. (Eds). Marcel Dekker, Inc., New York, p. 5–63.

12. Niki, E. 1996. Alpha-Tocopherol. In *Handbook of Antioxidants*, Cadenas, E. and Packer, L. (Eds). Marcel Dekker, New York, p. 3–25.

13. Rao, P.S., Liu, X.K., Das, D.K., Weinstein, G.S. and Tyras, D.H. 1991. Protection of ischemic heart from reperfusion injury by myo-inositol hexaphosphate, a natural antioxidant. *Ann. Thorac. Surg.* 52:908–912.

14. Kamp, D.W., Greenberger, M.J., Sbalchierro, J.S., Preusen, S.E. and Weitzman, S.A. 1998. Cigarette smoke augments asbestos-induced alveolar epithelial cell injury: role of free radicals. *Free Radic. Biol. Med.* 25:728–739.

15. Porres, J.M., Stahl, C.H., Cheng, W.H., Fu, Y., Roneker, K.R., Pond, W.G. and Lei, X.G. 1999. Dietary intrinsic phytate protects colon from lipid peroxidation in pigs with a moderately high dietary iron intake. *Proc. Soc. Exp. Biol. Med.* 221:80–86.

16. Zhou, J.R. and Erdman, J.W., Jr. 1995. Phytic acid in health and disease. *Crit. Rev. Food. Sci. Nutr.* 35:495–508.

17. Rimbach, G. and Pallauf, J. 1999. Effect of dietary phytate on magnesium bioavailability and liver oxidant status in growing rats. *Food Chem. Toxicol.* 37:37–45.

18. Kuo, C.F., Cheng, S. and Burgess, J.R. 1995. Deficiency of vitamin E and selenium enhances calcium-independent phospholipase A2 activity in rat lung and liver. *J. Nutr.* 125:1419–1429.

19. Burgess, J.R. and Kuo, C.-F. 1996. Increased calcium-independent phospholipase A2 activity in vitamin E and selenium-deficient rat lung, liver, and spleen cytosol is time-dependent and reversible. *J. Nutr. Biochem.* 7:366–374.

20. Navarro, F., Navas, P., Burgess, J.R., Bello, R.I., De Cabo, R., Arroyo, A. and Villalba, J.M. 1998. Vitamin E and selenium deficiency induces expression of the ubiquinone-dependent antioxidant system at the plasma membrane. *FASEB J* 12:1665–1673.

21. Navarro, F., Arroyo, A., Martin, S.F., Bello, R.I., de Cabo, R., Burgess, J.R., Navas, P. and Villalba, J.M. 1999. Protective role of ubiquinone in vitamin E and selenium-deficient plasma membranes. *Biofactors* 9:163–170.

22. Shamsuddin, A.M. 1995. Inositol phosphates have novel anticancer function. *J. Nutr.* 125: 725S–732S.

23. Shamsuddin, A.M. 1999. Metabolism and cellular functions of IP6: a review. *Anticancer Res* 19:3733–3736.

24. Singh, A. and Singh, S.P. 1998. Postnatal effect of smokeless tobacco on phytic acid or the butylated hydroxyanisole-modulated hepatic detoxication system and antioxidant defense mechanism in suckling neonates and lactating mice. *Cancer Lett.* 122:151–156.

25. Chang, M., Burgess, J.R., Scholz, R.W. and Reddy, C.C. 1990. The induction of specific rat liver glutathione S-transferase subunits under inadequate selenium nutrition causes an increase in prostaglandin F2 alpha formation. *J. Biol. Chem.* 265:5418–5423.

CHAPTER 12

Potential Use of Phytate as an Antioxidant in Cooked Meats

DAREN P. CORNFORTH

1. INTRODUCTION

TRADITIONALLY, phytate (*myo*-inositol hexakisphosphate; InsP$_6$) has been considered a food toxicant in animal feed and in human diets, due to its powerful chelating ability, resulting in lower bioavailability of zinc, calcium, and iron. More recently, however, phytate has been recognized as a natural seed antioxidant, capable of chelating ionic iron and preventing iron catalysis of lipid oxidation [1]. With recognition of the antioxidant properties of phytate, several groups have investigated the possible use of phytate as an antioxidant in muscle foods [2–4]. A major concern with phytate addition to foods is the possible impairment of mineral absorption. However, a number of studies now indicate that iron and zinc absorption is not impaired in meat-based diets containing phytate [5–7]. This chapter will review phytate as an antioxidant in meat systems, meat-phytate interactions on mineral absorption, and effects of phytate on product cost and acceptability.

2. MEAT COUNTERACTS THE EFFECTS OF PHYTATE ON MINERAL ABSORPTION

The inhibitory effects of phytate on iron absorption are effectively counteracted by the presence of ascorbic acid [8], meat consumption [5], or enzymatic dephytinization of bran [9] (see reviews by Torre et al. [10] and Carpenter and Mahoney [11]). The addition of beef, pork, or chicken to a cereal + phytate diet increased iron (Fe^{59}) absorption compared to controls with cereal and phytate but without meat [6]. Meat consumption also improves Zn absorption. Zinc bioavailability was about fourfold greater from beef than from a high-fiber breakfast cereal [7]. Ascorbic acid enhances iron absorption by reducing insoluble ferric iron to the more soluble ferrous form. In addition to stimulation of gastric acid secretion, meat factor(s) may chelate solubilized iron in the acid environment of the stomach, thereby maintaining iron solubility during intestinal digestion and absorption [11].

3. PHYTATE AS AN ANTIOXIDANT IN CHICKEN AND BEEF

Graf and coworkers [2,12] were the first to evaluate phytate as an antioxidant in a meat system. In cooked, minced chicken breast muscle +10% added water, phytic acid (1.5 mM, pH 6.0) was highly effective in reducing thiobarbituric acid (TBA) values (Figure 12.1) and warmed-over flavor (WOF) intensity (Figure 12.2), compared to controls [12]. Phytic acid (2 mM) was more effective than other antioxidants (2 mM ascorbic acid, BHT, or EDTA) for lowering TBA values in fresh beef homogenates incubated for 60 min at 37°C [13].

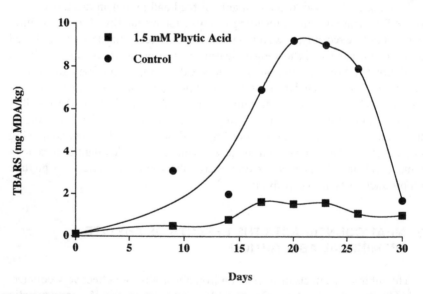

Figure 12.1 Effect of phytic acid on thiobarbituric acid reactive substances (TBARS) levels in cooked, minced chicken breast samples stored at 4°C in oxygen impermeable vacuum pouches. MDA = malonaldehyde. Reprinted with permission. (*Source*: Reference [12]).

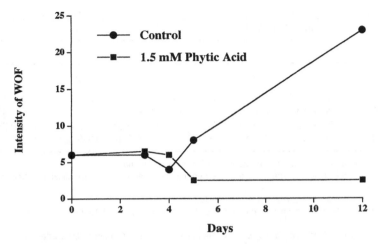

Figure 12.2 Effect of phytic acid on warmed-over flavor (WOF) generation in cooked, minced chicken breast samples stored at 4°C in oxygen impermeable pouches. Reprinted with permission. (*Source*: Reference [12]).

4. EFFECT OF PHYTATE ON COLOR, BIND STRENGTH, AND TBA VALUES

Lee et al. [14] compared sodium phytate (SPT) to traditional phosphates [sodium pyrophosphate (SPP) or sodium tripolyphosphate (STPP)] on characteristics of raw and cooked beef rolls. The rolls were formulated to contain 10% added water, 1% salt, and 0.5% phosphate as a percent of raw meat weight. On a millimolar basis, the 0.5% concentrations of SPT, SPP, and STPP were equivalent to 4.5, 11.2, and 13.6 mM, respectively. The uncooked rolls with SPT were redder, with higher Hunter color "a" values than controls or rolls containing STPP (Table 12.1) and lower levels of brown metmyoglobin (Table 12.2). Compared to controls, phytate and other phosphate treatments increased salt-soluble protein levels and raw meat pH and decreased the TBA value (Table 12.2). Phytate was apparently resistant to the effects of endogenous meat phosphatases, because orthophosphate levels were lower in rolls with phytate than in rolls with sodium tripolyphosphate or sodium pyrophosphate (Table 12.2).

After cooking, rolls with phytate or other phosphates had higher bind strength, cooked yield, and higher meat pH than controls (Table 12.3). All phosphate-treated rolls had lower TBA values than cooked controls. Phytate was slightly more effective than other phosphates for reduction of TBA values of cooked rolls. Among phosphate treatments, rolls with tripolyphosphate had the highest cooked yield. It was somewhat surprising that phytate was as effective as other phosphates for increasing bind strength of cooked rolls, compared to controls

TABLE 12.1.   Effect of Sodium Phytate Compared to Other Phosphates on Hunter Color Values of Raw and Cooked Beef Rolls.

Treatment	Raw			Cooked		
	L	A	B	L	A	B
Control	31 ± 2.2^a	13 ± 1.5^a	10 ± 1.0^a	45 ± 1.3^a	10 ± 1.1^a	10 ± 0.1^a
SPT	26 ± 1.6^b	17 ± 0.4^b	8 ± 0.3^b	43 ± 2.7^a	11 ± 0.1^a	9 ± 0.4^a
SPP	28 ± 1.8^b	16 ± 0.8^b	8 ± 0.6^b	43 ± 3.6^a	11 ± 0.3^a	10 ± 0.3^a
STPP	26 ± 1.7^b	15 ± 1.3^b	8 ± 0.4^b	44 ± 2.4^a	10 ± 0.9^a	9 ± 0.6^a

Data represent the mean ± standard deviation of three determinations.
[a–b] Means within columns sharing the same superscript letter were not significantly different at $p < 0.05$.
Hunter color: L = lightness; A = redness; B = yellowness.
Abbreviations: SPT = sodium phytate; SPP = sodium pyrophosphate; STPP = sodium tripolyphosphate.
(Reprinted from *Meat Science*, 50, Lee, B., Hendricks, D.G. and Cornforth, D.P., Effect of sodium phytate, sodium pyrophosphate, and sodium tripolyphosphate on physico-chemical characteristics of restructured beef, 273–283, Copyright 1998, with permission from Elsevier Science.)

without phosphates. Phosphates increase cooked yield and bind strength of cooked meats via two mechanisms. Phosphates raise meat pH, and there is a clear tendency for the meat yield and water-holding capacity to increase with increasing pH [15,16]. Second, phosphates increase bind strength through their ability to dissociate actomyosin into myosin and actin, thereby increasing protein extraction from post-rigor meats [17,18]. One mole of STPP hydrolyzes to one mole of orthophosphate and one mole of pyrophosphate due to the action of meat phosphatases [19]. The hydrolysis products of traditional phosphates contribute to actomyosin dissociation. Beef rolls with phytate had similar orthophosphate levels to controls (1167 versus 1148 ppm, respectively), indicating

TABLE 12.2.   Effect of Sodium Phytate Compared to Other Phosphates on Various Characteristics of Raw Restructured Beef Rolls after Storage for One Day at 2°C.

Treatment	MetMb (%)	Salt-soluble protein (mg/mL)	TBA number (mg MDA/kg meat)	Orthophosphate (ppm)	pH
Control	42 ± 3^a	6.8 ± 0.1^a	1.1 ± 0.10^a	1135 ± 27^a	5.3 ± 0.02^a
SPT	28 ± 1.5^b	$8.6 \pm 0.3^{b,c}$	0.3 ± 0.03^b	1127 ± 104^a	5.9 ± 0.05^c
SPP	29 ± 2.5^b	8.0 ± 0.3^b	0.2 ± 0.01^b	1817 ± 173^b	5.8 ± 0.04^b
STPP	36 ± 2.8^c	9.0 ± 0.2^c	0.2 ± 0.02^b	2206 ± 78^c	5.8 ± 0.05^b

Data represent the mean ± standard deviation of three determinations.
[a–c] Means within columns sharing the same superscript letter were not significantly different at $p < 0.05$.
Abbreviations: SPT = sodium phytate; SPP = sodium pyrophosphate; STPP = sodium tripolyphosphate; MetMb = metmyoglobin as % of total myoglobin; TBA = thiobarbituric acid; MDA = malonaldehyde.
(Reprinted from *Meat Science*, 50, Lee, B., Hendricks, D.G. and Cornforth, D.P., Effect of sodium phytate, sodium pyrophosphate, and sodium tripolyphosphate on physico-chemical characteristics of restructured beef, 273–283, Copyright 1998, with permission from Elsevier Science.)

TABLE 12.3. Effects of Sodium Phytate Compared to Other Phosphates on Various Characteristics of Cooked Beef Rolls.

Treatment	Bind Strength (g)	Cook Yield (%)	TBA number (mg MDA/kg meat)	Orthophosphate (ppm)	pH
Control	761 ± 148^a	77 ± 0.7^a	1.5 ± 0.08^a	1148 ± 45^a	5.5 ± 0.4^a
SPT	1878 ± 251^b	82 ± 0.3^b	0.5 ± 0.02^c	1167 ± 65^a	6.0 ± 0.5^c
SPP	1965 ± 224^b	84 ± 1.4^{bc}	0.7 ± 0.09^b	1657 ± 60^b	5.9 ± 0.3^b
STPP	1996 ± 416^b	86 ± 1.7^c	0.6 ± 0.10^b	1838 ± 136^c	5.8 ± 0.6^b

Data represent the mean ± standard deviation of three determinations.
[a–c] Means within columns sharing the same superscript letter were not significantly different at $p < 0.05$.
Abbreviations: SPT = sodium phytate; SPP = sodium pyrophosphate; STPP = sodium tripolyphosphate; TBA = thiobarbituric acid; MDA = malonaldehyde.
(Reprinted from *Meat Science*, 50, Lee, B., Hendricks, D.G. and Cornforth, D.P., Effect of sodium phytate, sodium pyrophosphate, and sodium tripolyphosphate on physico-chemical characteristics of restructured beef, 273–283, Copyright 1998, with permission from Elsevier Science.)

that endogenous phosphatases were unable to hydrolyze added phytate [4]. Thus, increased bind and cooked yield in rolls with phytate was primarily due to higher meat pH (6.0, phytate > 5.8, STPP > 5.5, controls) (Table 12.3).

Carnosine or phytate both increase color stability of a fresh beef homogenate during refrigerated storage [14]. After cooking, however, lipid oxidation and TBA values were much lower for samples with phytic acid than for controls or samples with added carnosine (Figure 12.3). After meat cookery, ionic iron

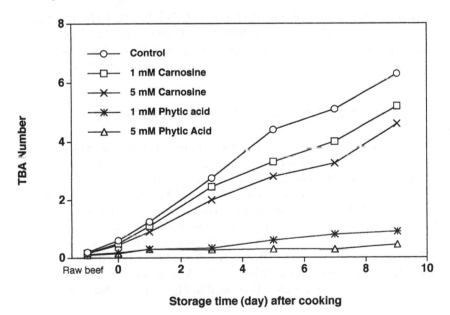

Figure 12.3 Effect of L-carnosine and phytic acid on TBA number in cooked beef during storage for 9 days at 4°C. Reprinted with permission. (*Source*: Reference [4]).

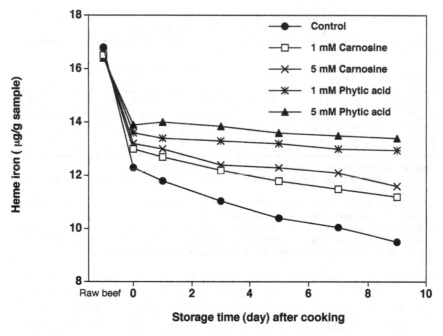

Figure 12.4 Effect of L-carnosine and phytic acid on heme iron content in cooked beef during storage for 9 days at 4°C. Reprinted with permission. (*Source*: Reference [4]).

levels are higher due to heme pigment oxidation [20]. Lee et al. [14] reported a highly negative correlation ($r = -0.92$) of heme iron content with lipid peroxidation in cooked beef. Phytate was significantly more effective than carnosine for inhibition of iron release from heme during cooking (Figure 12.4).

5. MECHANISM OF PHYTATE ANTIOXIDANT EFFECTS

Phytate inhibits lipid peroxidation by forming a complex in which all six coordination sites of Fe (III) are occupied by phytate [21]. In addition to binding, phytate strongly facilitates the oxidation of Fe (II) to Fe (III). Phytic acid at 0.1, 1, and 5 mM catalyzed oxidation of 100 μM Fe (II) by 71, 78, and 98%, respectively (Table 12.4). Both lipid and deoxyribose oxidation are catalyzed by ionic iron, and the resultant aldehydes may be quantitated by reaction with thiobarbituric acid (TBA). Using deoxyribose as a substrate, Lee and Hendricks [22] showed that phytate very effectively inhibited oxidation, as indicated by lower TBA values than controls (Table 12.5). Another indication of the nature of the phytate-iron interaction is illustrated in Figure 12.5, where phytic acid

TABLE 12.4.   Effects of Chelates on Oxidation of Fe (II) to Fe (III) and on Degradation of Deoxyribose in the Presence of 100 μM Fe (II).

Chelates	Fe (II), μM	% Oxidation vs. Control	TBARS, nmole/10 mM Deoxyribose	% Inhibition vs. Control
Control (buffer)	100		12.8 ± 1.2	
Desferroxamine, 0.1 mM	13.6 ± 1.5	86.4	4.5 ± 0.4	64.6
EDTA, 0.1 mM	0.2 ± 0.2	99.8	1.4 ± 0.2	89.1
Phytic acid				
0.1 mM	29.3 ± 1.9	70.7	3.5 ± 0.4	78.8
1.0 mM	22.7 ± 1.8	78.3	2.4 ± 0.3	81.2
5.0 mM	2.1 ± 0.7	97.9	1.5 ± 0.3	88.1

Data represent the mean ± standard deviation of three determinations.
TBARS = thiobarbituric acid reactive substances.
(*Source*: Reference [22]. Reprinted with Permission.)

inhibited the reduction of ferric iron by ascorbate in a dose-dependent manner. At 10 mM phytate, the high (98%) retention of iron in the ferric state indicates that phytate tightly bound ferric iron, rendering it unavailable for reaction with reductants such as ascorbate.

The relationship of iron oxidation state versus rate of lipid oxidation has been investigated by Minotti and Aust [23], who proposed a complex of $Fe(II)$-O_2-$Fe(III)$ as an initiating species for lipid oxidation. The rate of oxidation was found to be highest when the ratio of ferrous to ferric iron was one to one

TABLE 12.5.   Effect of Cooked Wild Rice on the Thiobarbituric Acid Values of Cooked Ground Beef Stored for 10 Days at 4°C.

Treatment	TBARS (ppm)
Beef control	3.5[a]
Beef + BHA (200 ppm)	0.3[b]
Beef + BHT (200 ppm)	0.5[b]
85% Beef + 15% cooked wild rice	1.8[b]
85% Beef + 15% water	3.5[a]

[a-b] Means sharing the same superscript letter were not significantly different at $p < 0.05$. Standard deviation = ±0.1 ppm.
TBARS values were based on the weight of the beef to eliminate the dilution effect of the rice.
Abbreviations: TBARS = thiobarbituric acid reactive substances; BHA = butylated hydroxyanisole; BHT = butylated hydroxytoluene.
(Reprinted with permission from Reference [34]. Copyright 1994 American Chemical Society.)

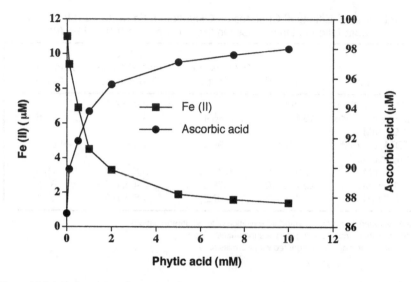

Figure 12.5 Inhibitory effect of phytic acid on reduction of 100 μM Fe (III) to Fe (II) by 100 μM ascorbic acid and the simultaneous oxidation of ascorbic acid in 35 mM HEPES buffer (pH 7.0) after standing for 25 min at 25°C. Reprinted with permission. (*Source*: Reference [22]).

[24–25]. The ability of chelators such as phytate to shift the ferrous-ferric ratio to 1:0 or 0:1 (as with phytate) facilitates their antioxidant effects [22].

6. COST AND REGULATORY STATUS OF PHYTATE IN MEAT PRODUCTS

Phytate is used as a food preservative in Japan. Phytate is not approved in the United States for use in meat products, because the USDA has not been petitioned to approve its use in a specific application. Cost is another drawback to commercial application of phytate in meats. At a current cost of 44 cents per gram (Sigma Chemical Company, St. Louis, MO), and at a use level of 0.5% of meat weight (the phosphate level permitted by the USDA) [26], phytate would cost 33 cents per pound raw meat. Obviously, phytate must be obtained more cheaply in order to justify its commercial application. One application where high phytate costs may be acceptable is in production of nitrite-free bacon or franks [27], using phytate rather than nitrite to inhibit rancid flavor development [28]. Nitrite is a precursor to formation of carcinogenic nitrosamines in fried bacon [29]. The USDA has reduced the permitted level of nitrite in cured products and now also requires use of 500 ppm ascorbate to lower residual nitrite levels after processing, virtually eliminating the potential for nitrosamine formation during frying (see review) [30]. However, there is a niche market for nitrite-free cured meats.

7. ANTIOXIDANT PROPERTIES OF WILD RICE IN COOKED MEAT PRODUCTS

A cheaper alternative to phytate is the addition of high-phytate grains to cooked ground meats. Textured soy proteins (TSP) or textured vegetable proteins (TVP) are economical extenders of ground beef patties [31]. Lower thiobarbituric acid (TBA) values were reported for ground beef patties extended with TSP [32], indicating possible antioxidant effects during storage. Addis and coworkers [3,33–35] at the University of Minnesota demonstrated the feasibility and acceptability of wild rice addition to ground beef patties and pork sausage [36]. In addition to a lowering of TBA values with the addition of wild rice (Table 12.5), panel evaluations indicated an actual preference for beef patties with 15 or 30% cooked wild rice, compared to all beef controls [33]. Whole-grain wild rice was a more effective antioxidant in ground beef patties than ground wild rice [3]. The antioxidant properties of wild rice kernels were due to the presence of phytic acid [34], as indicated by the identical HPLC chromatogram for wild rice extract versus phytic acid, and identical NMR phosphorus spectra of wild rice extract versus phytic acid hydrolyzate. The wild rice hull also contains antioxidant fractions, including anisole, vanillin, and *m*-hydroxybenzaldehyde [37].

Barley bran (1.5%) was also highly effective for prevention of oxidation in cooked ground beef patties [35]. Barley bran contains vitamin E and other tocotrienols, classified as Type I antioxidants. The phytate in wild rice is a Type II antioxidant [35]. Type I antioxidants are electron donors and act to interrupt the propagation step of lipid oxidation. Type II antioxidants, such as phytate, bind iron, preventing its participation in the initiation of lipid oxidation [38] and also preventing iron-catalyzed degradation of hydroperoxides to volatile aldehydes and ketones.

8. CONCLUSIONS

Through the mid 1980s, phytate was regarded strictly as a dietary inhibitor of mineral absorption. More recently, studies indicate that phytic acid is a natural antioxidant important for seed viability. Evidence also indicates that the inhibitory effects of phytate on mineral absorption are not seen in varied diets containing animal protein. Phytate may actually be beneficial as a dietary antioxidant in an animal protein diet. There may be nutritional advantages, or at least no disadvantage, to addition of phytate to meat products. Phytate retards oxidation and enhances fresh meat color stability. However, its primary application appears to be in cooked meats. Phytate can be as effective as traditional phosphates for enhancement of cooked yield, cohesion, and inhibition of lipid oxidation. The major obstacle to use of phytate in cooked meats is cost. Cheaper methods of phytate purification are needed before it can be cost competitive with traditional phosphates. However, phytate in cooked wild rice/ground beef blends has been shown to be a cost-effective method of phytate addition and highly acceptable by consumer panelists.

9. REFERENCES

1. Graf, E., Empson, K.L. and Eaton, J.W. 1987. Phytic acid: A natural antioxidant. *J. Biol. Chem.* 262:11647–11650.
2. Graf, E. and Panter, S.S. 1991. Inhibition of warmed-over flavor development by polyvalent cations. *J. Food Sci.* 56:1055–1058, 1067.
3. Johnson, M.H., Addis, P.B. and Epley, R.J. 1994. Rancidity in beef patties and reduction by wild rice. *J. Food Service Systems* 8:47–59.
4. Lee, B., Hendricks, D.G. and Cornforth, D.P. 1998. Antioxidant effects of carnosine and phytic acid in a model beef system. *J. Food Sci.* 63:394–398.
5. Zhang, D., Carpenter, C.E. and Mahoney, A.M. 1990. A mechanistic hypothesis for meat enhancement of iron absorption: stimulation of gastric secretions and iron chelation. *Nutr. Res.* 10:929.
6. Kim, Y., Carpenter, C.E. and Mahoney, A.W. 1993. Gastric acid production, iron status, and dietary phytate alter enhancement by meat of iron absorption in rats. *J. Nutr.* 123:940–946.
7. Zheng, J., Mason, J.B., Rosenberg, I.H. and Wood, R.J. 1993. Measurement of zinc bioavailability from beef and a ready-to-eat high-fiber breakfast cereal in humans: application of a whole gut lavage technique. *Amer. J. Clin. Nutr.* 58:902–907.
8. Siegenberg, D., Baynes, R.D., Bothwell, T.H., Macfarlane, B.J., Lamparelli, R.D., Car, N.G., MacPhail, P., Schmidt, U., Tal, A. and Mayet, F. 1991. Ascorbic acid prevents the dose dependent inhibitory effect of polyphenols and phytates on non-heme iron absorption. *Amer. J. Clin. Nutr.* 53:57–65.
9. Sandstrom, B. and Sandberg, A.S. 1992. Inhibitory effects of isolated inositol phosphates on zinc absorption in humans. *J. Trace Elem. Elec. Health Dis.* 6:99–103.
10. Torre, M., Rodriguez, A.R. and Saura-Calixto, F. 1991. Effects of dietary fiber and phytic acid on mineral availability. *Crit. Rev. Food Sci. Nutr.* 1:1–22.
11. Carpenter, C.E. and Mahoney, A.W. 1992. Contributions of heme and nonheme iron to human nutrition. *Crit. Rev. Food Sci. Nutr.* 31:333–367.
12. Empson, K.L., Labuza, T.P. and Graf, E. 1991. Phytic acid as a food antioxidant. *J. Food Sci.* 56:560–563.
13. Lee, B. and Hendricks, D.G. 1995. Phytic acid protective effect against beef round muscle lipid peroxidation. *J. Food Sci.* 60:241–244.
14. Lee, B., Hendricks, D.G. and Cornforth, D.P. 1998. Effect of sodium phytate, sodium pyrophosphate, and sodium tripolyphosphate on physico-chemical characteristics of restructured beef. *Meat Sci.* 50:273–283.
15. Thomsen, H.H. and Zeuthen, P. 1988. The influence of mechanically deboned meat and pH on the water-holding capacity and texture of emulsion type meat products. *Meat Sci.* 22:189–201.

16. Moiseev, I.V. and Cornforth, D.P. 1997. Sodium hydroxide and sodium tripolyphosphate effects on bind strength and sensory characteristics of restructured beef rolls. *Meat Sci.* 45:53–60.

17. Hamm, R. 1971. *Phosphates in Food Processing.* J.M. Demann and P. Melnychyn (Eds). Avi Publishing Co., Westport, CT. p. 65.

18. Theno, D.M., Siegel, D.G. and Schmidt, G.R. 1978. Meat massaging: Effects of salt and phosphate on the microstructure of binding junctions in sectioned and formed hams. *J. Food Sci.* 43:493–498.

19. Awad, M.K. 1968. Hydrolysis of polyphosphates added to meat. M.S. thesis, The University of Alberta, Edmonton, Alberta.

20. Igene, J.O., King, J.A., Pearson, A.M. and Gray, J.I. 1979. Influence of heme pigments, nitrite, and non-heme iron on the development of warmed-over flavor (WOF) in cooked meat. *J. Agric. Food Chem.* 27:838–842.

21. Graf, E. and Eaton, J.W. 1990. Antioxidant functions of phytic acid. *Free Radicals in Biol. Med.* 8:61–69.

22. Lee, B. and Hendricks, D.G. 1997. Metal-catalyzed oxidation of ascorbate, deoxyribose and linoleic acid as affected by phytic acid in a model system. *J. Food Sci.* 935–938, 984.

23. Minotti, G. and Aust, S.D. 1987. The role of iron in the initiation of lipid peroxidation. *Chem. Phys. Lipids* 44:191–208.

24. Minotti, G. and Aust, S.D. 1989. The role of iron in oxygen radical mediation of lipid peroxidation. *Chem. Biol. Interactions* 71:1–19.

25. Miller, D.M. and Aust, S.D. 1987. Studies of ascorbate dependent, iron-catalyzed lipid peroxidation. *Arch. Biochem. Biophys.* 271:113–119.

26. de Holl, J.C. 1981. Encyclopedia of Labelling Meat and Poultry Products. *Meat Plant Magazine*, p. 131.

27. O'Boyle, A.R., Rubin, L.J., Diosady, L.L., Aladin Kassam, N., Comer, F. and Brightwell, W. 1990. A nitrite-free curing system and its application to the production of weiners. *Food Technol.* 44:88.

28. Cornforth, D.P. 1998. Potential for use of phytate as an antioxidant in cooked meats. *IFT Muscle Foods Division Newsletter* (Fall) 25(1):4–5.

29. Crosby, N.T., Forman, J.K., Palframan, J.F. and Sawyer, R. 1972. Estimation of steam-volatile N-nitrosamines in foods at the 1 μg/kg level. *Nature* (London) 238:342–343.

30. Cornforth, D.P. 1996. Role of Nitric Oxide in Treatment of Foods, Ch. 8 in *Nitric Oxide: Principles and Actions*, J.R. Lancaster, Jr., Ed. Acad. Press, San Diego, CA, pp. 259–287.

31. McWatters, K.H. 1977. Performance of defatted peanut, soybean, and field pea meals as extenders in ground beef patties. *J. Food Sci.* 42:1492.

32. Kotula, A.W., Twigg, G.G. and Young, E.P. 1976. Evaluation of beef patties containing soy protein during 12-month frozen storage. *J. Food Sci.* 41:1142.

33. Minerich, P.L., Addis, P.B., Epley, R.J. and Bingham, C. 1991. Properties of wild rice/ground beef mixtures. *J. Food Sci.* 56:1154–1157.

34. Wu, K., Zhang, W., Addis, P.B., Epley, R.J., Salih, A.M. and Lehrfeld, J. 1994. Antioxidant properties of wild rice. *J. Agric. Food Chem.* 42:34–37.

35. Katsanidis, E., Addis, P.B., Epley, R.J. and Fulcher, R.G. 1997. Evaluation of the antioxidant properties of barley flour and wild rice in uncooked and precooked ground beef patties. *J. Food Service Systems* 10:9–22.

36. Rivera, J.A., Addis, P.B., Epley, R.J., Asamarai, A.M. and Breidenstein, B.B. 1996. Properties of wild rice/pork sausage blends. *J. Muscle Foods* 7:453–470.

37. Asamarai, A.M., Addis, P.B., Epley, R.J. and Krick, T.P. 1996. Wild rice hull antioxidants. *J. Agric. Food Chem.* 44:126–130.

38. Aust, S.D., Morehouse, L.A. and Thomas, C.E. 1985. Role of metals in oxygen radical reactions. *J. Free Radical Biol. Med.* 1:3–25.

CHAPTER 13

Phytate and Mineral Bioavailability

CONNIE M. WEAVER
SRIMATHI KANNAN

1. INTRODUCTION

FOR several decades, concerns have been raised about the role of phytic acid in reducing mineral bioavailability. Because dietary phytic acid is a ubiquitous plant constituent present in nuts, cereals, legumes, and oilseeds, current trends in food choices merit a reexamination of this issue. Recommendations for increasing consumption of cereals and grains as the foundation of the food guide pyramid by the U.S. Dietary Guidelines Committee has prompted one such trend. A second trend is that soy-containing foods are becoming increasingly popular in the United States due to intensified research on their health benefits. Increased consumption of snack foods with plant seeds including poppy seeds, sesame seeds, and pumpkin seeds, and granola mixes of nuts and dried foods that contain appreciable amounts of phytate is a third trend. An emerging trend is the interest of manufacturers and consumers in functional foods. Addition of antioxidants such as ascorbic acid or fructooligosaccharides to foods could have tremendous effects on mineral bioavailability that temper the effect of dietary phytate. Genetically modified crops with reduced phytate as discussed in another chapter in this book and still others with higher levels of micronutrients or absorption enhancers as reviewed by Frossard et al. [1] could substantially alter the current food supply.

In this chapter, the nature of the inhibitory effect of phytic acid on mineral bioavailability is reviewed as well as factors influencing the extention of the inhibition. A potential role of inositol phosphate hydrolysis products as enhancers of mineral absorption is discussed. The chapter is meant to be more of a perspective, especially in the context of other dietary enhancers and inhibitors, than a

211

comprehensive review. As such, examples are selected that studied endogenous phytate as opposed to the highly reactive sodium salt when possible.

2. PHYTATE-MINERAL INTERACTIONS

The effect of phytate on mineral bioavailability has been the focus of numerous research studies and the subject of extensive reviews [2–7]. Phytic acid has a potential for binding positively charged proteins, amino acids, and/or multivalent cations or minerals in foods (Figure 13.1). The resulting complexes are insoluble, difficult for humans to hydrolyze during digestion, and thus, typically are nutritionally less available for absorption. Phytate forms chelating conjugates with nutritionally important minerals such as calcium, magnesium, copper, iron (Fe^2 and Fe^{3+}), zinc, cobalt, and manganese. Solubility is a prerequisite for absorption of most minerals, although solubility at neutral pH has been shown to be less important for calcium absorption [8]. The chemical structure of phytic acid is indicative of strong chelating potential. Phytic acid has six strongly dissociated protons (pKs 1.1 to 2.1) and six weakly dissociated protons (pKs 4.6 to 10.0). The effect on minerals is observed through the formation of phytate-mineral (M) or peptide-mineral-phytate complexes. These complexes have stoichiometries of the M^+(n)-phytate type ($n =$ 1–6). Phytate forms a wide variety of insoluble salts with divalent and trivalent cations. Usually, the divalent cations (e.g., Zn^{2+}, Ca^{2+}, Mg^{2+}) form insoluble penta- and hexa-substituted salts. The insolubility of these complexes is regarded as the major reason for the reduced bioavailability of minerals due to diets high in phytic acid. When the complex includes peptides, bioavailability of proteins and enzymatic activity may be reduced [9]. Humans lack sufficient intestinal phytase to degrade the complexes. As much as 30–97% of the intake of phytic acid (0.3–3.7 g/d) may be undigested before it reaches the

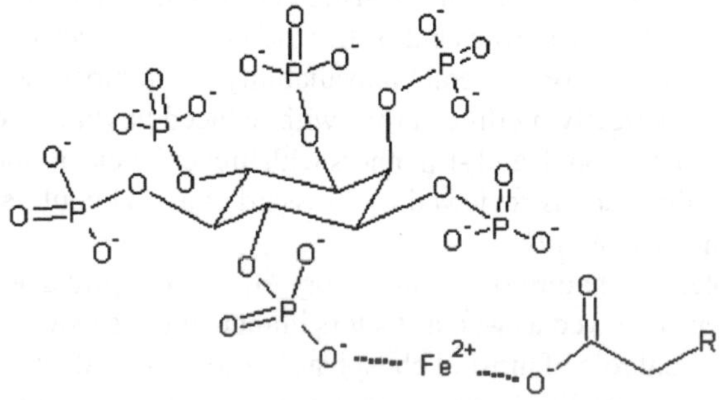

Figure 13.1 Phytate showing an example of an interaction with iron and a protein.

colon, and this phytic acid may have protective or adverse effects on colonic health.

Several factors determine the effect of phytate on mineral bioavailability: pH, size and valence of the mineral, mineral and phytate concentrations and ratios, and food matrix that includes the presence of enhancers and/or inhibitors. The effects of processing and degree of phosphorylation of phytate will be discussed in later sections.

2.1. pH

The effect of pH on mineral-phytate and protein-mineral-phyate interactions has been reviewed [10]. As the pH increases, and under sufficient phytate concentrations, phytic acid becomes more ionized and begins binding cations. pH also affects the charge of peptides. On the acidic side of the isoelectric pH of the peptide, the negatively charged carboxyl groups can react directly with the positively charged amino group. On the alkaline side of the isoelectric pH of the peptide where carboxy groups are negatively charged, binding occurs through positively charged mineral ions. This can occur in processing or during digestion. For example, at higher pH, zinc was more associated with phytate in soy protein isolates than at acidic pH (pH for soy protein is 4.5) [11]. A higher pH environment would typically occur in the intestine. However, Champagne and Phillippy [12] reported that high intraluminal gastric pH leads to the formation of calcium-zinc-phytate complexes as early as the stomach following the ingestion of soy protein isolate.

2.2. MINERAL TYPE

The relative binding strengths of different minerals to phytic acid vary greatly. Chelation strength increases with increasing atomic number of the mineral moving from the alkaline earth metals through transition metals in the periodic table. Vohra et al. [13], using titration curves of phytate as free acid in the presence of single cations, reported that phytate forms complexes with cations in the following descending order of strength:

$$Cu^{2+} > Zn^{2+} > Co^{2+} > Mn^{2+} > Fe^{3+} > Ca^{2+}$$

The decreasing order of stability of phytate-mineral complexes is as follows:

$$Zn^{2+} > Cu^{2+} > Ni^{2+} > Co^{2+} > Mn^{2+} > Ca^{2+}$$

Consequently, zinc is the essential mineral most affected by phytate. The presence of unhydrolyzed phytate in unleavened bread is considered to be responsible for zinc deficiency in the Middle East. Zinc deficiency is corrected by leavening [14] or by zinc supplementation to the phytate-rich cereal legume diet of Egyptians and Iranians [15,16].

2.3. MINERAL AND PHYTATE CONCENTRATIONS AND RATIOS

Phytate concentrations have to be sufficiently high to exert a substantial effect on mineral bioavailability in the diet. Harland and Oberleas [17] published the phytate values in approximately 200 foods. The inhibitory effect of phytate on mineral absorption is often dose responsive. The relationship between iron absorption and phytate content of cereals in young adults was strong at $r = -0.801$, $p < 0.02$ [18]. Genetically modified corn that reduced the phytate to one-third of the parent strain improved iron absorption by 50% [19]. In soy, the amount of phytate required to inhibit iron absorption is small and has to be reduced to concentrations of <10 mg/meal to substantially remove the inhibitory effect of phytate [20]. This may explain why iron absorption in rats and humans was not affected by a wide range in phytic acid concentration in soybeans produced hydroponically by manipulating phosphorus content of the nutrient solution because even the lowest concentration exceeded this threshold [21].

It is not clear whether phytate concentration per se or the ratio of phytate to minerals in the foodstuff or diet is dominant in influencing mineral bioavailability. It is possible that these factors have different impacts for different nutrients. The molar ratio of phytate to zinc has been reported to be a major factor in influencing bioavailability of zinc to rats from breakfast cereals [22]. Molar ratios of phytate:zinc >10 have been associated with zinc deficiency symptoms [23]. The phytate × calcium/zinc molar ratio was even a stronger predictor of zinc bioavailability from seeds than was the phytate-zinc molar ratio, especially at lower protein intakes [24,25]. Poor zinc availability in rats was observed when the phytate × Ca^{2+}/Zn^{2+} molar ratio exceeded 3.5 [24]. Similarly, absorption of ^{59}Fe in rats was decreased with a phytate:Fe^{2+} molar ratio above 14.2. A phytate:Ca^{2+} molar ratio >0.24 also reduces calcium bioavailability [26]. Thus, phytate:mineral molar ratios have been used to evaluate several minerals for predicted bioavailability.

Total phytate and phytate:mineral ratios have been directly compared for predicting mineral bioavailability for zinc and calcium. In rats, zinc retention was determined from wheat produced with a range of phytate concentrations [27]. Neither phytate Zn^{2+} nor phytate × Ca^{2+}/Zn^{2+} molar ratio improved the predictability of zinc retention over total phytate content ($r^2 = 0.625$–0.636 for all predictors). When human calcium fractional absorption from soybeans and wheat products with a range of phytate concentrations were used in correlation models, the phytate:calcium molar ratio only predicted 32% of the variation ($r^2 = 0.5694$, $p < 0.0001$), whereas, total phytate content predicted 70% of the variation ($r^2 = 0.8396$, $p < 0.001$) [28]. When the phytate concentration

TABLE 13.1. Fractional Calcium Absorption from Calcium Carbonate as Influenced by Calcium Load and Phytate:Calcium Molar Ratios in Healthy Premenopausal Women.[a]

Calcium Load (mmol)	Phytate Load (mmol)	Phytate:Ca Molar Ratio	Fractional Ca Absorption ($\overline{X} \pm$ SD)
0.5	0	—	0.77 ± 0.13
0.5	0.716	1.45	0.26 ± 0.08
1.175	0.716	0.61	0.22 ± 0.07
2.75	0.716	0.26	0.25 ± 0.06
6.425	0.716	0.11	0.22 ± 0.07
12.5	0	—	0.38 ± 0.07
15.5	0.716	0.05	0.20 ± 0.06

[a]Extruded wheat bran cereal provided the phytate.
Reference: Compiled data from Reference [29].

was held constant in a meal of 40 g extruded wheat bran cereal but the calcium load was increased with the addition of calcium carbonate, fractional calcium absorption was not altered [29]. Fractional calcium absorption was less in meals containing the wheat bran cereal than from $CaCO_3$ fed alone and did not exhibit the inverse relationship with calcium load as did $CaCO_3$ (Table 13.1). *In vitro* binding studies showed linear binding over seven orders of magnitude of calcium, suggesting that binding was unrelated to the phytate:calcium molar ratio.

There are several reasons that could explain why phytate:mineral molar ratios in a food or meal cannot reliably predict mineral bioavailability. There is not a fixed stoichiometric relationship between phytate and any particular mineral, because it depends on the presence of multiple ions, pH, temperature, and ionic strength. Furthermore, the intestinal milieu alters these factors. The pH alters along the digestive tract, and endogenous secretions can dilute the phytate:mineral ratios. The specific environment at the site of absorption determines absorption but cannot be readily measured.

2.4. FOOD MATRIX

The environment that surrounds phytate affects its interaction with minerals. This is true for both the foodstuff and the intestine and for minerals that enter the intestine from the diet or through endogenous secretions. The influence of pH was discussed previously, and the effect of processing will be discussed in a later section. Here, we will concentrate on the other constituents that may enhance or inhibit mineral absorption in the presence of phytate, and this effect can be substantial. For example, iron absorption from a meal can vary between 2 and 35%, depending on the presence of enhancers and inhibitors [30].

Fiber is a predominant plant constituent that has long been attributed to decreasing mineral absorption [31]. Dietary fiber is the nondigestible carbohydrate that provides bulk matrix to plants. Minerals entrapped in this matrix that cannot be freed during digestion can be excreted along with the fiber. To this extent, increasing digestibility of fiber can increase mineral absorption. Some fibers such as pectin and psyllium have negative charges that bind mineral ions *in vivo* [32]. However, their impact on *in vivo* calcium bioavailability at usual fiber intakes at least is negligible [33]. Substantial evidence now suggests that the phytate associated with fiber in seeds is more responsible for the reduction in mineral bioavailability. For example, phytate-rich extruded wheat bran cereal decreased calcium absorption in humans [34], whereas, purified cellulose, the major fiber in wheat bran, did not [33]. Wheat bran also decreases zinc absorption in rats, but the purified wheat bran fiber does not [35]. Purified fibers including cellulose and pectin did not influence iron or zinc absorption in humans [36]. Removal of phytates in wheat bran nearly removed the inhibition by bran of iron absorption [37]. The addition of $\sim$25 g of a low-phytate barley fiber did not exhibit adverse effects on calcium, magnesium, iron, or zinc retention in young women [38].

The influence of the nature of the protein in mineral-rich foods and diets has been the subject of much study. Generally, minerals are better absorbed from animal sources than plant sources. Zinc bioavailability to rats was greater from egg and chicken diets than from soy diets [39,40]. In addition, when the diet was mixed, zinc bioavailability was intermediate between the animal protein or plant protein given alone, suggesting formation of a common absorptive pool. This has also been demonstrated for iron [41]. For iron, this is partly attributable to the occurrence of iron in meat partially as heme iron which is not vulnerable to the inhibition of phytate as is nonheme iron. Some proteins have specific effects on binding minerals. Partially digested peptides from meat, especially those containing cysteine, enhance absorption of iron and zinc [42]. Peptides that contain serine phosphate, such as casein, reduce its absorption [20], or carboxylic acid groups as in soy protein bind iron. The albumin extract of white beans is very inhibitory to iron and zinc dialyzability, primarily due to its high phytic acid content and also to the protein [43]. However, upon digestion, the released cysteine enhances dialyzability of these minerals. Recall that binding of minerals by these protein residues is pH dependent.

An anomaly is the comparable bioavailability of calcium from many soy foods with milk despite the high phytate and oxalate content of soy [44], although calcium bioavailability from fortified soy milk is only $\sim$75% that of cow's milk [45]. This may be because calcium is not located in situ in combination with phytate and oxalate, or the calcium phytate complex was of the form of Ca_1-phytate or Ca_2-phytate, which would be more soluble than other Ca^{2+}-phytate complexes [46].

Differential mineral bioavailability also occurs within various plant protein sources. Over half of the iron in wheat occurs as monoferric phytate, which is highly soluble at neutral pH, unlike ferric phytate, and the iron appears to enter the common nonheme iron pool [47]. The chemical form of minerals in most plant foods has not been well characterized. A unique method was used to study the chemical form of iron in wheat. Wheat was grown hydroponically and intrinsically labeled with a stable isotope of iron with an absorption profile that could be identified by Mössbauer spectroscopy.

These examples from plant seeds contrast with mineral bioavailability from plant foods that are low in inhibitors. For example, calcium absorption is higher from Brassica vegetables than from milk [48,49]. Molybdenum absorption was greater from kale than from soybeans [50].

Enhancers of mineral absorption can counteract the deleterious effect of phytic acid. The classic example is ascorbic acid, often recommended with vegetarian diets, or in other cases where heme iron ingestion is low to increase absorption of iron [1].

A newer example is the effect of fructooligosaccharides on enhancing mineral absorption. Several studies have demonstrated their role in increasing calcium and magnesium absorption in rats [51] and humans [52]. Recently, inulin at 10% of diets in rats counteracted the deleterious effects of phytic acid alone at 0.7% in rats on apparent absorption of calcium (-31%), zinc (-62%), and iron (-48%) [53]. There is some evidence that vitamin A and β-carotene counter the inhibitory effect of phytate from cereal diets [54]. Both compounds improved solubility of iron *in vitro* when the pH was adjusted from 2 to 6. These authors suggested that at the pH of the intestine, iron liberated during digestion might be chelated to vitamin A or β-carotene, preventing insolubilization by phytate or polyphenols.

3. PHYTATE HYDROLYSIS

3.1. MINERAL BIOAVAILABILITY

As phosphate groups are progressively removed from the inositol hexaphosphate ($InsP_6$), the mineral binding strength decreases and solubility increases [55]. At phosphorylations ≥ 5, iron solubility was decreased [56], zinc absorption was suppressed [57], and calcium absorption to suckling rats was inhibited [58].

A series of inositol phosphates were tested for their ability to increase the solubility of iron—$InsP_3$ and $InsP_4$ with a 1,2,3-triphosphate grouping did so [59]. Furthermore, $Ins(1,2,3,6)P_4$, but not $Ins(1,2,5,6)P_4$, fed at the same levels significantly enhanced calcium absorption from calcium ascorbate to rats [60].

Ins(1,2,3,6)P_4 is also a more potent second messenger than Ins(1,2,5,6)P_4 and may be involved in gut uptake [61]. Thus, the degree of phosphorylation determines whether mineral absorption is enhanced, not affected or inhibited.

Hydrolysis of phytate can occur by nonenzymatic or enzymatic means. Nonenzymatic hydrolysis could occur during extrusion or thermal processing [62]. Phytate can also be hydrolyzed by phytase which may occur endogenous to cereals and uncooked green leafy vegetables or may be added exogenously.

3.2. PHYTASE

Phytase catalyzes the stepwise hydrolysis of phytate to phosphate and inositol. Commercial sources of phytase are derived from wheat bran or microbial or fungal sources and are used in pretreatment of feed or during food fermentation. Phytase is added to swine and poultry rations to improve phytate phosphorus utilization. This can lessen the environmental concern of phosphorus water pollution from unabsorbed phytate. Fermentation of food with yeast, germination of seeds, treatment of foods with phytase, or removal of phytase with acid plus salt and ultrafiltering can improve mineral absorption (Table 13.2). Leavening during bread making improves mineral absorption. The impact of added phytase on mineral bioavailability can depend on the source. As shown in Table 13.2, microbial phytase removed all the inhibition of phytate on iron absorption, whereas, wheat phytase was ineffective. *A. niger* phytase has pH optima at 2.0 and 6.0, so hydrolysis could occur in the stomach, whereas, the pH optimum for wheat phytase is 5.0.

TABLE 13.2. The Influence of Phytase on Mineral Absorption in Humans.

Mineral	Test Food	Absorption (%)	Reference
Calcium	Wheat cookies	64.8 ± 8.7	[34]
	Wheat bread	70.3 ± 10.8[a]	
	Pinto bean, untreated	23.1 ± 5.3	[63]
	Pinto bean, phytase treated	31.8 ± 7.1[a]	
Iron	Wheat rolls + bran	10.4 ± 2.5	[64]
	Wheat rolls + autoclaved phytase-deactivated bran	10.3 ± 2.64	
	Bran	14.3 ± 2.6	
	Bran + microbial phytase	26.1 ± 3.8	
Iron	Soy protein isolate	1.36	[20]
	Soy protein + microbial phytase	5.48	
	Soy protein isolate + acid-salt-reduced phytate	4.17	

[a]Significantly different at $p < 0.05$ from unhydrolyzed food with the same protein source.

4. PROCESSING

Processing conditions that result in hydrolysis of phytate can improve mineral absorption as discussed above. Other processing conditions that could influence the effect of phytate on mineral absorption include pH and physical removal.

If the pH of the product is adjusted to minimize complexation of minerals with phytic acid and peptides, mineral absorption can be improved. The pH-dependent interaction of phytate, zinc, and soy protein has been studied *in vitro* [11]. Away from the isoelectric pH of the protein, 4.5, binding of ^{65}Zn was greatest. This explains why zinc from neutralized, isolated soy protein is less than from the acid precipitated isolate that is not adjusted to neutral pH [65].

Cooking does not improve dialyzability of minerals from legumes [44]. ^{65}Zn association with soy proteins was unaffected by autoclaving [11]. An adverse effect of cooking is deactivation of endogenous phytase. Genetic engineering has recently produced a heat-stable phytase that has 90% of its original activity after 20 minutes at 100°C [66].

Milling or other processing steps that lead to the removal of phytate can improve mineral absorption [67]. However, minerals are concentrated in the aleurone layer of cereals that is removed during the refining of cereals. Phytic acid is associated with protein bodies in legumes and is more difficult to remove by physical means.

5. ADAPTATION TO HIGH PHYTATE DIET

The potential effect of adapting to a high-phytate diet on mineral bioavailability has been tested in animals through intervention studies and in humans through population studies. No intestinal adaptation to a high-phytate diet was found for iron. Weanling rats adapted to either soy or wheat flour diets did not have improved ability to absorb iron from a soy test meal compared to rats adapted to chicken or casein diets [68]. In fact, rats adapted to casein had superior ^{59}Fe retention. Neither were vegans adapted to high-phytate diets able to absorb iron from wheat rolls with or without added bran better than a control group [69].

In contrast to iron, whole body retention of ^{65}Zn in rats adapted to soy did not decrease with age as for rats adapted to casein [70]. Retention of zinc from soy was initially higher if rats were adapted to casein, but by 56 days of adaptation, retention of zinc from a soy test meal was similar in rats adapted to soy or casein.

6. CONCLUSIONS

The literature suggests that if the essential minerals in a diet are present in adequate concentrations and in reasonable ratios with respect to one another, and to phytate, no reason exists for nutritional concern [71]. However, many populations might ingest inadequate quantities of calcium, iron, and zinc and may be marginally deficient in magnesium. Thus, concern continues about the impact of dietary phytate upon the mineral status of certain vulnerable segments of the population including children, teenagers, pregnant women, and the elderly. For the first three groups, representing periods of prolific growth, the phytate/mineral ratios are critical. In the final group, the baby-boomers, inadequate nutrient intakes coupled with use of over-the-counter medications may further compromise mineral status. Still others will strive to increase phytate consumption even as a supplement because of the health benefits [72]. The changing face of the food supply through genetic modification, fortification, creation of functional foods, and increased use of supplements will affect mineral nutriture. It will be complicated to sort out in the various subpopulations until we have better measures of mineral status as we currently have for iron.

7. REFERENCES

1. Frossard, E., Bucher, M., Mächler, F., Mozafar, A. and Hurrell, R. 2000. Potential for increasing the content and bioavailability of Fe, Zn, and Ca in plants for human nutrition, *J. Food Agric.* 80:861–879.
2. Harland, B.F. 1989. Dietary fibre and mineral bioavailability, *Nutr. Res. Rev.*, 2:133–147.
3. Harland, B.R. and Morris, E.R. 1995. Phytate: A good or bad food component? *Nutr. Res.* 15:733–754.
4. Reddy, N.R., Pierson, M.D., Sathe, S.K. and Salunkhe, D.K. 1989. *Phytates in Cereals and Legumes.* Boca Raton, FL, CRC Press. pp. 80–110.
5. Plaami, S. 1997. Myoinositol phosphates: Analysis, content in foods and effect in nutrition, *Lebensm. Wiss a Tech.* 30:633–647.
6. Zhou, J.R. and Erdman Jr., J.W. 1995. Phytic acid in health and disease, *Crit. Rev. Food Sci. Nutr.* 35:495–508.
7. Paullauf, J., Pietsch, M. and Reimbach, G. 1998. Dietary phytate reduces magnesium bioavailability in growing rats, *Nutr. Res.* 18:1029–1037.
8. Heaney, R.P., Recker, R.R. and Weaver, C.M. 1990. Absorbability of calcium sources. The limited role of solubility, *Calcif. Tissue Intl.* 46:300–304.
9. Deshpande, S.S. and Cheryan, M. 1984. Effects of phytic acid, divalent cations, and their interactions on α-amylase activity, *J. Food Sci.* 49:516–519.
10. Champagne, E.T. and Phillippy, B.Q. 1989. Effects of pH on calcium, zinc, and phytate solubilities and complexes following *in vitro* digestions of soy protein isolate, *J. Food Sci.* 54:587–592.
11. Khan, A., Weaver, C.M. and Sathe, S. 1990. Association of zinc with soy proteins as affected by heat and pH, *J. Food Sci.* 55:263–264.
12. Champagne, E.T. and Phillippy, B.Q. 1989. Effects of pH on calcium, zinc, and phytate solubilities and complexes following *in vitro* digestions of soy protein isolate, *J Food Sci.* 54:587–592.
13. Vohra, P., Gray, G.A. and Kratzer, F.H. 1965. Phytic acid-metal complexes, *Proc. Soc. Exp. Biol.* 120:447–454.
14. Nävert, B., Sandström, B. and Aderblad, A. 1985. Reduction of the phytate content of bran by leavening the bread and its effect on zinc absorption in man, *Brit. J. Nutr.* 53:47–53.

15. Prasad, A.S., Miale, A., Jr., Farid, Z., Sanstead, H.H., Schulert, A.R. and Darby, W.J. 1963. Biochemical studies on dwarfism, hypogonadism, and anemia, *Arch. Int. Med.* 111:407–428.
16. Reinhold., J.G., Nasr, K., Lahimgarzadeh, A. and Hedayati, H. 1973. Effects of purified phytate and phytate rich bread upon metabolism zinc, calcium, phosphorus, and nitrogen in man, *Lancet* 1:283–288.
17. Harland, B.F. and Oberleas, D. 1987. Phytate in foods, *Wrld. Rev. Nutr. Diet* 52:235–259.
18. Cook, J.D., Reddy, M.B., Burri, J., Juillerat, M.A. and Hurrell, R.F. 1997. The influence of different cereal grains on iron absorption from infant cereal foods, *Am. J. Clin. Nutr.* 65:964–969.
19. Mendoza, C., Viteri, F.E., Lonnerdal, B., Young, K.A., Raboy, V. and Brown, K.H. 1998. Effect of genetically modified, low-phytic acid maize on absorption of iron from tortillas, *Amer. J. Clin. Nutr.* 68:1123–1127.
20. Hurrell, R.F., Juillerat, M.-A., Reddy, M.B., Lynch, S.R., Dassenko, S.A. and Cook, J.D. 1992. Soy protein, phytate, and iron absorption in humans, *Amer. J. Clin. Nutr.* 56:573–578.
21. Beard, J.L., Weaver, C.M., Lynch, S., Johnson, C.D., Dassenko, S. and Cook, J.D. 1988. The effect of soybean phosphate and phytate content on iron bioavailability, *Nutr. Res.* 8:345–352.
22. Morris, E.R. and Ellis, R. 1981. Phytate-zinc molar ratio of breakfast cereals and bioavailability of zinc to rats, *Cereal Chem.* 58:363–366.
23. Morris, E.R. and Ellis, R. 1980. Bioavailability to rats of iron and zinc in wheat bran: response to low-phytate bran and effect of the phytate/zinc molar ratio, *J. Nutr.* 110:2000–2010.
24. Fordyce, E.S., Forbes, R.M., Fobbins, K.R. and Erdman, J.W. 1987. Phytate × calcium/zinc molar ratios: are they predictive of zinc bioavailability? *J. Food Sci.* 52:440–444.
25. Davies, N.T., Carswell, A.J.P. and Mills, C.F. 1986. The effect of variation in dietary calcium intake on the phytate-zinc interaction in rats. In: *Trace Element Metabolism in Man and Animals (TEMA V)*, pp. 456–457.
26. Morris, E.R. and Ellis, R. 1985. Bioavailaiblity of dietary calcium-effect of phytate on adult men consuming nonvegetarian diets. In: *Nutritional Bioavailability of Calcium*, Kies, C., Ed.; ACS Symposium Series 275: American Chemical Society: Washington, DC, p. 63.
27. Saha, P.R., Weaver, C.M. and Mason, A.C. 1994. Mineral bioavailability in rats from intrinsically labeled whole wheat flour of various phytate levels, *J. Agric. Food Chem.* 42:2531–2535.
28. Weaver, C.M., Martin, B.R. and Heaney, R.P. 1991. Calcium absorption from foods. In: *Nutritional Aspects of Osteoporosis*, Burckhardt, P. and Heaney, R.P., Ed., Serono Symposia Publications from Raven Press 85:133–137.
29. Weaver, C.M., Heaney, R.P., Teegarden, D. and Hinders, S.M. 1996. Wheat bran abolishes the inverse relationship between calcium load size and absorption fraction in women, *J. Nutr.* 126:303–307.
30. Monsen, E.R., Hallberg, L., Layrisse, M., Hegsted, D.M., Cook, J.D., Merz, W. and Finds, C.A. 1978. Estimation of available dietary iron, *Amer. J. Clin. Nutr.* 31:134–141.
31. Widdowson, E.M. and McCance, R.A. 1942. Mineral metabolism of healthy adults on white and brown bread dietaries, *J. Physiol.* 101:44–85.
32. James, W.P.T., Branch, W.J. and Southgate, D.A.T. 1978. Calcium binding by dietary fibre, *Lancet* 1:638.
33. Heaney, R.P. and Weaver, C.M. 1995. Effect of psyllium on absorption of co-ingested calcium, *J. Amer. Ger. Soc.* 43:1–3.
34. Weaver, C.M., Heaney, R.P., Martin, B.R. and Fitzsimmons, M.L. 1991. Human calcium absorption from whole wheat products, *J. Nutr.* 121:1769–1775.
35. Davies, N.T., Hristic, V. and Flett, A.A. 1977. Phytate rather than fibre in bran as the major determinant of zinc availability to rats, *Nutr. Rep. Inter.* 15:207.
36. Rossander, L., Sandberg, A.-S. and Sandström, B. 1992. The influence of dietary fiber on mineral absorption and utilization. In: *Dietary Fiber—A Component of Food: Nutritional Function in Health and Disease*, Schweizer, T.F. and Edwards, C.A., Eds. London: Springer Verlag, pp. 197–216.

37. Hallberg, L., Rossander, L. and Skanberg, A.-B. 1987. Phytates and the inhibitory effect of bran on iron absorption in man, *Amer. J. Clin. Nutr.* 45:988–996.

38. Wisker, E., Nage, R., Tanudjaja, T.K. and Feldman, W. 1991. Calcium, magnesium, zinc, and iron balances in young women: effects of a low-phytate barley-fiber concentrate, *Amer. J. Clin. Nutr.* 54:553–559.

39. Stuart, M.A., Ketelsen, S.M., Weaver, C.M. and Erdman, Jr., J.W. 1986. Bioavailability of zinc to rats as affected by protein source and previous dietary intake, *J. Nutr.* 116:1423–1431.

40. Meyer, N.R., Stuart, M.A. and Weaver, C.M. 1983. Bioavailability of zinc from defatted soy flour, soy hulls, and whole eggs as determined by intrinsic and extrinsic labeling techniques, *J. Nutr.* 113:1255–1264.

41. Hallberg, L. 1974. The pool concept in food iron absorption and some of its implications, *Proc. Nutr. Soc.* 33:285–291.

42. Taylor, P.G., Martinez-Torres, C., Ramono, E.L. and Layrisse, M. 1986. The effect of cysteine-containing peptides released during meat digestion on iron absorption in humans, *Amer. J. Clin. Nutr.* 43:68–71.

43. Lombardi-Boccia, G., Schlemer, U., Cappelloni, M. and Di Lullo, G. 1998. The inhibitory effect of albumin extracts from white beans (Phaseolus vulgaris L.) on *in vitro* iron and zinc dialysability: role of phytic acid, *Food Chem.* 63:1–7.

44. Heaney, R.P., Weaver, C.M. and Fitzsimmons, M.L. 1991. Soybean phytate content: effect on calcium absorption, *Amer. J. Clin. Nutr.* 53:745–747.

45. Heaney, R.P., Dowell, M.S., Rafferty, K. and Bierman, J. 2000. Bioavailability of the calcium in fortified soy imitation milk with some observations on methods, *Amer. J. Clin. Nutr.* 71:1166–1169.

46. Graf, E. 1983. Calcium binding to phytic acid, *J. Agric. Food Chem.* 31:851–855.

47. Morris, E.R. and Ellis, R. 1976. Isolation of monoferric phytate from wheat bran and its biological value as an iron source to the rat, *J. Nutr.* 106:753–760.

48. Heaney, R.P., Weaver, C.M., Hinders, S.M., Martin, B. and Packard, P.T. 1993. Absorbability of calcium from Brassica vegetables: broccoli, bok choy, and kale, *J. Food Sci.* 59:1378–1380.

49. Weaver, C.M., Heaney, R.P., Nickel, K.P. and Packard, P.I. 1997. Calcium bioavailability from high oxalate vegetables: Chinese vegetables, sweet potatoes and rhubarb, *J. Food Sci.* 62:524–525.

50. Turnlund, J.R., Weaver, C.M., Kim, S.K., Keyes, W.R., Gizaw, Y., Thompson, K.H. and Peiffer, G.L. 1999. Molybdenum absorption and utilization in humans from soy and kale intrinsically labeled with stable isotopes of molybdenum, *Amer. J. Clin. Nutr.* 69:1217–1223.

51. Ohta, A., Ohtsuki, M., Baba, S., Takizawa, T., Adachi, T. and Kimura, S. 1995. Effects of fructo-oligosaccharides on the absorption of iron, calcium and magnesium in iron deficient anemic rats, *J. Nutr. Sci. Vitamin* 41:281–291.

52. Coudray, C., Bellager, J., Castiglia-Delavaud, C., Rémésy, C., Vermorel, M. and Rayssiguier, Y. 1997. Effect of soluble or partly soluble dietary fibres supplementation on absorption and balance of calcium, magnesium, iron and zinc in healthy young men, *Eur. J. Clin. Nutr.* 51: 375–380.

53. Lopez, H.W., Coudray, C., Levart-Verny, M.-A., Feillet-Coudray, C., Demmqué, C. and Rémésy, C. 2000. Fructooligosaccharides enhance mineral apparent absorption and counteract the deleterious effects of phytic acid on mineral homeostasis in rats, *J. Nutr. Biochem.* 111:500–508.

54. Garcia-Casal, M.N., Layrisse, M., Solano, L., Barón, M.A., Arguello, F., Llovera, D., Ramírez, J., Leets, I. and Tropper, E. 1998. Vitamin A and β-carotene can improve nonheme iron absorption from rice, wheat and corn by humans, *J. Nutr.* 128:646–650.

55. Jackman, R.H. and Black, C.A. 1951. Solubility of iron, aluminum, calcium and magnesium inositol phosphates at different pH values, *Soil. Sci.* 72:179–186.

56. Sandberg, A.-S., Carlsson, N.-G. and Svanberg, U. 1989. Effects of inositol Tri-, Tetra-, and Hexaphosphates on *in vitro* estimation of iron availability, *J. Food Sci.* 54:159–161.

57. Sandström, B. and Sandberg, A.-S. 1992. Inhibitory effects of isolated inositol phosphates on zinc absorption in humans, *Trace Elements and Electrolytes in Health and Disease*, 6:99–103.

58. Lonnerdal, B., Sandberg, A.S., Sandstrom, B. and Kunz, C. 1989. Inhibitory effects of phytic acid and other inositol phosphates on zinc and calcium absortion in suckling rats, *J. Nutr.* 119:211–214.

59. Phillippy, B.Q. and Graf, E. 1997. Antioxidant functions of inositol 1,2,3-triphosphate and inositol 1,2,3,6-tetrakisphosphate, *Free Rad. Biol. Med.* 22:939–946.

60. Shen, X., Weaver, C.M., Kempa-Steczko, A., Martin, B.R., Phillippy, B.Q. and Heaney, R.P. 1998. An inositol phosphate as a calcium absorption enhancer in rats, *J. Nutr. Biochem.* 9:298–301.

61. Delisle, S., Radenberg, T., Wintermantel, M.R., Tietz, C., Parys, J.B., Pittet, D., Welsh, J. and Mayr, G.W. 1994. Second messenger specificity of the inositol trisphophate receptor: Reappraisal based on novel inositol phosphates, *Amer. J. Physiol.* 226:C429–C436.

62. Phillippy, B.Q., White, K.D., Johnson, M.R., Tao, S.H. and Fox, M.R.S. 1987. Preparation of inositol phosphates from sodium phytate by enzymatic and nonenzymatic hydrolysis, *Ann. Biochem.* 162:115–121.

63. Weaver, C.M., Heaney, R.P., Proulx, W.R., Hinders, S.M. and Packard, P.T. 1993. Absorbability of calcium from common beans, *J. Food Sci.* 58:1401–1403.

64. Sandberg, A.-S., Hulthen, L.R. and Turk, M. 1996. Dietary *Aspergillus niger* phytase increases iron absorption in humans, *J. Nutr.* 126:476–480.

65. Kettelsen, S.M., Stuart, M., Weaver, C.M., Forbes, R.M. and Erdman Jr., J.W. 1984. Bioavailability of zinc from defatted soy flour, acid-precipitated soy concentrate and neutralized soy concentrate as determined by intrinsic and bioavailability to rats from defatted flour, *J. Nutr.* 114:1035–1042.

66. Pasamontes, L., Hacker, M., Wyss, M., Tessier, M. and van Loon, A.P. 1997. Gene cloning, purification, and characterization of a heat-stable phytase from the fungus *Aspergillus fumigatus*, *Appl. Environ. Microbiol.* 63:1696–1700.

67. Weaver, C.M., Mason, A.C. and Hamaker, B.R. 2000. Food Uses. Ch. 3. In: *Designing Crops for Added Value*. C.F. Murphy and D.M. Peterson, Eds. Published by the American Society of Agronomy, Crop Science Society of American, and Soil Science Society of America. Agronomy Monograph 40, pp. 21–55.

68. Johnson, C.D. and Weaver, C.M. 1983. Effect of previous diets on iron absorption from an intrinsically labeled soy flour testmeal, *Nutr. Reports. Intl.* 28:1129–1135.

69. Brune, M., Rossander, L. and Halberg, L. 1989. Iron absorption: no intestinal adaptation to a high-phytate diet, *Amer. J. Clin. Nutr.* 49:542–545.

70. Khan, A. and Weaver, C.M. 1989. Bioavailability of zinc to rats from soybeans and casein as affected by protein source and length of adaptation, *Nutr. Res.* 9:327–336.

71. Kelsay, J. 1987. Effects of fiber, phytic acid, and oxalic acid in the diet on mineral bioavailability, *Amer. J. Gastroenterol.* 82:983–986.

72. Thompson, L.U. 1989. Nutritional and physiological effects of phytic acid. In: *Food Proteins*. J.E. Kensella and V.G. Soucie, Eds. Champaign, IL: American Oil Chemists Society, pp. 410–431.

CHAPTER 14

Role of Phytic Acid in Cancer and Other Diseases

MAZDA JENAB
LILIAN U. THOMPSON

1. INTRODUCTION

PHYTIC acid (PA, InsP$_6$), is found in cereals, legumes, nuts and oilseeds, constitutes up to 1–5% of the weight of cereals or oilseeds and serves as the chief storage form of phosphorus [1–4]. PA has long been considered an antinutrient [5], mainly due to its ability to bind with many divalent cations, proteins and starch and to consequently reduce their bioavailability [4,6–8]. This binding ability is attributed to PA's highly negatively charged phosphorylated structure. To combat PA's antinutritive effects, particularly in nutritionally compromised populations, many ways of removing it from foods have been suggested [9–11]. However, it has also been suggested by many studies that consumption of PA may convey some beneficial health effects [12–17]. This chapter will discuss *in vitro* and *in vivo* studies on the effect of PA on the risk of cancer and other chronic diseases and some of its mechanisms of action.

2. CANCER

As will be discussed in more detail below, PA has been shown to be protective of a variety of cancers in many different *in vitro* and *in vivo* models. However, comparison of these studies is complicated by differences in the method of PA supplementation. A majority of the studies on PA and cancer have used either 1–2% pure PA supplemented to a low-fiber diet or added to the drinking water, while a limited number of studies used cereal brans as a source of naturally occurring PA. Which method of PA supplementation is more effective is not clear. Because animals eat various amounts of diet and drink varying volumes of water, those fed, for example, 1% PA in the diet, may be receiving different amounts of PA than animals consuming drinking water supplemented with the same percentage of PA, particularly because the addition of PA to the drinking water may also require adjusting the pH of the water to make it more palatable for the animals [18]. This issue is further complicated by the fact that few of the studies on PA and cancer provided details of the total amounts of PA, food or drinking water consumed. Thus, it is very difficult to judge which amount of PA and delivery via which medium is most effective for cancer prevention or treatment. Nonetheless, as will be discussed below, it is clear that PA, at moderate levels of intake, is having a protective effect on various cancers.

3. COLON CANCER

3.1. *IN VITRO* STUDIES–PURE PA

Purified PA has been shown in *in vitro* studies to inhibit cell growth and increase cell differentiation and maturation of HT-29 human colon carcinoma cells [19,20]. In these studies, PA caused a reversion of malignant phenotype leading to a decrease in expression of tumor markers. Also, in HT-29 human colon carcinoma cells, upregulation of the expression of p53, a tumor suppressor gene, and p21$^{WAF1/CIP1}$, a growth inhibitor, has been shown to be effected by pure PA [21]. This suggests that the effects of PA on the growth and maturation of these cells may involve the direct modulation of genes.

3.2. *IN VIVO* STUDIES–PURE PA IN DRINKING WATER

In *in vivo* studies, pure PA given in the drinking water has been shown to reduce the rate of colonic cell proliferation, an early biomarker of colon cancer risk, at early (four weeks) [22] and late (36–40 weeks) [23,24] time points (Table 14.1).

It has also been shown to reduce various colon tumor parameters when given at either the initiation [14,25] or promotional stages [24,26] of colon carcinogenesis (Table 14.1). When administered to azoxymethane (AOM)-treated rats, up to five months post-initiation, PA (2%) significantly reduced colon tumor number, size and mitotic rate, when compared to the control group [24], suggesting that PA can have anticancer effects at both the initiation and promotion stages of colon cancer. Also, using 2% PA provided in the drinking water, Pretlow et al. [26] showed a reduced number of colon tumors and tumor volume (Table 14.1).

Pretlow et al. [26] also observed a significant increase in large aberrant crypt foci (ACF) with time in the control group versus the PA treated group, suggesting that PA decreased the growth of large ACF, which are thought to be more likely to progress toward tumors. The total number of ACF, however, was

TABLE 14.1. Summary of Studies on the Effect of Endogenous or Exogenous Phytic Acid (PA) on Colon Tumorigenesis and Its Early Risk Markers.

Parameters Measured	Endogenous PA in Wheat Bran	Pure Exogenous PA provided in		Effect	Amount of PA	Reference
		Diet	Drinking Water			
Colon tumor						
Number		•		↓	0.25%	[33]
			•	↓	1.0%	[23]
			•	↓	1.0, 2.0%	[14]
			•	↓	2.0%	[24]
			•	↓	2.0%	[26]
Frequency*		•✿		↓	0.4%	[45]
			•	↓	0.1%, 1.0%	[25]
			•	↓	1.0, 2.0%	[14]
Volume/size		•✿		↓	0.4%	[45]
			•	↓	0.1%, 1.0%	[25]
			•	↓	2.0%	[24]
			•	↓	2.0%	[26]
ACF parameters						
Number		•		↓	1.0%	[17]
		•		↓	1.0%, 2.0%	[33]
		•		↓	2.0%	[17]
			•	↓	2.0%	[26]
Number of SIM	•	•		↓	1.0%	[17]
Degree of LA	•	•		↓	1.0%	[17]
Cell proliferation		•		↓	0.6%, 1.2%, 2.0%	[13]
		•		↓	1.2%	[15]
	•	•		↓	1.0%	[17]
			•	↓	2.0%	[22]
Mitotic rate			•	↓	1.0%	[23]
			•	↓	1.0, 2.0%	[14]
			•	↓	2.0%	[24]
Apoptosis	•	•		↑	1.0%	[17]
			•	NS	2.0%	[22]
Differentiation	•	•		↑	1.0%	[17]

Abbreviations: NS = not significant; some studies are repeated in different sections; *also includes tumor incidence and multiplicity, ✿significant effect observed with pure PA only in combination with 2% wheat bran oil; arrows pointing down indicate a lowering or protective effect; arrows pointing up indicate an increase.

not significantly different between the treatment and control groups. ACF are thought to be valid pre-neoplastic markers of colon cancer risk [27], because they have been shown to have increased proliferative activity in comparison to neighboring normal crypts [28,29]. They have also been shown to have numerous genetic mutations such as in the p53 tumor suppressor gene [30] and in the k-ras oncogene [31].

3.3. *IN VIVO* STUDIES–PURE PA IN DIET

In *in vivo* studies, pure PA supplemented to the diet has been shown to reduce the rate of colonic cell proliferation, an early biomarker of colon cancer risk, after 2–14 weeks of treatment [13–15,17] (Table 14.1). Similar to PA supplemented in the drinking water, PA given in the diet has also been shown to be colon cancer protective [32,33] (Table 14.1). Shivapurkar et al. [32] evaluated the chemoprotective effect of PA (2%) in a dual organ (mammary and colon) cancer model and found that PA (2%) significantly decreased the number of colon tumors after 30 weeks of treatment. Nelson et al. [33] found that PA supplemented to a high-iron diet, significantly decreased the number of colon tumors and eliminated the colon cancer promoting effects of iron.

Dietary PA has also been shown to reduce various ACF parameters [17,32,34]. Shivapurkar et al. [32] showed that in addition to reducing the number of colon tumors, 2% PA significantly reduced the total number of ACF and number of larger ACF (those with multiplicity of four or more) after nine weeks, while Challa et al. [34] found that both 1% and 2% PA reduced the number of ACF in various colon sections after 13 weeks of treatment. Challa et al. [34] also showed an interesting synergistic effect between PA and green tea on ACF reduction, suggesting that PA can interact with other phytochemicals to provide an enhanced protective effect. In the study of Jenab and Thompson [17], PA significantly decreased the number of sialomucin-producing ACF (SIM ACF) (Figure 14.1) in the distal colon of rats fed 1% PA supplied in the diet. Sialomucins are abnormal for the distal colon of rats and humans and are found in a majority of colon tumors. Because SIM ACF are thought to be more advanced toward tumorigenesis [29], their inhibition by PA provides strong evidence for its colon cancer protective role. In the same study, dietary PA also significantly reduced the rate of cell proliferation [17] while increasing the rate of cellular apoptosis and differentiation [35] in the normal colon. These data suggest that PA can influence the growth kinetics of normal colon cells as well as affect the growth of abnormal cells in ACF and cells in tumors.

3.4. CLINICAL STUDIES

Only a few clinical studies have been conducted on the role of PA in colon cancer. Because iron has been shown *in vitro* to be strongly bound by PA [36,37]

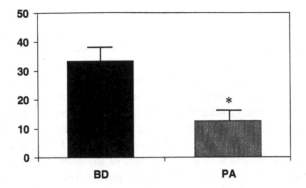

Figure 14.1 The effect of phytic acid (PA) on sialomucin producing aberrant crypt foci in the whole colon of rats. Values are means ± SEM; $n = 12$ rats per group; BD = basal diet control group; PA = 1% pure phytic acid group. There is a significant difference between the groups, $p < 0.05$. (Adapted from Reference [17].)

and to be a promoter of colon carcinogenesis in epidemiological [38–41] and animal studies [33], Owen et al. [42] determined if there was a correlation between fecal PA and fecal iron and rate of cell proliferation in adenoma patients. Although they observed a strong correlation between fecal PA and iron, they did not see any relationship between fecal PA and the rate of cell proliferation, a key early biomarker of colon cancer. This study is only preliminary, and its results may be influenced by other factors such as the presence of adenomas affecting colorectal cell kinetics [43], matched time of fecal collection and colonic sampling for the rate of cell proliferation, whether the patients were maintained on the same diets as they were at the time of fecal collection or how fecal PA levels relate to dietary PA intake. Any effects of dietary PA on cell proliferation may require long-term exposure to constant levels of dietary PA, and therefore, clinical experiments in this area will require long-term consumption of controlled diets with known levels of PA and minerals. Fecal PA may not necessarily be reflective of the rate of cell proliferation, because it is essentially unreactive as it is complexed with other dietary components. It is possible that colonic cell proliferation may be influenced by PA that is degraded within the colon to lower inositol phosphates or PA that is absorbed by the colonocytes.

3.5. PA SUPPLEMENTATION–ENDOGENOUS
VERSUS EXOGENOUS

Because wheat bran is one of the richest dietary sources of PA, some studies [17,44,45], have used it in dietary models of PA intake. However, the use of wheat bran as a source of PA compared to addition of pure PA to a purified animal diet or drinking water brings out the issue of endogenous versus exogenous sources of PA. Due to its chemical properties, pure exogenous PA in the diet can interact with other food components. For example, it may form complexes with the mineral, protein or starch components in the diet [12]. Pure PA given in the drinking water may also interact with other food components provided that it is consumed concurrently with diet. If the animal consumes its water and food at different time periods, then the interaction of the pure PA in the drinking water with other components of the diet may be minimized. However, endogenous PA present within the matrix of a high-fiber food source such as wheat bran, may not be able to interact with other dietary components or even the colonic mucosa, because it may already be tightly bound to the fiber, proteins, starch or minerals. Thus, the extent of the anticancer effects of PA may be related to whether it is given as exogenous, pure PA, in the diet or drinking water, or provided as a natural component (endogenous) of a high-fiber food such as wheat bran. Although the literature contains suggestions that supplementation of a diet with pure PA, for the purpose of cancer risk reduction, may be better than eating fiber-rich foods [16,46], it must be noted that a high-fiber diet will likely provide many more phytochemicals than just PA. These other phytochemicals may also be protective of a variety of cancers as well as diseases other than cancer. Therefore, it is still more practical to eat diets containing high-fiber foods than a refined diet supplemented with just one phytochemical. Nonetheless, to date, very little comparison has been made between endogenous and exogenous PA and their effects on early markers of colon carcinogenesis [17] or colon tumorigenesis [45].

The study of Jenab and Thompson [17] differentiated the role of endogenous PA present within the matrix of 25% wheat bran versus exogenous pure PA (1.0%; equivalent to the amount of PA in the 25% wheat bran diet) added to a low-fiber diet on aberrant crypt foci (ACF) and sialomucin producing ACF, early biomarkers of colon cancer risk. To test the effect of endogenous PA, they compared the results of the 25% wheat bran group to a group of animals fed 25% dephytinized wheat bran (wheat bran with the endogenous PA removed), 1.0% pure PA or 25% dephytinized wheat bran plus 1.0% pure PA. In this study, since food intake did not differ among the treatment groups, the actual intake of pure exogenous PA was equivalent to that of the endogenous PA in wheat bran. Thus, any variation in results may not be due to differences in level of PA intake. The exogenous PA significantly reduced the number and size of ACF, while all the treatment groups significantly reduced the num-ber of sialomucin-producing ACF versus the control diet. These data suggest that although endogenous and exogenous PA are both effective, exogenous PA may be more so. In addition, Jenab and Thompson [17] showed that although all the treatment diets significantly reduced the labeling index of cell prolif-eration in the top 40% of the crypt versus the control diet, the wheat bran with its endogenous PA was significantly more effective than the other diets (Figure 14.2). This suggests that the endogenous PA is playing a more important role than the wheat bran fiber because the removal of PA from the wheat bran

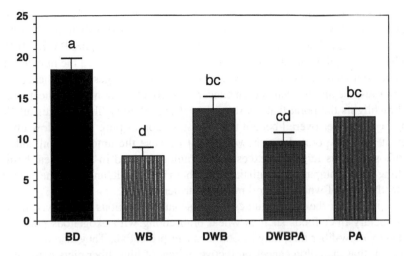

Figure 14.2 Labeling index in the top 40% of the crypt in the colon of rats. Values are means ± SEM; $n = 12$ rats per group; BD = basal diet control group; WB = 25% wheat bran; DWB = 25% dephytinized wheat bran; DWBPA = 25% dephytinized wheat bran plus 1% pure PA; PA = 1% pure phytic acid group. Values with different superscripts are significantly different, $p < 0.05$. (Adapted from Reference [17].)

caused an increase in the rate of cell proliferation. In addition, the endogenous PA may be more effective than the exogenous PA, because although the rate of cell proliferation with the 1% PA diet was significantly lower relative to the control group, it was significantly higher than the wheat bran group. This differential effect of endogenous and exogenous PA suggests that although both are effective, they may be acting through different mechanisms.

In a long-term colon tumorigenesis study, Reddy et al. [45] found that the effect of a 10% wheat bran diet did not differ significantly from a 10% wheat bran diet with its oil or endogenous PA components removed, with or without addition of 0.4% exogenous PA. Many of the other colon carcinogenesis studies on wheat bran and PA discussed above used levels of PA greater than 0.4% (Table 14.1) and levels of wheat bran greater than 10% [47–51]. Thus, the use of higher levels of wheat bran and PA supplementation in this study may have produced more positive results. Nonetheless, the results suggest that PA may not be a strong factor in the protective effects of wheat bran, since removal of endogenous PA and addition of exogenous PA did not significantly affect colon tumorigenesis.

However, the most important finding of the study by Reddy et al. [45] is that removal of both the oil and PA components of wheat bran caused a significant increase in the number of adenocarcinomas, suggesting that the colon cancer protective effects of wheat bran may be associated with a combination of its

PA and other components such as its oil. In the Reddy et al. [45] study, wheat bran oil (at a level of 2% which is an amount in excess to that found in 10% wheat bran) added either alone or in combination with 0.4% pure PA to a 10% dephytinized and defatted wheat bran diet significantly reduced the number of adenocarcinomas versus the 10% wheat bran diet. This suggests that wheat bran oil may also be an important colon cancer protective component of wheat bran and highlights the point that consumption of high dietary fiber sources of PA may be of greater overall benefit than consumption of pure PA in a low-fiber diet. It must be pointed out, however, that because the amount of oil added in this study was largely in excess of the amount found in 10% wheat bran, making direct comparisons with the wheat bran is difficult, and thus, the cancer protective role of wheat bran oil must be further studied.

The results of the above studies suggest that endogenous PA derived from high dietary fiber foods such as wheat bran, along with exogenous pure PA added to a low-fiber diet, may be colon cancer protective. They also highlight the point that the colon cancer protective effects of high-fiber diets may stem from more than just one important component.

4. MAMMARY CANCER

In vitro, pure PA added to estrogen receptor positive MCF-7 and estrogen receptor negative MDA-MB-231 human breast cancer cells has been shown to inhibit their growth and increase their cell differentiation and maturation [52].

Table 14.2 provides a summary of animal studies on the effect of endogenous or exogenous PA on mammary cancer and its early risk markers. In animal studies, dietary PA supplementation (1.2%) reduced cell proliferation in the mammary gland [15]. The reductions were stronger when the PA was added to diets supplemented with high levels of iron and calcium, suggesting that PA was binding these cations and inhibiting their promotive effects. In mammary tumorigenesis studies, PA supplementation was initially shown to bring about a slight, nonsignificant decrease [53]. However, subsequent experiments [16,54] showed that pure PA, supplemented in the drinking water, effectively decreased tumor incidence in a dimethylbenzanthracene (DMBA) mammary tumor model. Since then, similar effects have been observed with pure PA provided in the diet [32,55].

The debate over endogenous versus exogenous PA discussed above for models of colon cancer has been addressed to an extent for models of mammary cancer by Vucenik et al. [16], who showed that levels of wheat bran up to 20% were ineffective in reducing rat mammary tumorigenesis, while PA (0.4%) added in pure form to the drinking water caused a significant reduction in the tumor incidence. However, the researchers [16] indicated that the level of pure exogenous PA in the drinking water of one group was matched to the level of endogenous PA present in the 20% wheat bran diet of the other group, it is

TABLE 14.2. Summary of Studies on the Effect of Endogenous or Exogenous Phytic Acid (PA) on Various Cancers and Early Risk Markers.

Parameters Measured	Endogenous PA in Wheat Bran	Pure Exogenous PA provided in		Effect	Amount of PA	Reference
		Diet	Drinking Water			
Mammary cell proliferation	•			↓	1.2%	[15]
Mammary tumor						
Number			•	NS	15 mM	[53]
		•		↓	2.0%	[32]
Frequency*			•	↓$^\Phi$	15 mM	[54]
	•		•	↓^{f}	0.4%	[44]
Size$^\lambda$		•		↓	2.0%	[55]
			•	↓	2.0%	[129]
Soft tissue tumor		•		↓	8.9%	[61]
Skin tumor			•	↓	2.0%	[63]
Liver tumor			•	↓	2.0%	[129]
Bladder tumor		•		↑	2.0%	[64]
		•		↑$^\xi$	2.0%	[130]
			•	↑$^\xi$	2.0%	[129]
Multi-organ tumorigenesis		•		NS	1.0%	[131]

Abbreviations: NS = not significant; *also includes tumor incidence and multiplicity; $^\Phi$significant effect only when PA combined with 15mM inositol; ^{f}endogenous PA in wheat bran not shown to have any significant effects; $^\lambda$also includes tumor diameter; $^\xi$significant effect only with sodium salt of PA; arrows pointing down indicate a lowering or protective effect; arrows pointing up indicate an increase.

unclear how much PA the two groups consumed or whether these levels were comparable. Thus, in this case, it is very difficult to compare the effects of the endogenous PA in wheat bran to exogenous pure PA given in the drinking water or to judge one as being more effective than the other.

5. OTHER CANCERS

In vitro, purified PA at various concentrations has been shown to decrease the growth and increase the differentiation of K562 human erythroleukemia cell lines [56], human PC-3 prostate adenocarcinoma [57] and human HepG2 liver cancer [58] cells. In addition, PA has also been shown to decrease the growth of human rhabdomyosarcoma [59] cells in culture.

In vivo, the growth of HepG2 liver cancer cells injected in immunodeficient mice was inhibited by pure PA (40 mg/kg of body weight) given via the drinking

TABLE 14.3. Summary of Studies Using Nondietary Modes of Phytic Acid (PA) Supplementation.

Parameters Measured	Mode of PA Administration	Effect	Amount of PA	Reference
Liver tumor weight	Intratumoral injection	↓	40 mg/kg of body weight	[60]
Soft tissue tumor Incidence	Peritumoral injection	↓	40 mg/kg of body weight	[59]
Growth	Peritumoral injection	↓	40 mg/kg of body weight	[59]
	i.p. injection	↓	0.25%	[62]

Arrows pointing down indicate a lowering or protective effect.

water [60] (Table 14.2). Furthermore, immunodeficient mice, not treated with PA but inoculated with HepG2 cells that were pretreated with PA, developed substantially less tumors than mice inoculated with untreated cells [60]. PA (40 mg/kg of body weight) has been shown to inhibit human rhabdomyosarcoma growth when injected directly into the tumor in mice [59] (Table 14.3), while PA treatment, at 8.9% given in the diet, has reduced the growth of rat fibrosarcoma [61] (Table 14.2) and, at a level of 0.25% delivered by direct i.p. injection, subcutaneously transplanted mice fibrosarcoma cells as well as the number of established pulmonary metastases from those cells [62] (Table 14.3). PA (2%) has also been protective against skin 2-step papillomas [63] when given in the drinking water and has inhibited hepatocellular carcinomas in the liver and putative preneoplastic lesions in the pancreas [64] when administered via the diet (Table 14.2).

6. POSSIBLE MECHANISMS OF ANTICANCER EFFECTS

Because PA is ubiquitous to every mammalian cell, it is not surprising that it has a cancer protective effect on different tissues, in different experimental models and under various conditions. However, the mechanisms of PA action are not clear and are open to conjecture.

Many of the proposed mechanisms of PA action are related to its structure and its physical properties (Figure 14.3). For example, PA may potentially bind important proteins and enzymes, either within the cell or in the digestive tract. Within the cell, binding of enzymes may alter the cell's growth characteristics. Although PA has been shown in vitro to inhibit the activity of several cellular enzymes such as serine/threonine protein phosphatases [65], the potential effect of this on individual cells or in vivo is not known. The mechanisms of PA

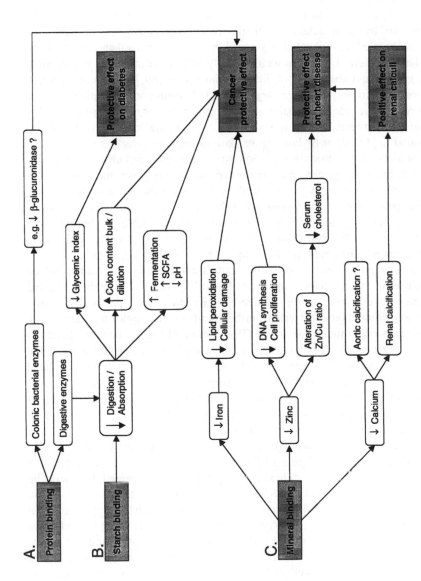

Figure 14.3 Possible mechanisms of phytic acid action.

action may well depend on whether PA enters the cell intact or is reformed intracellularly. PA has been shown to be rapidly absorbed and converted to lower inositol phosphates in various murine and human cells *in vitro* [66] and in rats [67]. However, the potential exists that hydrolysis products of dietary PA, hydrolyzed by either dietary, intestinal or bacterial phytases, may also enter the cell and be used for intracellular production of inositol phosphates and PA [68].

The inhibition of enzymes within the digestive tract may lead to inhibition of digestion and absorption of dietary components [69,70] [Figure 14.3(A)]. PA may bind to enzymes necessary for starch digestion or, alternatively, it may also reduce the rate of digestion and absorption of starches [Figure 14.3(B)] either by hydrogen binding to starch [12,71], binding to proteins that starch is bound to [12,70] or by binding amylase or enzyme cofactors such as Ca^{2+} [4]. Studies suggest that PA may slow starch digestion and absorption *in vivo* [72,73]. Such undigested and unabsorbed starch may reach the colon where it may either contribute to fecal bulk and increase the dilution of potential carcinogens, or it may be fermented to short-chain fatty acids (SCFA), which may subsequently decrease the colonic pH [Figure 14.3(B)]. The increased production of SCFA, particularly butyrate, may play a protective role in colon carcinogenesis. Butyrate has been shown in several *in vitro* studies to slow the growth rates of human colorectal cancer cell lines [74,75] and to induce apoptosis in human colon cancer cell lines [76]. In *in vivo* studies, rats fed butyrate pellets were shown to have reduced rates of apoptosis [77], while increased production of butyrate from dietary fiber consumption has been shown to suppress colon tumor formation [78]. Decreased pH has been suggested to be protective of colon carcinogenesis [79] by possibly causing alterations in the metabolic activity of colonic flora [80], altering bile acid metabolism [81] and inhibiting ammonia production and absorption [82]. Thus, PA consumption as part of the diet may potentially lead to starch malabsorption resulting in colon content dilution, increased butyrate and decreased pH, which could be potentially protective of colon carcinogenesis.

Also due to its structure, PA can chelate polyvalent cations, such as iron, calcium and zinc [83] [Figure 14.3(C)]. These minerals are required for vital cellular functions, and their chelation may affect cellular growth kinetics. The chelation of minerals vital to DNA synthesis and cell growth may be a factor in studies that have shown a decrease in cell proliferation with PA supplementation [13,17]. This may also be a mechanism in the protective effects of PA on human rhabdomyosarcoma [59] and HepG2 liver cancer cells in athymic mice, where PA was injected directly into the tumor. In fact, studies have shown that PA can reduce increased colonic cell proliferation [15] and inhibit colon tumorigenesis [33] induced by the addition of minerals such as iron to the diet.

Iron may play a role in oxidative damage [84] and catalysis of hydroxyl radical production [85]. Such radicals may play a role in the etiology of colon carcino-genesis by causing either direct cellular and genetic damage or by promoting the conversion and oxidation of procarcinogens to carcinogens, if they are generated in fecal material close to colonic epithelium. *In vitro*, the addition of PA to a superoxide radical-generating system has been shown to inhibit the hydroxyl radical formation [2,86], inhibit iron-catalyzed lipid peroxidation [3] and inhibit the production of reactive oxygen species [87].

Because the iron necessary to catalyze hydroxyl radical production may be diet derived [88], its chelation by PA may inhibit hydroxyl radical production and subsequent oxidative damage [Figure 14.3(C)]. In fact, individuals consuming less dietary fiber, and presumably less PA, have been shown to produce significantly more hydroxyl radicals [89]. Dietary PA is capable of binding iron in the digestive tract. A correlation has been observed between fecal PA and iron content [42,87], suggesting that increased dietary PA consumption may contribute to decreased hydroxyl radical production and oxidative damage. Although Porres et al. [90] have shown that breakdown of endogenous PA in a corn-soy diet by addition of phytase enzyme can cause an increase in colonic lipid peroxidation, more research is required in this area of PA action, particularly in light of observations by Rimbach and Pallauf [86] showing no effect of PA on liver oxidant or antioxidant status even in a high iron situation.

Since PA is ubiquitous to all mammalian cells, PA from the diet may be absorbed and possibly converted to lower inositol phosphates ($InsP_{1-5}$) [66,67]. The PA or lower InsPs formed from it may participate in cellular inositol phosphate pools. In fact, *in vitro*, PA has been shown to be rapidly absorbed and to be converted to lower inositol phosphates in various murine and human cells [66]. Similar results have been shown *in vivo* [19]. Because InsPs may act as secondary messengers responsible for vital cellular functions, the modulation of the levels of these InsPs by dietary PA may play a role in alteration of cellular growth kinetics in a variety of tissues.

This role may be potentiated by the ability of PA or lower InsPs to modulate cellular calcium levels. For example, inositol triphosphate ($InsP_3$), is a second messenger capable of causing mobilization of intracellular Ca^{2+} and alteration of cellular calcium concentration [91]. The potential of $InsP_3$ and possibly other InsPs [92] to modulate cellular calcium concentration may also allow them to modulate some isoforms of protein kinase C (PKC), a family of calcium-dependent enzymes involved in many cellular processes such as cell proliferation and differentiation [93]. The activity of PKC may also be affected by diacylglycerol (DAG). DAG, along with $InsP_3$, are produced from the breakdown of phosphatidylinositol-4,5-bisphosphate (PIP_2) [94]. Thus, if dietary PA can influence intracellular $InsP_3$ levels and modulate PIP_2 breakdown, it may also alter DAG production, affecting PKC activity and manipulating cellular processes such as proliferation, differentiation and apoptosis. In fact, relatively new data suggest that PA may induce G1 phase cell cycle arrest via PKC δ activation and affect the levels of cyclin D1, cyclin E, cyclin-dependent kinase 2 and 4, retinoblastoma protein and cyclin-dependent kinase inhibitor p27 in various human breast cancer cell lines [95]. This study shows that the cellular effects of PA are very complicated, and that PA, or its breakdown products, may be intimately involved in many cellular processes. Certainly, these effects of PA require much further and in-depth study.

Intracellularly, PA may also be affecting the activity of certain enzymes, such as phosphatidylinositol-3-kinase, vital to cell growth processes [96]. In fact, this enzyme has been implicated in the growth promotion of a number of cancer cell lines [97,98], and its activity is shown to be higher in colorectal tumors [99]. Within the cell, PA, or lower InsPs produced from it, may also act to upregulate p53, involved in some apoptotic pathways, and p21[WAF-1/CIP1] gene expression, whose protein product is growth inhibitory and results in G1 cell cycle arrest by binding PCNA [21]. This may explain the ability of PA to reduce cell proliferation in vivo as observed via a decrease in PCNA levels [17].

The potential for the effects of PA to be modulated by lower InsPs or myo-inositol, the parent compound of PA, produced from its breakdown, have been observed in some studies. Myo-inositol alone has been shown to protect against both mammary [54] and lung cancers [100]. Shamsuddin et al. [23] have observed that the colon cancer protective effects of PA are more pronounced in the presence of myo-inositol. The Shamsuddin group has made similar observations in metastatic and mammary models of cancer [53,54]. In these studies, the PA + myo-inositol groups showed greater decrease of tumor burden along with lesser number and multiplicity of tumors than animals fed either the PA or inositol alone.

Potential modulation of lower InsP and cellular secondary messenger levels by PA suggest that PA may directly affect cellular processes such as cell proliferation and apoptosis. In fact, a number of in vitro [20,21,67] and in vivo studies [13–15,17,22,25] have shown that PA can reduce the rate of colonic cell proliferation. Jenab and Thompson [17] have recently shown that both the endogenous PA in 25% wheat bran and exogenous PA (1%) added either to dephytinized wheat bran or to a low-fiber diet can increase the colonic rate of apoptosis and the degree of differentiation along with the previously observed decrease in the rate of cell proliferation. Because these measures are important indicators of colon cancer risk and tumor development [101], they provide a mechanism whereby endogenous and exogenous PA may be protective of early biomarkers of colon cancer risk and suggest that part of the beneficial effects of PA can be modulated through alterations in cell proliferation, apoptosis and differentiation.

PA may also exert its effects by affecting the immune system. Enhancement of natural killer cell activity has been shown in carcinogen-treated mice fed with PA and in vitro when splenocytes from normal mice were treated with PA [102].

It is clear that PA, whether from a high dietary fiber source or in its pure form, along with its breakdown products, may modulate many cellular and physiological processes. However, the extent of this modulation and how it can be harnessed to prevent or treat various cancers is not yet clear and should be the focus of intense study in the future. PA, perhaps via many of these same potential mechanisms, may also affect disease states other than cancer. These are discussed below.

7. HEART DISEASE

A high zinc/copper ratio has been linked to hypercholesterolemia [103]. Thus, by virtue of its ability to bind these divalent cations and potentially alter their balance and availability [7,8], PA may also affect serum cholesterol levels [Figure 14.3(C)]. In fact, some foods rich in PA, such as bengal gram beans, have been shown to have hypocholesterolemic effects [104]. Although cereal fiber has been associated with decreased coronary heart disease risk [105] and wheat fiber left over from the amylolytic digestion of wheat flakes, which may contain some PA, have been suggested to favorably affect serum cholesterol [106], wheat bran, a rich source of PA, has not been shown to reduce blood lipid levels [107,108]. A reduction of serum cholesterol as a result of PA supplementation has been shown in animal studies [109–111]. Sharma [110] showed that the addition of 0.2% PA to a high-cholesterol diet in hypercholesterolemic rats reduced serum cholesterol along with serum triglycerides. Jariwalla et al. [111] tested the effect of a much higher amount of PA on serum lipids in hyperlipidemic rats fed a high-cholesterol diet. They fed groups of rats either a control diet with 6% saturated fat or control diet supplemented with 0.6% cholesterol for 11 weeks. In addition, two other groups of rats received one or the other of these diets for five weeks and were then fed the same diet as before, supplemented with 8.3% PA. They found that increased consumption of cholesterol led to an increase in total serum cholesterol and serum zinc/copper ratio. The addition of 8.3% PA to the control and the high-cholesterol diet significantly reduced the elevated total cholesterol and triglyceride levels. However, the zinc/copper ratio was only significantly lowered by the addition of PA to the high-cholesterol diet. Despite the high level of PA, the researchers [11] did not note any adverse effects on serum mineral levels. However, because others have shown that lesser amounts of PA can lead to renal [112] and urinary bladder [64] papillomas, the safety and efficacy of such high levels of PA intake must be further scrutinized.

The ability of PA to chelate iron and possibly reduce its free radical generating potential and subsequent lipid peroxidative damage [Figure 14.3(C)] may also protect the heart from ischemic and reperfusion injury [113]. Rao et al. [113] intravenously injected rats with saline or PA at levels up to 15 mg per 100 g of BW. Shortly after, the hearts were excised, and reperfusion injury was induced *in vitro*. The higher levels of injected PA resulted in lessened reperfusion injury by significantly reducing creatine kinase release, decreasing lipid peroxidation and enhancing coronary flow and recovery of ventricular function versus the control group.

It has also been suggested that the calcium-binding ability of PA may help to reduce the level of calcium in soft tissues prone to calcification [114]. Seely [114] suggests that calcification of the aorta may lead to loss of elasticity and development of hypertension. It is argued that PA may bind excess calcium, reduce the level of aortic calcification and improve hypertension. Although hydrolysates of PA have been shown to reduce aortic calcifications in the rat [115], the effect on hypertension is still unclear.

It is clear that the role of PA, particularly dietary PA, in the prevention of heart disease is still open to much conjecture, and thus, a greater amount of emphasis must be placed on research in this area.

8. DIABETES

PA may play an important role in the inhibition of starch digestion and absorption [116] [Figure 14.3(B)]. It can bind to starch either by hydrogen bonding, indirectly via proteins it is bound to [70,116], or by inhibiting the proper digestion and hence absorption of starch by binding amylase or enzyme cofactors such as Ca^{2+} [4]. In fact, PA has been shown *in vitro* to inhibit amylase activity under a variety of circumstances [117,118]. Yoon et al. [72] have shown that increasing levels of PA intake from cereal and legume foods are negatively correlated with the glycemic index, suggesting that PA can slow dietary starch digestion and absorption. In addition, an increased rate of digestion and blood glucose response seen upon dephytinization of navy bean flour was reversed by re-addition of PA [73], again suggesting a role for PA in starch digestion and absorption. Similar effects have been observed *in vitro*, where the addition of PA (2%) to wheat starch significantly reduced the rate of starch digestion [72].

Naturally, the extent and severity of these interactions depends on the ratio of starch to PA and the overall PA content of the diet. The role of PA in starch absorption may also depend on the interaction of PA with other dietary compounds. For example, the inhibitory effects of PA on starch digestibility and absorption have been reversed by the addition of calcium, which is preferentially chelated by the PA. The PA is then unable to bind to or interact with starches in the ingested food [107]. Thus, the effects of pure PA added to a diet may be different from the same amount of PA present within the matrix of a foodstuff.

Any starch that is not digested and absorbed in the small intestine would probably lessen the glucose absorption at that site, thus decreasing the glycemic effect and the required insulin response. This is of consequence not only in diabetes, but also in heart disease. Because reductions in plasma glucose and insulin may lead to lower hepatic lipid synthesis, the hypoglycemic effects of PA may also contribute to its hypolipidemic effects described above. In fact, PA

(0.5%) has been shown to significantly decrease sucrose-induced increases in total hepatic lipids, serum triacylglycerols and phospholipids [119]. The hypoglycemic effects of PA may also be important in light of the insulin hypothesis of colon carcinogenesis [120–122], whereby lesser insulin response may equate to decreased tumor growth promotion. The slowing of digestion and absorption of starch by PA allows the starch to reach the colon, where it may increase fecal bulk or be fermented to short-chain fatty acids (SCFA), thereby decreasing pH. All of these factors may also play a role in the observed colon cancer protective effects of dietary PA (Table 14.1).

9. OTHER DISEASES

By virtue of its mineral-binding abilities, PA may also aid in the prevention of renal calculi [Figure 14.3(C)]. PA has been shown *in vitro* to best prevent brushite, a form of calcium phosphate and component of renal calculi, crystallization and precipitation [123]. In rats, it has been observed that PA treatment can reduce the number of ethylene-glycol-induced calcifications and total calcium amount in the kidney [124]. Grases et al. [125] have also shown that the propensity of the AIN-76A rat diet to cause renal calculi is due to a lack of dietary PA. In humans, it has been shown that individuals prone to stone formation had significantly lower urinary PA levels than those not prone to stone formation [126]. Also in humans, Ohkawa et al. [127] have shown that consumption of rice bran, rich in PA, for periods ranging from one to three years can reduce kidney stone formation. Thus, it appears that PA can effectively inhibit renal crystallization of calcium salts. Because urinary PA levels are directly related to dietary PA intake [128], a sufficient intake of PA is necessary to inhibit renal calculi formation.

10. CONCLUSIONS

From numerous *in vitro* and *in vivo* studies, it is clear that purified exogenous PA, given either in the drinking water or the diet, as well as endogenous PA, as a phytochemical component of wheat bran, have cancer protective effects on a variety of tissues. It is also apparent that pure PA may play a role in reducing serum cholesterol and lipids and thus reduce the risk of heart disease, while both exogenous and endogenous PA may have hypoglycemic effects and thus be of consequence in diabetes. PA consumption may also be beneficial for those suffering from buildup of renal calculi. Although only a limited number of studies have been conducted on the role of endogenous versus exogenous PA, it appears that endogenous PA may be, in part, responsible for the cancer protective effects of high-fiber foods, particularly wheat bran, a rich source of PA, while exogenous PA is also very effective. However, the effects of PA observed in animal studies need to be validated in humans, so well-designed, prospective clinical studies are necessary. Several mechanisms have been suggested for the disease protective effects of PA, including its ability to bind starch, proteins, enzymes, and minerals such as the pro-oxidant iron, its potential participation in cellular inositol phosphate pools and its involvement in signal transduction, cell signaling cascades and gene expression. PA is a major phytochemical in many high dietary fiber foods, which may, in part, be responsible for the disease risk reduction effect attributed to many of these foods. However, additional research is necessary to better characterize the effects of PA on various diseases and to further compare the effectiveness of endogenous versus exogenous PA, particularly in human populations.

11. REFERENCES

1. Reddy, N.R., Sathe, S., and Salunkhe, D.K. 1982. Phytates in legumes and cereals. *Adv. Food Res.* 28:1–92.
2. Graf, E. and Eaton, J.W. 1984. Effects of phytate on mineral bioavailability in mice. *J. Nutr.* 114:1192–1198.
3. Graf, E., Empson, K.L., and Eaton, J.W. 1987. Phytic acid: a natural antioxidant. *J. Biol. Chem.* 262:11647–11650.
4. Rickard, S.E. and Thompson, L.U. 1997. Interactions and biological effects of phytic acid, in *Antinutrients and Phytochemicals in Food*, Shahidi, F., Ed., American Chemical Society, Washington, D.C., pp. 294–312.
5. Cheryan, M. 1980. Phytic acid interactions in food systems. *Crit. Rev. Food Sci. Nutr.* 13:297–335.
6. Brune, M., Rossander, L., and Hallberg, L. 1989. Iron absorption: no intestinal adaptation to a high-phytate diet. *Am. J. Clin. Nutr.* 49:542–545.
7. Torre, M., Rodriguez, A.R., and Saura-Calixto, F. 1991. Effects of dietary fiber and phytic acid on mineral bioavailability. *Crit. Rev. Food. Sci. Nutr.* 1:1–22.
8. Zhou, J.R. and Erdman, J.W. 1995. Phytic acid in health and disease. *Crit. Rev. Food Sci. Nutr.* 35:495–508.
9. Morris, E.R. and Ellis, R. 1980. Bioavailability to rats of iron and zinc in wheat bran: response to low-phytate bran and effect of the phytate/zinc molar ratio. *J. Nutr.* 110:2000–2010.
10. Sathe, S.K. and Salunkhe, D.K. 1984. Technology of removal of unwanted components of dry beans. *Crit. Rev. Food Sci. Nutr.* 21:263–287.
11. Thompson, L.U. and Cho, Y.S. 1984. Effect of acylation upon extraction of nitrogen, minerals and phytic acid in rapeseed flour and protein concentrate. *J. Food Sci.* 49:771–774.
12. Thompson, L.U. 1995. Phytic acid and other nutrients: are they partly responsible for health benefits of high fiber foods, in *Dietary Fiber in Health and Disease*, Kritchevsky, D. and Bonfield, C., Eds., Egan Press, New York, pp. 305–317.
13. Nielsen, B.K., Thompson, L.U., and Bird, R.P. 1987. Effect of phytic acid on colonic epithelial cell proliferation. *Cancer Lett.* 37:317–325.
14. Shamsuddin, A.M., Elsayed, A.M., and Ullah, A. 1988. Suppression of large intestinal cancer in F-344 rats by inositol hexaphosphate. *Carcinogenesis* 9:577–583.
15. Thompson, L.U. and Zhang, L. 1991. Phytic acid and minerals: effect on early markers of risk for mammary and colon carcinogenesis. *Carcinogenesis* 12:2041–2045.

16. Vucenik, I., Yang, G.Y., and Shamsuddin, A.M. 1997. Comparison of pure inositol hexaphosphate and high–bran diet in the prevention of DMBA-induced rat mammary carcinogenesis. *Nutr. Cancer* 28:7–13.
17. Jenab, M. and Thompson, L.U. 1998. The influence of phytic acid in wheat bran on early biomarkers of colon carcinogenesis. *Carcinogenesis* 19:1087–1092.
18. Shamsuddin, A.M. 1995. Inositol phosphates have novel anticancer function. *J. Nutr.* 125:725S–732S.
19. Sakamoto, K., Venkatraman, G., and Shamsuddin, A.M. 1993. Growth inhibition and differentiation of HT–29 cells *in vitro* by inositol hexaphosphate (phytic acid). *Carcinogenesis* 14:1815–1819.
20. Yang, G.Y. and Shamsuddin, A.M. 1995. IP6-induced growth inhibition and differentiation of HT-29 human colon cancer cells: involvement of intracellular inositol phosphates. *Anticancer Res.* 15:2479–2487.
21. Saied, I.T. and Shamsuddin, A.M. 1998. Up-regulation of the tumour suppressor gene p53 and WAF1 gene expression by IP6 in HT-29 human colon carcinoma cell line. *Anticancer Res.* 18:1479–1484.
22. Corpet, D.E., Tache, S., and Peiffer, G. 1997. Colon tumour promotion: is it a selection process? Effects of cholate, phytate and food restriction in rats on proliferation and apoptosis in normal and aberrant crypts. *Cancer Lett.* 114:135–138.
23. Shamsuddin, A.M., Ullah, A., and Chakravarthy, A. 1989. Inositol and inositol hexaphosphate suppress cell proliferation and tumor formation in CD-1 mice. *Carcinogenesis* 10:1461–1463.
24. Shamsuddin, A.M. and Ullah, A. 1989. Inositol hexaphosphate inhibits large intestinal cancer in F344 rats 5 months after induction by azoxymethane. *Carcinogenesis* 10:625–626.
25. Ullah, A. and Shamsuddin, A.M. 1990. Dose-dependent inhibition of large intestinal cancer by inositol hexaphosphate in F344 rats. *Carcinogenesis* 11:2219–2222.
26. Pretlow, T.P., O'Riordan, M.A., Somich, G.A., Amini, S.B., and Pretlow, T.G. 1992. Aberrant crypts correlate with tumour incidence in F344 rats treated with azoxymethane and phytate. *Carcinogenesis* 13:1509–1512.
27. Bird, R.P. 1995. Role of aberrant crypt foci in understanding the pathogenesis of colon cancer. *Cancer Lett.* 93:55–71.
28. Pretlow, T.P., Cheyer, C., and O'Riordan, M.A. 1994. Aberrant crypt foci and colon tumors in F344 rats have similar increases in proliferative activity. *Int. J. Cancer* 56:599–602.
29. Jenab, M. and Thompson, L.U. 2001. Sialomucin production in aberrant crypt foci relates to degree of dysplasia and rate of cell proliferation. *Cancer Lett.* 165:19–25.
30. Stopera, S.A. and Bird, R.P. 1993. Immunohistochemical demonstration of mutant p53 tumor suppressor gene product in aberrant crypt foci. *Cytobios.* 73:73–88.
31. Stopera, S.A., Murphy, L.C., and Bird, R.P. 1992. Evidence for a ras gene mutation in azoxymethane-induced colonic aberrant crypts in Sprague-Dawley rats: earliest recognizable precursor lesions of experimental colon cancer. *Carcinogenesis* 13:2081 2085.
32. Shivapurkar, N., Tang, Z.C., Frost, A., and Alabaster, O. 1996. A rapid dual organ rat carcinogenesis bioassay for evaluating the chemoprevention of breast and colon cancer. *Cancer Lett.* 100:169–179.
33. Nelson, R.L., Yoo, S.J., Tanure, G., Andrianopoulos, G., and Misumi, A 1989. The effect of iron on experimental colorectal carcinogenesis. *Anticancer Res.* 9:1477–1482.
34. Challa, A., Rao, D.R., and Reddy, B.S. 1997. Interactive suppression of aberrant crypt foci induced by azoxymethane in rat colon by phytic acid and green tea. *Carcinogenesis* 18:2023–2026.
35. Jenab, M. and Thompson, L.U. 2000. Phytic acid in wheat bran affects colon morphology, cell differentiation and apoptosis. *Carcinogenesis* 21:1547–1552.
36. Graf, E. and Eaton, J.W. 1990. Antioxidant functions of phytic acid. *Free Radical Biol. Med.* 8:61–69.

37. Graf, E. and Eaton, J.W. 1993. Suppression of colonic cancer by dietary phytic acid. *Nutr. Cancer* 19:11–19.
38. Stevens, R.G., Beasely, R.P., and Blumberg, B.S. 1986. Iron binding proteins and risk of cancer in Taiwan. *J. Nat. Cancer Inst.* 76:605–610.
39. Stevens, R.G., Jones, D.Y., Micozzi, M.S., and Taylor, P.R. 1988. Body iron stores and the risk of cancer. *N. Eng. J. Med.* 319:1047–1052.
40. Selby, J.V. and Friedman, G.D. 1988. Epidemiologic evidence of an association between body iron stores and risk of cancer. *Int. J. Cancer* 41:677–682.
41. Wurzelmann, J.I., Silver, A., Schreinemachers, D.M, Sandler, R.S., and Everson, R.B. 1996. Iron intake and risk of colon cancer. *Cancer Epidemiol. Biomark. Prev.* 5:503–507.
42. Owen, R.W., Weisgerber, U.M., Spiegelhalder, B., and Bartsch, H. 1996. Faecal phytic acid and its relation to other putative markers of risk for colorectal cancer. *Gut* 38:591–597.
43. Risio, M., Lipkin, M., Candelaresi, G., Bertone, A., Coverlizza, S., and Rossini, F.P. 1991. Correlations between rectal mucosa cell proliferation and the clinical and pathological features of non-familial neoplasia of the large intestine. *Cancer Res.* 51:1917–1921.
44. Vucenik, I., Yang, G.Y., and Shamsuddin, A.M. 1997. Comparison of pure inositol hexaphosphate and high-bran diet in the prevention of DMBA-induced rat mammary carcinogenesis. *Nutr. Cancer* 28:7–13.
45. Reddy, B.S., Hirose, Y., Cohen, L.A., Simi, B., Cooma, I., and Rao, C.V. 2000. Preventive potential of wheat bran fractions against experimental colon carcinogenesis: implications for human colon cancer prevention. *Cancer Res.* 60:4792–4797.
46. Shamsuddin, A.M. and Vucenik, I. 1999. Mammary tumor inhibition by IP6: a review. *Anticancer Res.* 19:3671–3674.
47. Wilson, R.B., Hutcheson, D.P., and Wideman, L. 1977. Dimethylhydrazine induced colon tumors in rats fed diets containing beef fat or corn oil with or without wheat bran. *Amer. J. Clin. Nutr.* 30:176–181.
48. Chen, W.F., Patchevsky, A.S., and Goldsmith, H.S. 1978. Colonic protection from dimethyl-hydrazine by a high fiber diet. *Surg. Gyn. Obs.* 147:503–506.
49. Reddy, B.S. and Mori, H. 1981. Effect of dietary wheat bran and dehydrated citrus fiber on 3,2-dimethyl-4-aminobiphenyl induced intestinal carcinogenesis in F344 rats. *Carcinogenesis* 2:21–25.
50. Barnes, D.S., Clapp, N.K., Scott, D.A., Oberst, D.L., and Berry, S.G. 1983. Effects of wheat, rice, corn and soybean bran on 1,2-dimethylhydrazine-induced large bowel tumourigenesis in F344 rats. *Nutr. Cancer* 5:1–9.
51. Maziya-Dixon, B.B., Klopfenstein, C.F., and Leipold, H.W. 1994. Protective effects of hard red wheat versus hard white wheats in chemically induced colon cancer in CF1 mice. *Cereal Chem.* 71:359–363.
52. Shamsuddin, A.M., Yang, G.Y., and Vucenik, I. 1996. Novel anticancer function of IP6: growth inhibition and differentiation of human mammary cancer cell lines *in vitro*. *Anticancer Res.* 16:3287–3292.
53. Vucenik, I., Sakamoto, K., Bansal, M., and Shamsuddin, A.M. 1993. Inhibition of rat mammary carcinogenesis by inositol hexaphosphate (phytic acid). A pilot study. *Cancer Lett.* 75:95–101.
54. Vucenik, I., Yang, G.Y., and Shamsuddin, A.M. 1995. Inositol hexaphosphate and inositol inhibit DMBA-induced rat mammary cancer. *Carcinogenesis* 16:1055–1058.
55. Hirose, M., Hoshiya, T., Akagi, K., Futakuchi, M., and Ito, N. 1994. Inhibition of mammary gland carcinogenesis by green tea catechins and other naturally occurring antioxidants in female Sprague-Dawley rats pretreated with 7,12-dimethylbenz[alpha]anthracene. *Cancer Lett.* 83:149–156.
56. Shamsuddin, A.M., Baten, A., and Lalwani, N.D. 1992. Effects of inositol hexaphosphate on growth and differentiation in K-562 erythroleukemia cell line. *Cancer Lett.* 64:195–202.

57. Shamsuddin, A.M. and Yang, G.Y. 1995. Inositol hexaphosphate inhibits growth and induces differentiation of PC-3 human prostate cancer cells. *Carcinogenesis* 16:1975–1979.
58. Vucenik, I., Tantivejkul, K., Zhang, Z.S., Cole, K.E., Saied, I., and Shamsuddin, A.M. 1998. IP6 in treatment of liver cancer. I. IP6 inhibits growth and reverses transformed phenotype in HepG2 human liver cancer cell line. *Anticancer Res.* 18:4083–4090.
59. Vucenik, I., Kalebic, T., Tantivejkul, K., and Shamsuddin, A.M. 1998. Novel anticancer function of inositol hexaphosphate: inhibition of human rhabdomyosarcoma *in vitro* and *in vivo*. *Anticancer Res.* 18:1377–1384.
60. Vucenik, I., Zhang, Z.S., and Shamsuddin, A.M. 1998. IP6 in treatment of liver cancer. II. Intra-tumoural injection of IP6 regresses pre-existing human liver cancer xenotransplanted in nude mice. *Anticancer Res.* 18:4091–4096.
61. Jariwalla, R.J., Sabin, R., Lawson, S., Bloch, D.A., Prender, M., Andrews, V., and Herman, Z.S. 1988. Effect of dietary phytic acid (phytate) on the incidence and growth rate of tumors promoted in Fisher rats by magnesium supplement. *Nutr. Res.* 8:813–827.
62. Vucenik, I., Tomazic, V.J., Fabian, D., and Shamsuddin, A.M. 1992. Antitumour activity of phytic acid (inositol hexaphosphate) in murine transplanted and metastatic fibrosarcoma. A pilot study. *Cancer Lett.* 65:9–13.
63. Ishikawa, T., Nakatsuru, Y., Zarkovic, M., and Shamsuddin, A.M. 1999. Inhibition of skin cancer by IP6 *in vivo*: initiation-promotion model. *Anticancer Res.* 19:3749–3752.
64. Hirose, M., Ozaki, K., Takaba, K., Fukushima, S., Shirai, T., and Ito, N. 1991. Modifying effects of the naturally occurring antioxidants gamma oryzanol, phytic acid, tannic acid and n-tritriacontane-16, 18-dione in a rat wide-spectrum organ carcinogenesis model. *Carcinogenesis* 12:1917–1921.
65. Larsson, O., Barker, C.J., Sjoholm, A., Carlqvist, H., Michell, R.H., Bertorello, A., Nilsson, T., Honkanen, R.E., Mayr, G.W., Zwiler, J., and Berggren, P.O. 1997. Inhibition of phosphatases and increased Ca2+ channel activity by inositol hexakisphosphate. *Science* 278:471–474.
66. Vucenik, I. and Shamsuddin, A.M. 1994. [^{3}H]Inositol Hexaphosphate (Phytic Acid) is rapidly absorbed and metabolized by murine and human malignant cells *in vitro*. *J. Nutr.* 124:861–868.
67. Sakamoto, K., Vucenik, I., and Shamsuddin, A.M. 1993. [3H]Phytic acid (inositol hexaphosphate) is absorbed and distributed to various tissues in rats. *J. Nutr.* 123:713–720.
68. Shears, S.B. 1998. The versatility of inositol phosphates as cellular signals. *Biochim. Biophys. Acta* 1436:49–67.
69. Singh, M. and Krikorian, A.D. 1982. Inhibition of trypsin activity *in vitro* by phytate. *J. Agric. Food Chem.* 30:799–800.
70. Thompson, L.U. 1986. Phytic acid: a factor influencing starch digestibility and blood glucose response, in *Phytic Acid: Chemistry and Applications*, Graf, E. Ed., Pilatus Press, Minneapolis, MN, pp. 173–194.
71. Gupta, M., Khetarpaul, N., and Chauhan, B.M. 1992. Rabadi fermentation of wheat: changes in phytic acid content and *in vitro* digestibility. *Plant Foods Hum. Nutr.* 42:109–116.
72. Yoon, J.H., Thompson, L.U., and Jenkins, D.J.A. 1983. The effect of phytic acid on *in vitro* rate of starch digestibility and blood glucose response. *Amer. J. Clin. Nutr.* 38:835–842.
73. Thompson, L.U., Button, C.L., and Jenkins, D.J.A. 1987. Phytic acid and calcium affect the *in vitro* rate of navy bean starch digestion and blood glucose response in humans. *Am. J. Clin. Nutr.* 46:467–473.
74. Coradini, D., Pellizzaro, C., Marimpietri, D., Abolafio, G., and Daidone, M.G. 2000. Sodium butyrate modulates cell cycle-related proteins in HT29 human colonic adenocarcinoma cells. *Cell Prolif.* 33:139–146.
75. Basson, M.D., Turowski, G.A., Rashid, Z., Hong, F., and Madri, J.A. 1996. Regulation of human colonic cell line proliferation and phenotype by sodium butyrate. *Dig. Dis. Sci.* 41:1989–1993.

76. Hague, A., Manning, A.M., Hanlon, K.A., Huschtscha, L.I., Hart, D., and Paraskeva, C. 1993. Sodium butyrate induces apoptosis in human colonic tumour cell lines in a p53-independent pathway: implications for the possible role of dietary fiber in the prevention of large bowel cancer. *Int. J. Cancer* 55:498–505.

77. Caderni, G., Luceri, C., Lancioni, L., Tessitore, L., and Dolara, P. 1998. Slow-release pellets of sodium butyrate increase apoptosis in the colon of rats treated with azoxymethane, without affecting aberrant crypt foci and colonic proliferation. *Nutr. Cancer* 30:175–181.

78. McIntyre, A., Gibson, P.R., and Young, G.P. 1993. Butyrate production from dietary fibre and protection against large bowel cancer in a rat model. *Gut* 34:386–391.

79. Newmark, H.L. and Lupton, J.R. 1990. Determinants and consequences of colonic luminal pH: implications for colon cancer. *Nutr. Cancer* 14:161–173.

80. Mallett, A.K., Bearne, C.A., and Rowland, I.R. 1989. The influence of incubation pH on the activity of rat and human gut flora enzymes. *J. Appl. Bacteriol.* 66:433–437.

81. Thornton, J.R. 1981. High colonic pH promotes colorectal cancer. *Lancet* 1:1081–1083.

82. Clinton, S.K., Dieterich, M., Bostwick, D.G., Olson, L.M., Montag, A.G., and Michelassi, F. 1987. The effects of ammonia on N-methyl-N-nitrosoguanidine induced colon carcinogenesis and ras oncogene expression. *FASEB J.* 46:585–588.

83. Graf, E. and Eaton, J.W. 1990. Antioxidant functions of phytic acid. *Free Radical Biol. Med.* 8:61–69.

84. Nelson, R.L. 1992. Dietary iron and colorectal cancer risk. *Free Rad. Bio. Med.* 12:161–168.

85. Graf, E. and Eaton, J.W. 1993. Suppression of colonic cancer by dietary phytic acid. *Nutr. Cancer* 19:11–19.

86. Rimbach, G. and Pallauf, J. 1998. Phytic acid inhibits free radical formation *in vitro* but does not affect liver oxidant or antioxidant status in growing rats. *J. Nutr.* 128:1950–1955.

87. Owen, R.W., Spiegelhalder, B., and Bartsch, H. 1998. Phytate, reactive oxygen species and colorectal cancer. *Eur. J. Cancer Prev. (Suppl)* 2:S41–54.

88. Babbs, C.F. 1990. Free radicals and the etiology of colon cancer. *Free Rad. Biol. Med.* 8:191–200.

89. Erhardt, J.G., Lim, S.S., Bode, J.C., and Bode, C. 1997. A diet rich in fat and poor in dietary fiber increases the *in vitro* formation of reactive oxygen species in human feces. *J. Nutr.* 127:706–709.

90. Porres, J.M., Stahl, C.H., Cheng, W.H., Fu, Y., Roneker, K.R., Pond, W.G., and Lei, X.G. 1999. Dietary intrinsic phytate protects colon from lipid peroxidation in pigs with a moderately high dietary iron intake. *Proc. Soc. Exp. Biol. Med.* 221:80–86.

91. Menniti, F.S., Oliver, K.G., Putney, J.W. Jr., and Shears, S.B. 1993. Inositol phosphates and cell signaling: new views of InsP5 and InsP6. *Trends Biochem. Sci.* 18:53–56.

92. Berridge, M.J. and Irvine, R.F. 1989. Inositol phosphates and cell signalling. *Nature* (London) 341:197–205.

93. Davidson, L.A., Jiang, Y.H., Derr, J.N., Aukema, H.M., Lupton, J.R., and Chapkin, R.S. 1994. Protein kinase C isoforms in human and rat colonic mucosa. *Arch. Biochem. Biophys.* 312:547–553.

94. Berridge, M.J. 1981. Phosphatidyl glycerol hydrolysis: a multifunctional transducing mechanism. *Mol. Cell. Endocrinol.* 24:115–140.

95. Vucenik. I., Tantivejkul, K., Anderson, L.M., and Ramljak, D. 2001. Inositol hexaphosphate may exert its antiproliferative effects by affecting the proteins involved in the regulation of the G1/S transition of the cell cycle (Abstract) *J. Nutr.* 131:193S.

96. Huang, C., Ma, W.Y., Hecht, S., and Dong, Z. 1997. Inositol hexaphosphate inhibits cell transformation and activator protein 1 activation by targeting phosphatidyl-inositol-3'kinase. *Cancer Res.* 57:2873–2878.

97. Huang, C., Ma, W.Y., and Dong, Z. 1996. Requirement of phosphatidylinositol-3-kinase in epidermal growth factor induced AP-1 transactivation and transformation in JB6 P+ cells. *Mol. Cell. Biol.* 16:6427–6435.

98. Huang, C., Schmid, P.C., Ma, W.Y., and Schmid, H.H.O. 1997. Phosphatidylinositol-3-kinase is necessary for 12-0-tetradecanoylphorbol-13-acetate-induced cell transformation and AP-1 activation. *J. Biol. Chem.* 272:4187–4194.

99. Phillips, W.A., St. Clair, F., Munday, A.D., Thomas, R.I.S., and Mitchell, C.A. 1998. Increased levels of phosphatidyl-3-kinase activity in colorectal tumours. *Cancer* 83:41–47.

100. Wattenberg, L.W. and Estensen, R.D. 1996. Chemopreventive effects of myoinositol and dexamethasone on benzo[a]pyrene and 4-(methylnitrosoamino)-1(3-pyridyl)-butanone-induced pulmonary carcinogenesis in female A/J mice. *Cancer Res.* 56:5132–5135.

101. Chang, W.C.L., Chapkin, R.S., and Lupton, J.R. 1997. Predictive value of proliferation, differentiation and apoptosis as intermediate markers for colon tumourigenesis. *Carcinogenesis* 18:721–730.

102. Baten, A., Ullah, A., Tomazic, V.J., and Shamsuddin, A.M. 1989. Inositol-phosphate-induced enhancement of natural killer cell activity correlates with tumor suppression. *Carcinogenesis* 10:1595–1598.

103. Klevay, L.M. 1975. Coronary heart disease: the zinc/copper hypothesis. *Amer. J. Clin. Nutr.* 28:764–774.

104. Mathur, K.S., Singhal S.S., and Sharma R.D. 1964. Effect of Bengal gram on experimentally induced high levels of cholesterol in tissue and serum in albino rats. *J. Nutr.* 84:201–204.

105. Wolk, A., Manson, J.E., Stampfer, M.J., Colditz, G.A., Hu, F.B., Speizer, F.E., Hennekens, C.H., and Willett, W.C. 1999. Long term intake of dietary fiber and decreased risk of coronary heart disease among women. *J. Amer. Med. Assoc.* 281:1998–2004.

106. Vuksan, V., Jenkins, D.J., Vidgen, E., Ransom, T.P., Ng, M.K., Culhane, C.T., and O'Connor, D. 1999. A novel source of wheat fiber and protein: effects on fecal bulk and serum lipids. *Amer. J. Clin. Nutr.* 69:226–230.

107. Thompson, L.U. 1989. Nutritional and physiological effects of phytic acid, in *Food Proteins*, Kinsella, J.E. and Soucie, W.G., Eds,. American Oil Chemists Society, Champaign, IL, pp. 410–431.

108. Jenkins, D.J., Kendall, C.W., Vuksan, V., Augustin, L.S., Mehling, C., Parker, T., Vidgen, E., Lee, B., Faulkner, D., Seyler, H., Josse, R., Leiter, L.A., Connelly, P.W., and Fulgoni, V. 1999. Effect of wheat bran on serum lipids: influence of particle size and wheat protein. *J. Amer. Coll. Nutr.* 18:159–165.

109. Klevay, L.M. 1977. Elements of ischemic heart disease. *Perspect Biol Med.* 20:186–92.

110. Sharma, R.D. 1980. Effect of hydroxy acids on hypercholesterolaemia in rats. *Atherosclerosis* 37:463–468.

111. Jariwalla, R.J., Sabin, R., Lawson, S., and Herman, Z. 1990. Lowering of serum cholesterol and triglycerides and modulation of divalent cations by dietary phytate. *J. Applied Nutr.* 42:18–28.

112. Hiasa, Y., Kitahori, Y., Morimoto, J., Konishi, N., Nakaoka, S., and Nishioka, H. 1992. Carcinogenicity study in rats of phytic acid "daiichi", a natural food additive. *Food Chem. Toxicol.* 30:117–125.

113. Rao, P.S., Liu, X., Das, D.K., Weinstein, G.S., and Tyras, D.H. 1991. Protection of ischemic heart from reperfusion injury by myo-inositol hexaphosphate, a natural antioxidant. *Ann. Thorac. Surgery* 52:908–912.

114. Seely, S. 1991. Is calcium excess in western diet a major cause of arterial disease? *Int. J. Cardiol.* 33:191–198.

115. Van den Berg, C.J., Hill L.F., and Stanbury S.W. 1972. Inositol phosphates and phytic acid as inhibitors of biological calcification in the rat. *Clin. Sci.* 43:377–383.

116. Thompson, L.U. 1993. Potential health benefits and problems associated with antinutrients in foods. *Food Res. Intl.* 26:131–149.

117. Deshpande, S.S. and Cheryan, M. 1984. Effects of phytic acid, divalent cations and their interactions on alpha-amylase activity. *J. Food Sci.* 49:516–519.

118. Thompson, L.U. and Yoon, J.H. 1984. Starch digestibility as affected by polyphenols and phytic acid. *J. Food Sci.* 49:1228–1229.
119. Katayama, T. 1995. Effect of dietary sodium phytate on the hepatic and serum levels of lipids and on the hepatic activities of NADPH-generating enzymes in rats fed on sucrose. *Biosci. Biotechnol. Biochem.* 59:1159–1160.
120. McKeown-Eyssen, G. 1994. Epidemiology of colorectal cancer revisited: are serum triglycerides and/or plasma glucose associated with risk? *Cancer Epidemiol. Biomarkers Prev.* 3:687–695.
121. Giovannucci, E. 1995. Insulin and colon cancer. *Cancer Causes Control* 6:164–179.
122. Bruce, W.R. and Corpet, D.E. 1996. The colonic protein fermentation and insulin resistance hypothesis for colon cancer etiology: experimental tests using precursor lesions. *Eur. J. Cancer Prev.* 2:41–47.
123. Grases, F., Ramis, M., and Costa-Bauza, A. 2000. Effects of phytate and pyrophosphate on brushite and hydroxyapatite crystallization. Comparison with the action of other polyphosphates. *Urol. Res.* 28:136–140.
124. Grases, F., Garcia-Gonzalez, R., Torres, J.J., and Llobera, A. 1998. Effects of phytic acid on renal stone formation in rats. *Scand. J. Urol. Nephrol.* 32:261–265.
125. Grases, F., Prieto, R.M., Simonet, B.M., and March, J.G. 2000. Phytate prevents tissue calcifications in female rats. *Biofactors* 11:171–177.
126. Grases, F., March, J.G., Prieto, R.M., Simonet, B.M., Costa-Bauza, A., Garcia-Raja, A., and Conte, A. 2000. Urinary phytate in calcium oxalate stone formers and healthy people—dietary effects on phytate excretion. *Scand. J. Urol. Nephrol.* 34:162–164.
127. Ohkawa, T., Ebisuno, S., Kitagawa, M., Morimoto, S., Miyazaki, Y., and Yasukawa, S. 1984. Rice bran treatment for patients with hypercalciuric stones: experimental and clinical studies. *J. Urol.* 132:1140–1145.
128. Grases, F., Simonet, B.M., March, J.G., and Prieto, R.M. 2000. Inositol hexakisphosphate in urine: the relationship between oral intake and urinary excretion. *Brit. J. Urol. Intl.* 85:138–142.
129. Hirose, M., Fukushima, S., Imaida, K., Ito, N., and Shirai, T. 1999. Modifying effects of phytic acid and gamma-oryzanol on the promotion stage of rat carcinogenesis. *Anticancer Res.* 19:3665–3670.
130. Takaba, K., Hirose, M., Ogawa, K., Hakoi, K., and Fukushima, S. 1994. Modification of n-butyl-n-(4-hydroxybutyl)nitrosamine-initiated urinary bladder carcinogenesis in rats by phytic acid and its salts. *Food Chem. Toxicol.* 32:499–503.
131. Takaba, K., Hirose, M., Yoshida, Y., Kimura, J., Ito, N., and Shirai, T. 1997. Effects of n-tritriacontane-16,18-dione, curcumin, chlorphyllin, dihydroguaiaretic acid, tannic acid and phytic acid on the initiation stage in a rat multi-organ carcinogenesis model. *Cancer Lett.* 113:39–46.

Index